Hadronic Matter

" An Overview "

Edited by

Paul F. Kisak

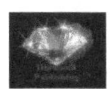

Virginia, USA

Visit our website at: https://www.createspace.com/5817066

Printed in The United States of America

First Trade Edition: 2015
10 9 8 7 6 5 4 3 2 1

Black & White on White paper
349 pages

ISBN-13: 978-1518729553
ISBN-10: 151872955X

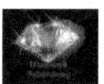

Virginia, USA

Contents

Chapter 1

Hadron

In particle physics, a **hadron** ◀))i /ˈhædrɒn/ (Greek: ἁδρός, *hadrós*, "stout, thick") is a composite particle made of quarks held together by the strong force (in a similar way as molecules are held together by the electromagnetic force).

Hadrons are categorized into two families: baryons, made of three quarks, and mesons, made of one quark and one antiquark. Protons and neutrons are examples of baryons; pions are an example of a meson. Hadrons containing more than three valence quarks (exotic hadrons) have been discovered in recent years. A tetraquark state (an exotic meson), named the Z(4430)⁻, was discovered in 2007 by the Belle Collaboration [1] and confirmed as a resonance in 2014 by the LHCb collaboration.[2] Two pentaquark states (exotic baryons), named P+
c(4380) and P+
c(4450), were discovered in 2015 by the LHCb collaboration.[3] There are several more exotic hadron candidates, and other colour-singlet quark combinations may also exist.

Of the hadrons, protons are stable, and neutrons bound within atomic nuclei are stable. Other hadrons are unstable under ordinary conditions; free neutrons decay with a half-life of about 611 seconds. Experimentally, hadron physics is studied by colliding protons or nuclei of heavy elements such as lead, and detecting the debris in the produced particle showers.

1.1 Etymology

The term "hadron" was introduced by Lev B. Okun in a plenary talk at the 1962 International Conference on High Energy Physics.[4] In this talk he said:

> Not withstanding the fact that this report deals with weak interactions, we shall frequently have to speak of strongly interacting particles. These particles pose not only numerous scientific problems, but also a terminological problem. The point is that "strongly interacting particles" is a very clumsy term which does not yield itself to the formation of an adjective. For this reason, to take but one instance, decays into strongly interacting particles are called non-leptonic. This definition is not exact because "non-leptonic" may also signify "photonic". In this report I shall call strongly interacting particles "hadrons", and the corresponding decays "hadronic" (the Greek ἁδρός signifies "large", "massive", in contrast to λεπτός which means "small", "light"). I hope that this terminology will prove to be convenient. — Lev B. Okun, 1962

1.2 Properties

According to the quark model,[5] the properties of hadrons are primarily determined by their so-called *valence quarks*. For example, a proton is composed of two up quarks (each with electric charge $+\frac{2}{3}$, for a total of $+\frac{4}{3}$ together) and one down quark (with electric charge $-\frac{1}{3}$). Adding these together yields the proton charge of +1. Although quarks also carry color charge, hadrons must have zero total color charge because of a phenomenon called color confinement. That is, hadrons must be "colorless" or "white". These are the simplest of the two ways: three quarks of different colors, or a quark of one color and an antiquark carrying the corresponding anticolor. Hadrons with the first arrangement are called baryons, and those with the second arrangement are mesons.

Hadrons, however, are not composed of just three or two quarks, because of the strength of the strong force. More accurately, strong force gluons have enough energy (E) to have resonances composed of massive (m) quarks ($E > mc^2$) . Thus, virtual quarks and antiquarks, in a 1:1 ratio, form the majority of massive particles inside a hadron. The two or three quarks are the excess of quarks vs. antiquarks in hadrons, and vice versa in anti-hadrons. Because the virtual quarks are not stable wave packets (quanta), but irregular and transient phenomena, it is not meaningful to ask which quark is real and which virtual; only the excess is apparent from the outside. Massless virtual gluons compose the numerical majority of particles inside hadrons.

Like all subatomic particles, hadrons are assigned quantum numbers corresponding to the representations of the Poincaré group: $J^{PC}(m)$, where J is the spin quantum number, P the intrinsic parity (or P-parity), and C, the charge conjugation (or C-parity), and the particle's mass, m. Note that the mass of a hadron has very little to do with the mass of its valence quarks; rather, due to mass–energy equivalence, most of the mass comes from the large amount of energy associated with the strong interaction. Hadrons may also carry flavor quantum numbers such as isospin (or G parity), and strangeness. All quarks carry an additive, conserved quantum number called a baryon number (B), which is $+\frac{1}{3}$ for quarks and $-\frac{1}{3}$ for antiquarks. This means that baryons (groups of three quarks) have $B = 1$ whereas mesons have $B = 0$.

Hadrons have excited states known as resonances. Each ground state hadron may have several excited states; several hundreds of resonances have been observed in particle physics experiments. Resonances decay extremely quickly (within about 10^{-24} seconds) via the strong nuclear force.

In other phases of matter the hadrons may disappear. For example, at very high temperature and high pressure, unless there are sufficiently many flavors of quarks, the theory of quantum chromodynamics (QCD) predicts that quarks and gluons will no longer be confined within hadrons, "because the strength of the strong interaction diminishes with energy". This property, which is known as asymptotic freedom, has been experimentally confirmed in the energy range between 1 GeV (gigaelectronvolt) and 1 TeV (teraelectronvolt).[6]

All free hadrons except the proton (and antiproton) are unstable.

1.3 Baryons

Main article: Baryon

All known baryons are made of three valence quarks, so they are fermions, *i.e.*, they have odd half-integer spin, because they have an odd number of quarks. As quarks possess baryon number $B = \frac{1}{3}$, baryons have baryon number $B = 1$. The best-known baryons are the proton and the neutron.

One can hypothesise baryons with further quark-antiquark pairs in addition to their three quarks. Hypothetical baryons with one extra quark-antiquark pair (5 quarks in all) are called pentaquarks. As of August 2015, there are two known pentaquarks, P+
c(4380) and P+
c(4450), both discovered in 2015 by the LHCb collaboration.[3]

Each type of baryon has a corresponding antiparticle (antibaryon) in which quarks are replaced by their corresponding antiquarks. For example, just as a proton is made of two up-quarks and one down-quark, its corresponding antiparticle, the antiproton, is made of two up-antiquarks and one down-antiquark.

1.4 Mesons

Main article: Meson

Mesons are hadrons composed of a quark-antiquark pair. They are bosons, meaning they have integral spin, *i.e.*, 0, 1, or −1, as they have an even number of quarks. They have baryon number $B = 0$. Examples of mesons commonly produced in particle physics experiments include pions and kaons. Pions also play a role in holding atomic nuclei together via the residual strong force.

In principle, mesons with more than one quark-antiquark pair may exist; a hypothetical meson with two pairs is called a tetraquark. Several tetraquark candidates were found in the 2000s, but their status is under debate.[7] Several other hypothetical "exotic" mesons lie outside the quark model of classification. These include glueballs and hybrid mesons (mesons bound by excited gluons).

1.5   See also

- Hadronization, the formation of hadrons out of quarks and gluons

- Large Hadron Collider (LHC)

- List of particles

- Standard model

- Subatomic particles

- Hadron therapy, a.k.a. particle therapy

- Exotic hadrons

1.6   References

[1] Choi, S.-K.; Belle Collaboration; et al. (2007). "Observation of a resonance-like structure in the π±Ψ′ mass distribution in exclusive B→Kπ±Ψ′ decays". *Physical Review Letters* **100** (14). arXiv:0708.1790. Bibcode:2008PhRvL.100n2001C. doi:10.1103/PhysRevLett.100.142001.

[2] LHCb collaboration (2014): Observation of the resonant character of the Z(4430)⁻ state

[3] R. Aaij et al. (LHCb collaboration) (2015). "Observation of J/ψp resonances consistent with pentaquark states in Λ0
b→J/ψK−
p decays". *Physical Review Letters* **115** (7). doi:10.1103/PhysRevLett.115.072001.

[4] Lev B. Okun (1962). "The Theory of Weak Interaction". *Proceedings of 1962 International Conference on High-Energy Physics at CERN*. Geneva. p. 845. Bibcode:1962hep..conf..845O.

[5] C. Amsler *et al.* (Particle Data Group) (2008). "Review of Particle Physics – Quark Model" (PDF). *Physics Letters B* **667**: 1. Bibcode:2008PhLB..667....1P. doi:10.1016/j.physletb.2008.07.018.

[6] S. Bethke (2007). "Experimental tests of asymptotic freedom". *Progress in Particle and Nuclear Physics* **58** (2): 351. arXiv:hep-ex/0606035. Bibcode:2007PrPNP..58..351B. doi:10.1016/j.ppnp.2006.06.001.

[7] Mysterious Subatomic Particle May Represent Exotic New Form of Matter

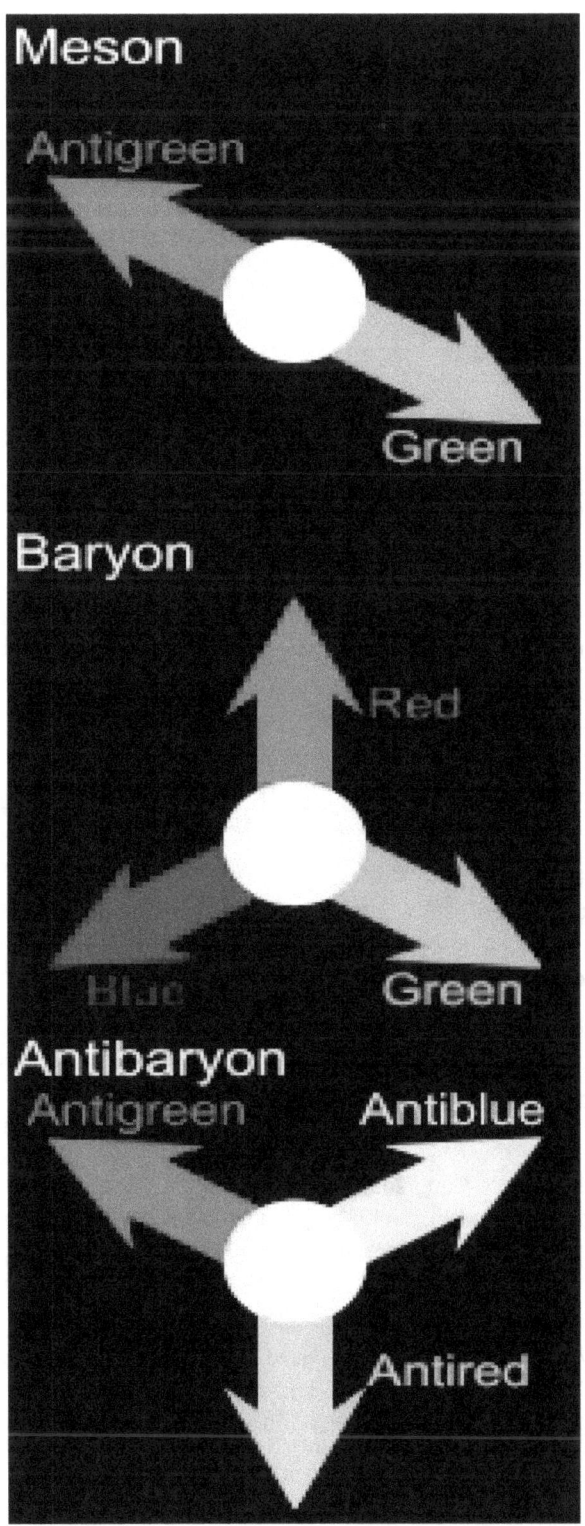

All types of hadrons have zero total color charge. (three examples shown)

Chapter 2

Hadronization

In particle physics, **hadronization** (or **hadronisation**) is the process of the formation of hadrons out of quarks and gluons. This occurs after high-energy collisions in a particle collider in which free quarks or gluons are created. Due to postulated colour confinement, these cannot exist individually. In the Standard Model they combine with quarks and antiquarks spontaneously created from the vacuum to form hadrons. The QCD (Quantum Chromodynamics) of the hadronization process are not yet fully understood, but are modeled and parameterized in a number of phenomenological studies, including the Lund string model and in various long-range QCD approximation schemes.[1][2][3]

The tight cone of particles created by the hadronization of a single quark is called a jet. In particle detectors, jets are observed rather than quarks, whose existence must be inferred. The models and approximation schemes and their predicted Jet hadronization, or **fragmentation**, have been extensively compared with measurement in a number of high energy particle physics experiments; e.g. TASSO,[4] OPAL,[5] H1.[6]

Hadronization also occurred shortly after the Big Bang when the quark–gluon plasma cooled to the temperature below which free quarks and gluons cannot exist (about 170 MeV). The quarks and gluons then combined into hadrons.

A top quark, however, has a mean lifetime of 5×10^{-25} seconds, which is shorter than the time scale at which the strong force of QCD acts, so a top quark decays before it can hadronize, allowing physicists to observe a "bare quark."[7] Thus, they have not been observed as components of any observed hadron, while all other quarks have been observed only as components of hadrons.

2.1 Hadronization simulation and models

Hadronization can be explored using Monte Carlo simulation. After the particle shower has terminated, partons with virtualities on the order of the cut off scale remain. From this point on, the parton is in the low momentum transfer, long-distance regime in which non-perturbative effects become important. The most dominant of these effects is hadronization, which converts partons into observable hadrons. No exact theory for hadronization is known but there are two successful models for parameterization.

The scale at which partons are given to the hadronization is fixed by the Shower Monte Carlo program. Hadronization models typically start at some predefined scale of their own. This can cause significant issue if not set up properly within the Shower Monte Carlo. Common choices of Shower Monte Carlo are PYTHIA and HERWIG. Each of these correspond to one of the two parameterization models.

2.2 References

[1] Yu. L. Dokshitzer, V. A. Khoze, A. H. Mueller and S. I. Troyan, *Basics of Perturbative QCD* Editions Frontieres (1991)

[2] A. Bassetto, M. Ciafaloni, G. Marchesini and A. H. Mueller, *Nucl. Phys.* 207B (1982) 189

[3] A. H. Mueller, *Phys. Lett.* 104B (1981) 161

[4] TASSO Collaboration, W. Braunschweig et al., *Zeit. Phys.* C47 (1990) 187

[5] OPAL Collaboration, M.Z. Akrawy et al., *Phys. Lett.* 247B (1990) 617.

[6] H1 Collaboration, S. Aid et al., "A Study of the fragmentation of quarks in e- p collisions at HERA." *Nucl.Phys.B* 445:3-24,1995.

[7] Abazov, et al., "Evidence for the Production of Single Top Quarks", Fermilab-Pub08/056-E (2008)

- Greco, V.; Ko, C. M.; Lévai, P. (2003). "Parton Coalescence and the Antiproton/Pion Anomaly at RHIC". *Physical Review Letters* **90**(20): 202302. arXiv:nucl-th/0301093. Bibcode:2003PhRvL..90t2302G..PMID12785885.

- Fries, R. J.; Müller, B.; Nonaka, C.; Bass, SA (2003). "Hadronization in Heavy-Ion Collisions: Recombination and Fragmentation of Partons Hadronization in Heavy-Ion Collisions". *Physical Review Letters* **90** (20): 202303. arXiv:nucl-th/0301087. Bibcode:2003PhRvL..90t2303F. doi:10.1103/PhysRevLett.90.202303. PMID 12785886.

Chapter 3

Subatomic particle

In the physical sciences, **subatomic particles** are particles much smaller than atoms.[1] There are two types of subatomic particles: elementary particles, which according to current theories are not made of other particles; and *composite* particles.[2] Particle physics and nuclear physics study these particles and how they interact.[3]

In particle physics, the concept of a particle is one of several concepts inherited from classical physics. But it also reflects the modern understanding that at the quantum scale matter and energy behave very differently from what much of everyday experience would lead us to expect.

The idea of a particle underwent serious rethinking when experiments showed that light could behave like a stream of particles (called photons) as well as exhibit wave-like properties. This led to the new concept of wave–particle duality to reflect that quantum-scale "particles" behave like both particles and waves (also known as wavicles). Another new concept, the uncertainty principle, states that some of their properties taken together, such as their simultaneous position and momentum, cannot be measured exactly.[4] In more recent times, wave–particle duality has been shown to apply not only to photons but to increasingly massive particles as well.[5]

Interactions of particles in the framework of quantum field theory are understood as creation and annihilation of *quanta* of corresponding fundamental interactions. This blends particle physics with field theory.

3.1 Classification

3.1.1 By statistics

Main article: Spin–statistics theorem
 Any subatomic particle, like any particle in the 3-dimensional space that obeys laws of quantum mechanics, can be either a boson (an integer spin) or a fermion (a half-integer spin).

3.1.2 By composition

The elementary particles of the Standard Model include:[6]

- Six "flavors" of quarks: up, down, bottom, top, strange, and charm;

- Six types of leptons: electron, electron neutrino, muon, muon neutrino, tau, tau neutrino;

- Twelve gauge bosons (force carriers): the photon of electromagnetism, the three W and Z bosons of the weak force, and the eight gluons of the strong force;

- The Higgs boson.

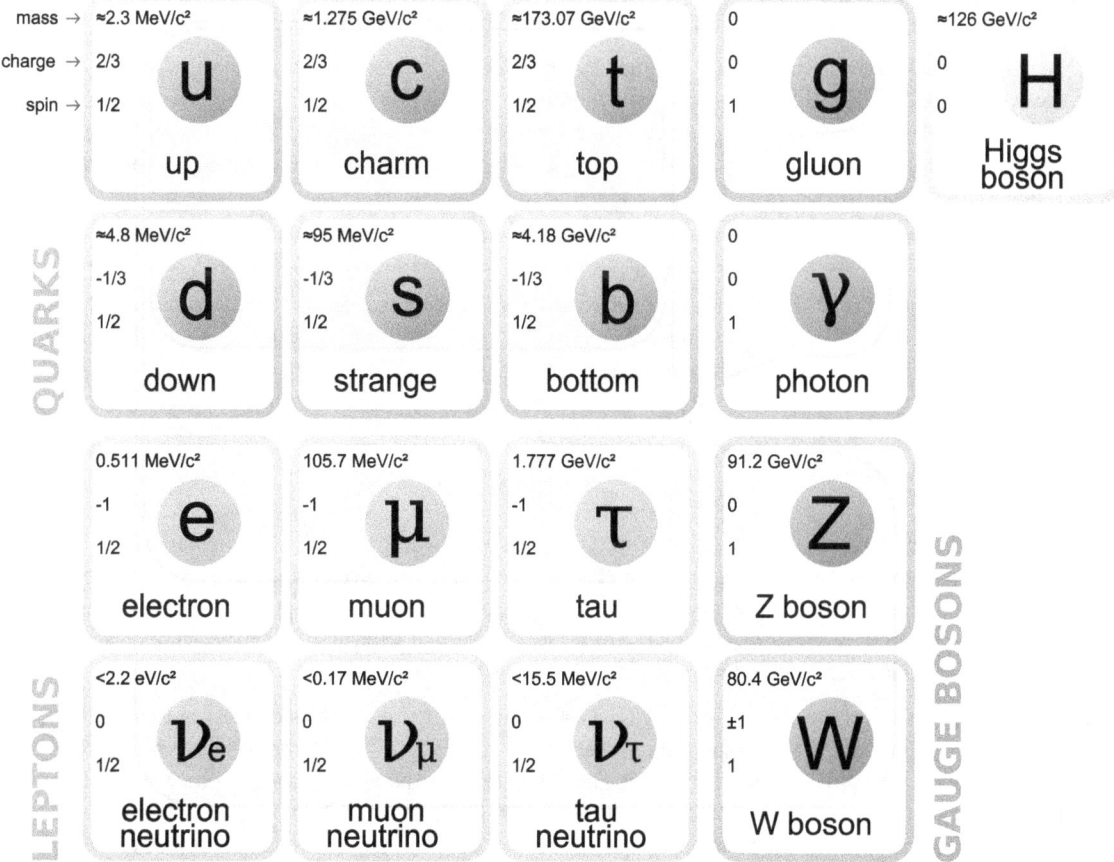

The Standard Model classification of particles

Various extensions of the Standard Model predict the existence of an elementary graviton particle and many other elementary particles.

Composite subatomic particles (such as protons or atomic nuclei) are bound states of two or more elementary particles. For example, a proton is made of two up quarks and one down quark, while the atomic nucleus of helium-4 is composed of two protons and two neutrons. The neutron is made of two down quarks and one up quark. Composite particles include all hadrons: these include baryons (such as protons and neutrons) and mesons (such as pions and kaons).

3.1.3 By mass

In special relativity, the energy of a particle at rest equals its mass times the speed of light squared ($E = mc^2$). That is, mass can be expressed in terms of energy and vice versa. If a particle has a frame of reference where it lies at rest, then it has a positive rest mass and is referred to as *massive*.

All composite particles are massive. Baryons (meaning "heavy") tend to have greater mass than mesons (meaning "intermediate"), which in turn tend to be heavier than leptons (meaning "lightweight"), but the heaviest lepton (the tau particle) is heavier than the two lightest flavours of baryons (nucleons). It is also certain that any particle with an electric charge is massive.

All massless particles (particles whose invariant mass is zero) are elementary. These include the photon and gluon, although the latter cannot be isolated.

3.2   Other properties

Through the work of Albert Einstein, Louis de Broglie, and many others, current scientific theory holds that *all* particles also have a wave nature.[7] This has been verified not only for elementary particles but also for compound particles like atoms and even molecules. In fact, according to traditional formulations of non-relativistic quantum mechanics, wave–particle duality applies to all objects, even macroscopic ones; although the wave properties of macroscopic objects cannot be detected due to their small wavelengths.[8]

Interactions between particles have been scrutinized for many centuries, and a few simple laws underpin how particles behave in collisions and interactions. The most fundamental of these are the laws of conservation of energy and conservation of momentum, which let us make calculations of particle interactions on scales of magnitude that range from stars to quarks.[9] These are the prerequisite basics of Newtonian mechanics, a series of statements and equations in *Philosophiae Naturalis Principia Mathematica*, originally published in 1687.

3.3   Dividing an atom

The negatively charged electron has a mass equal to $1/1836$ of that of a hydrogen atom. The remainder of the hydrogen atom's mass comes from the positively charged proton. The atomic number of an element is the number of protons in its nucleus. Neutrons are neutral particles having a mass slightly greater than that of the proton. Different isotopes of the same element contain the same number of protons but differing numbers of neutrons. The mass number of an isotope is the total number of nucleons (neutrons and protons collectively).

Chemistry concerns itself with how electron sharing binds atoms into structures such as crystals and molecules. Nuclear physics deals with how protons and neutrons arrange themselves in nuclei. The study of subatomic particles, atoms and molecules, and their structure and interactions, requires quantum mechanics. Analyzing processes that change the numbers and types of particles requires quantum field theory. The study of subatomic particles *per se* is called particle physics. The term *high-energy physics* is nearly synonymous to "particle physics" since creation of particles requires high energies: it occurs only as a result of cosmic rays, or in particle accelerators. Particle phenomenology systematizes the knowledge about subatomic particles obtained from these experiments.

3.4   History

Main articles: History of subatomic physics and Timeline of particle discoveries

The term "*subatomic* particle" is largely a retronym of 1960s made to distinguish a big number of baryons and mesons (that comprise hadrons) from particles that are now thought to be truly elementary. Before that hadrons were usually classified as "elementary" because their composition was unknown.

A list of important discoveries follows:

3.5   See also

- *Atom: Journey Across the Subatomic Cosmos* (book)

- *Atom: An Odyssey from the Big Bang to Life on Earth...and Beyond* (book)

- CPT invariance

- Dark Matter

- Hot spot effect in subatomic physics

- List of fictional elements, materials, isotopes and atomic particles

- List of particles

- Poincaré symmetry

- Ylem

3.6   References

[1] "Subatomic particles". NTD. Retrieved 5 June 2012.

[2] Bolonkin, Alexander (2011). *Universe, Human Immortality and Future Human Evaluation*. Elsevier. p. 25. ISBN.

[3] Fritzsch, Harald (2005). *Elementary Particles*. World Scientific. pp. 11–20. ISBN 978-981-256-141-1.

[4] Heisenberg, W. (1927), "Über den anschaulichen Inhalt der quantentheoretischen Kinematik und Mechanik", *Zeitschrift für Physik* (in German) **43** (3–4): 172–198, Bibcode:1927ZPhy...43..172H, doi:10.1007/BF01397280.

[5] Arndt, Markus; Nairz, Olaf; Vos-Andreae, Julian; Keller, Claudia; Van Der Zouw, Gerbrand; Zeilinger, Anton (2000). "Wave-particle duality of C60 molecules". *Nature* **401** (6754): 680–682. Bibcode:1999Natur.401..680A. doi:10.1038/44348. PMID 18494170.

[6] Cottingham, W. N.; Greenwood, D. A. (2007). *An introduction to the standard model of particle physics*. Cambridge University Press. p. 1. ISBN 978-0-521-85249-4.

[7] Walter Greiner (2001). *Quantum Mechanics: An Introduction*. Springer. p. 29. ISBN 3-540-67458-6.

[8] R. Eisberg & R. Resnick (1985). *Quantum Physics of Atoms, Molecules, Solids, Nuclei, and Particles* (2nd ed.). John Wiley & Sons. pp. 59–60. ISBN 0-471-87373-X. For both large and small wavelengths, both matter and radiation have both particle and wave aspects. [...] But the wave aspects of their motion become more difficult to observe as their wavelengths become shorter. [...] For ordinary macroscopic particles the mass is so large that the momentum is always sufficiently large to make the de Broglie wavelength small enough to be beyond the range of experimental detection, and classical mechanics reigns supreme.

[9] Isaac Newton (1687). Newton's Laws of Motion (*Philosophiae Naturalis Principia Mathematica*)

[10] Klemperer, Otto (1959). *Electron Physics: The Physics of the Free Electron*. Academic Press.

[11] Some sources such as The Strange Quark indicate 1947.

[12] http://press.web.cern.ch/press-releases/2014/06/cern-experiments-report-new-higgs-boson-measurements

3.7   Further reading

General readers

- Feynman, R.P. & Weinberg, S. (1987). *Elementary Particles and the Laws of Physics: The 1986 Dirac Memorial Lectures*. Cambridge Univ. Press.

- Brian Greene (1999). *The Elegant Universe*. W.W. Norton & Company. ISBN 0-393-05858-1.

- Oerter, Robert (2006). *The Theory of Almost Everything: The Standard Model, the Unsung Triumph of Modern Physics*. Plume.

- Schumm, Bruce A. (2004). *Deep Down Things: The Breathtaking Beauty of Particle Physics*. Johns Hopkins University Press. ISBN 0-8018-7971-X.

- Martinus Veltman (2003). *Facts and Mysteries in Elementary Particle Physics*. World Scientific. ISBN 981-238-149-X.

Textbooks

- Coughlan, G. D., J. E. Dodd, and B. M. Gripaios (2006). *The Ideas of Particle Physics: An Introduction for Scientists*, 3rd ed. Cambridge Univ. Press. An undergraduate text for those not majoring in physics.

- Griffiths, David J. (1987). *Introduction to Elementary Particles*. Wiley, John & Sons, Inc. ISBN 0-471-60386-4.

- Kane, Gordon L. (1987). *Modern Elementary Particle Physics*. Perseus Books. ISBN 0-201-11749-5.

3.8 External links

- particleadventure.org: The Standard Model.

- cpepweb.org: Particle chart.

- University of California: Particle Data Group.

- Annotated Physics Encyclopædia: Quantum Field Theory.

- Jose Galvez: Chapter 1 Electrodynamics (pdf).

Chapter 4

Standard Model

This article is about the Standard Model of particle physics. For other uses, see Standard model (disambiguation).

This article is a non-mathematical general overview of the Standard Model. For a mathematical description, see the article Standard Model (mathematical formulation).

For the Standard Model of Big Bang cosmology, Lambda-CDM model.

The **Standard Model** of particle physics is a theory concerning the electromagnetic, weak, and strong nuclear inter-

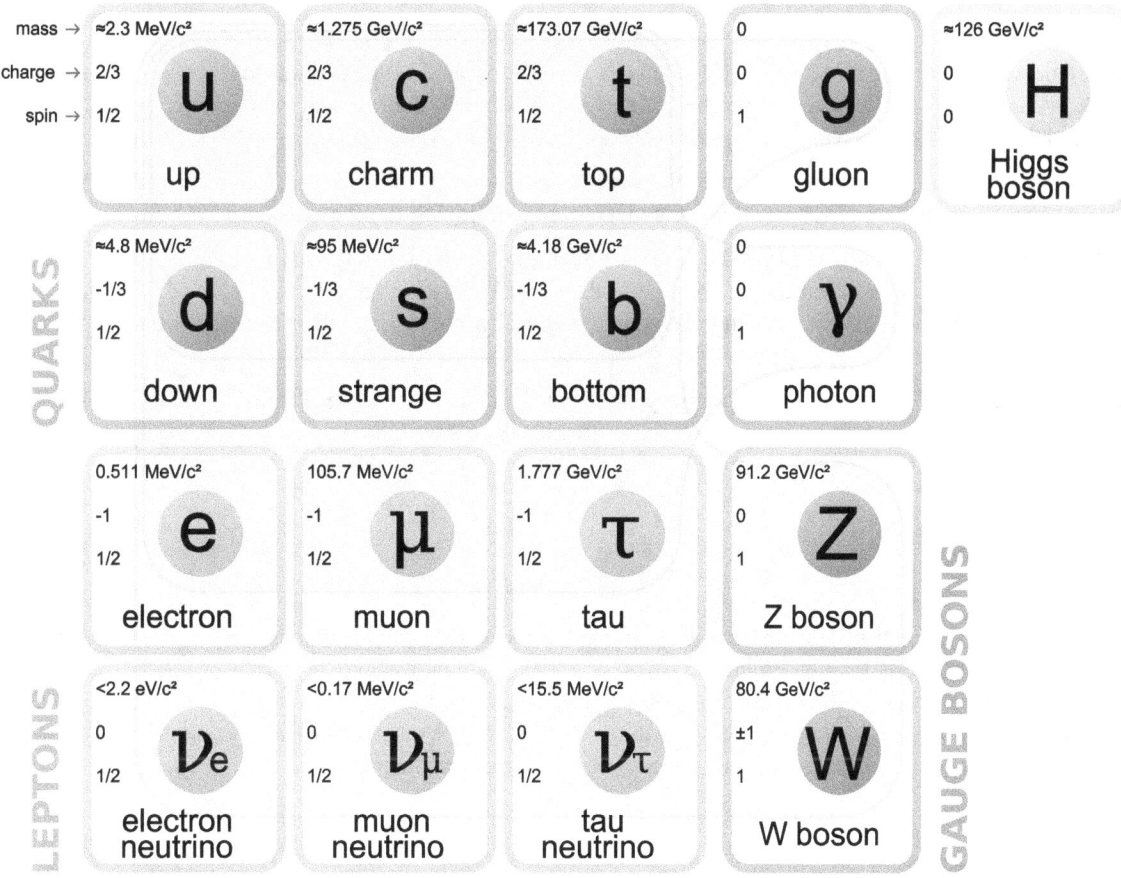

The Standard Model of elementary particles (more schematic depiction), with the three generations of matter, gauge bosons in the fourth column, and the Higgs boson in the fifth.

actions, as well as classifying all the subatomic particles known. It was developed throughout the latter half of the 20th century, as a collaborative effort of scientists around the world.[1] The current formulation was finalized in the mid-1970s upon experimental confirmation of the existence of quarks. Since then, discoveries of the top quark (1995), the tau neutrino (2000), and more recently the Higgs boson (2013), have given further credence to the Standard Model. Because of its success in explaining a wide variety of experimental results, the Standard Model is sometimes regarded as a "theory of almost everything".

Although the Standard Model is believed to be theoretically self-consistent[2] and has demonstrated huge and continued successes in providing experimental predictions, it does leave some phenomena unexplained and it falls short of being a complete theory of fundamental interactions. It does not incorporate the full theory of gravitation[3] as described by general relativity, or account for the accelerating expansion of the universe (as possibly described by dark energy). The model does not contain any viable dark matter particle that possesses all of the required properties deduced from observational cosmology. It also does not incorporate neutrino oscillations (and their non-zero masses).

The development of the Standard Model was driven by theoretical and experimental particle physicists alike. For theorists, the Standard Model is a paradigm of a quantum field theory, which exhibits a wide range of physics including spontaneous symmetry breaking, anomalies, non-perturbative behavior, etc. It is used as a basis for building more exotic models that incorporate hypothetical particles, extra dimensions, and elaborate symmetries (such as supersymmetry) in an attempt to explain experimental results at variance with the Standard Model, such as the existence of dark matter and neutrino oscillations.

4.1 Historical background

The first step towards the Standard Model was Sheldon Glashow's discovery in 1961 of a way to combine the electromagnetic and weak interactions.[4] In 1967 Steven Weinberg[5] and Abdus Salam[6] incorporated the Higgs mechanism[7][8][9] into Glashow's electroweak theory, giving it its modern form.

The Higgs mechanism is believed to give rise to the masses of all the elementary particles in the Standard Model. This includes the masses of the W and Z bosons, and the masses of the fermions, i.e. the quarks and leptons.

After the neutral weak currents caused by Z boson exchange were discovered at CERN in 1973,[10][11][12][13] the electroweak theory became widely accepted and Glashow, Salam, and Weinberg shared the 1979 Nobel Prize in Physics for discovering it. The W and Z bosons were discovered experimentally in 1981, and their masses were found to be as the Standard Model predicted.

The theory of the strong interaction, to which many contributed, acquired its modern form around 1973–74, when experiments confirmed that the hadrons were composed of fractionally charged quarks.

4.2 Overview

At present, matter and energy are best understood in terms of the kinematics and interactions of elementary particles. To date, physics has reduced the laws governing the behavior and interaction of all known forms of matter and energy to a small set of fundamental laws and theories. A major goal of physics is to find the "common ground" that would unite all of these theories into one integrated theory of everything, of which all the other known laws would be special cases, and from which the behavior of all matter and energy could be derived (at least in principle).[14]

4.3 Particle content

The Standard Model includes members of several classes of elementary particles (fermions, gauge bosons, and the Higgs boson), which in turn can be distinguished by other characteristics, such as color charge.

4.3.1 Fermions

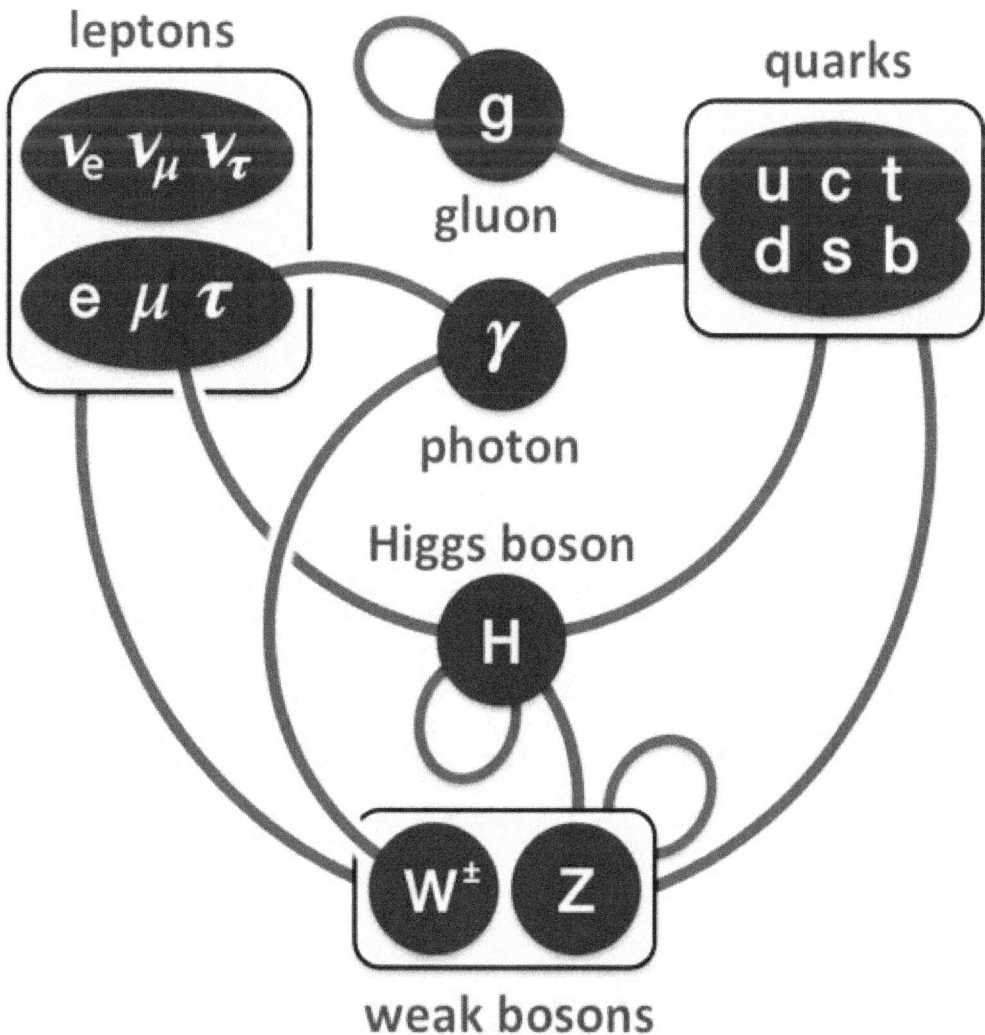

Summary of interactions between particles described by the Standard Model.

The Standard Model includes 12 elementary particles of spin-½ known as fermions. According to the spin-statistics theorem, fermions respect the Pauli exclusion principle. Each fermion has a corresponding antiparticle.

The fermions of the Standard Model are classified according to how they interact (or equivalently, by what charges they carry). There are six quarks (up, down, charm, strange, top, bottom), and six leptons (electron, electron neutrino, muon, muon neutrino, tau, tau neutrino). Pairs from each classification are grouped together to form a generation, with corresponding particles exhibiting similar physical behavior (see table).

The defining property of the quarks is that they carry color charge, and hence, interact via the strong interaction. A phenomenon called color confinement results in quarks being very strongly bound to one another, forming color-neutral composite particles (hadrons) containing either a quark and an antiquark (mesons) or three quarks (baryons). The familiar proton and the neutron are the two baryons having the smallest mass. Quarks also carry electric charge and weak isospin. Hence they interact with other fermions both electromagnetically and via the weak interaction.

The remaining six fermions do not carry colour charge and are called leptons. The three neutrinos do not carry electric charge either, so their motion is directly influenced only by the weak nuclear force, which makes them notoriously difficult to detect. However, by virtue of carrying an electric charge, the electron, muon, and tau all interact electromagnetically.

Each member of a generation has greater mass than the corresponding particles of lower generations. The first generation charged particles do not decay; hence all ordinary (baryonic) matter is made of such particles. Specifically, all atoms consist of electrons orbiting around atomic nuclei, ultimately constituted of up and down quarks. Second and third generation charged particles, on the other hand, decay with very short half lives, and are observed only in very high-energy environments. Neutrinos of all generations also do not decay, and pervade the universe, but rarely interact with baryonic matter.

4.3.2 Gauge bosons

In the Standard Model, gauge bosons are defined as force carriers that mediate the strong, weak, and electromagnetic fundamental interactions.

Interactions in physics are the ways that particles influence other particles. At a macroscopic level, electromagnetism allows particles to interact with one another via electric and magnetic fields, and gravitation allows particles with mass to attract one another in accordance with Einstein's theory of general relativity. The Standard Model explains such forces as resulting from matter particles exchanging other particles, generally referred to as *force mediating particles*. When a force-mediating particle is exchanged, at a macroscopic level the effect is equivalent to a force influencing both of them, and the particle is therefore said to have *mediated* (i.e., been the agent of) that force. The Feynman diagram calculations, which are a graphical representation of the perturbation theory approximation, invoke "force mediating particles", and when applied to analyze high-energy scattering experiments are in reasonable agreement with the data. However, perturbation theory (and with it the concept of a "force-mediating particle") fails in other situations. These include low-energy quantum chromodynamics, bound states, and solitons.

The gauge bosons of the Standard Model all have spin (as do matter particles). The value of the spin is 1, making them bosons. As a result, they do not follow the Pauli exclusion principle that constrains fermions: thus bosons (e.g. photons) do not have a theoretical limit on their spatial density (number per volume). The different types of gauge bosons are described below.

- Photons mediate the electromagnetic force between electrically charged particles. The photon is massless and is well-described by the theory of quantum electrodynamics.

- The W+, W−, and Z gauge bosons mediate the weak interactions between particles of different flavors (all quarks and leptons). They are massive, with the Z being more massive than the W±. The weak interactions involving the W± exclusively act on *left-handed* particles and *right-handed* antiparticles. Furthermore, the W± carries an electric charge of +1 and −1 and couples to the electromagnetic interaction. The electrically neutral Z boson interacts with both left-handed particles and antiparticles. These three gauge bosons along with the photons are grouped together, as collectively mediating the electroweak interaction.

- The eight gluons mediate the strong interactions between color charged particles (the quarks). Gluons are massless. The eightfold multiplicity of gluons is labeled by a combination of color and anticolor charge (e.g. red–antigreen).[nb 1] Because the gluons have an effective color charge, they can also interact among themselves. The gluons and their interactions are described by the theory of quantum chromodynamics.

The interactions between all the particles described by the Standard Model are summarized by the diagrams on the right of this section.

4.3.3 Higgs boson

Main article: Higgs boson

Standard Model Interactions
(Forces Mediated by Gauge Bosons)

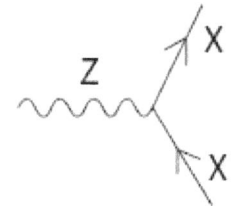

X is any fermion in
the Standard Model.

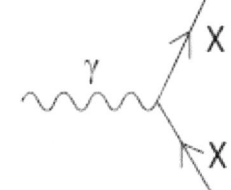

X is electrically charged.

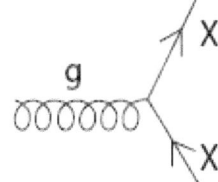

X is any quark.

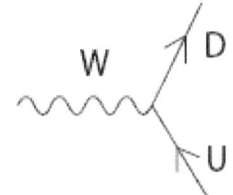

U is a up-type quark;
D is a down-type quark.

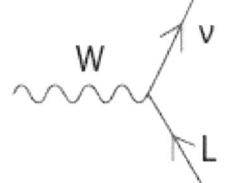

L is a lepton and ν is the
corresponding neutrino.

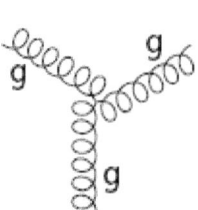

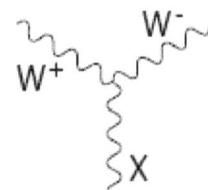

X is a photon or Z-boson.

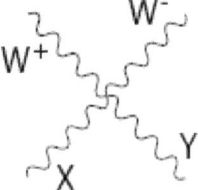

X and Y are any two
electroweak bosons such
that charge is conserved.

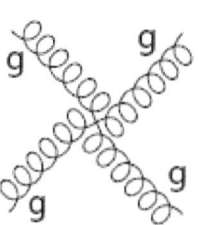

The above interactions form the basis of the standard model. Feynman diagrams in the standard model are built from these vertices. Modifications involving Higgs boson interactions and neutrino oscillations are omitted. The charge of the W bosons is dictated by the fermions they interact with; the conjugate of each listed vertex (i.e. reversing the direction of arrows) is also allowed.

The Higgs particle is a massive scalar elementary particle theorized by Robert Brout, François Englert, Peter Higgs, Gerald Guralnik, C. R. Hagen, and Tom Kibble in 1964 (see 1964 PRL symmetry breaking papers) and is a key building block in the Standard Model.[7][8][9][15] It has no intrinsic spin, and for that reason is classified as a boson (like the gauge bosons, which have integer spin).

The Higgs boson plays a unique role in the Standard Model, by explaining why the other elementary particles, except the photon and gluon, are massive. In particular, the Higgs boson explains why the photon has no mass, while the W and Z bosons are very heavy. Elementary particle masses, and the differences between electromagnetism (mediated by the photon) and the weak force (mediated by the W and Z bosons), are critical to many aspects of the structure of microscopic (and hence macroscopic) matter. In electroweak theory, the Higgs boson generates the masses of the leptons (electron, muon, and tau) and quarks. As the Higgs boson is massive, it must interact with itself.

Because the Higgs boson is a very massive particle and also decays almost immediately when created, only a very high-energy particle accelerator can observe and record it. Experiments to confirm and determine the nature of the Higgs boson using the Large Hadron Collider (LHC) at CERN began in early 2010, and were performed at Fermilab's Tevatron until its closure in late 2011. Mathematical consistency of the Standard Model requires that any mechanism capable of generating the masses of elementary particles become visible at energies above 1.4 TeV;[16] therefore, the LHC (designed to collide two 7 to 8 TeV proton beams) was built to answer the question of whether the Higgs boson actually exists.[17]

On 4 July 2012, the two main experiments at the LHC (ATLAS and CMS) both reported independently that they found a new particle with a mass of about 125 GeV/c^2 (about 133 proton masses, on the order of 10^{-25} kg), which is "consistent with the Higgs boson." Although it has several properties similar to the predicted "simplest" Higgs,[18] they acknowledged that further work would be needed to conclude that it is indeed the Higgs boson, and exactly which version of the Standard Model Higgs is best supported if confirmed.[19][20][21][22][23]

On 14 March 2013 the Higgs Boson was tentatively confirmed to exist.[24]

4.3.4   Total particle count

Counting particles by a rule that distinguishes between particles and their corresponding antiparticles, and among the many color states of quarks and gluons, gives a total of 61 elementary particles.[25]

4.4   Theoretical aspects

Main article: Standard Model (mathematical formulation)

4.4.1   Construction of the Standard Model Lagrangian

Technically, quantum field theory provides the mathematical framework for the Standard Model, in which a Lagrangian controls the dynamics and kinematics of the theory. Each kind of particle is described in terms of a dynamical field that pervades space-time. The construction of the Standard Model proceeds following the modern method of constructing most field theories: by first postulating a set of symmetries of the system, and then by writing down the most general renormalizable Lagrangian from its particle (field) content that observes these symmetries.

The global Poincaré symmetry is postulated for all relativistic quantum field theories. It consists of the familiar translational symmetry, rotational symmetry and the inertial reference frame invariance central to the theory of special relativity. The local SU(3)×SU(2)×U(1) gauge symmetry is an internal symmetry that essentially defines the Standard Model. Roughly, the three factors of the gauge symmetry give rise to the three fundamental interactions. The fields fall into different representations of the various symmetry groups of the Standard Model (see table). Upon writing the most general Lagrangian, one finds that the dynamics depend on 19 parameters, whose numerical values are established by experiment. The parameters are summarized in the table above (note: with the Higgs mass is at 125 GeV, the Higgs self-coupling strength $\lambda \sim 1/8$).

Quantum chromodynamics sector

Main article: Quantum chromodynamics

The quantum chromodynamics (QCD) sector defines the interactions between quarks and gluons, with SU(3) symmetry, generated by T^a. Since leptons do not interact with gluons, they are not affected by this sector. The Dirac Lagrangian of the quarks coupled to the gluon fields is given by

$$\mathcal{L}_{QCD} = i\overline{U}(\partial_\mu - ig_s G_\mu^a T^a)\gamma^\mu U + i\overline{D}(\partial_\mu - ig_s G_\mu^a T^a)\gamma^\mu D.$$

G_μ^a is the SU(3) gauge field containing the gluons, γ^μ are the Dirac matrices, D and U are the Dirac spinors associated with up- and down-type quarks, and g_s is the strong coupling constant.

Electroweak sector

Main article: Electroweak interaction

The electroweak sector is a Yang–Mills gauge theory with the simple symmetry group U(1)×SU(2)L,

$$\mathcal{L}_{\text{EW}} = \sum_\psi \bar\psi \gamma^\mu \left(i\partial_\mu - g'\frac{1}{2} Y_{\text{W}} B_\mu - g\frac{1}{2}\vec\tau_{\text{L}}\vec W_\mu \right) \psi$$

where $B\mu$ is the U(1) gauge field; YW is the weak hypercharge—the generator of the U(1) group; $\vec W_\mu$ is the three-component SU(2) gauge field; $\vec\tau_{\text{L}}$ are the Pauli matrices—infinitesimal generators of the SU(2) group. The subscript L indicates that they only act on left fermions; g' and g are coupling constants.

Higgs sector

Main article: Higgs mechanism

In the Standard Model, the Higgs field is a complex scalar of the group SU(2)L:

$$\varphi = \frac{1}{\sqrt 2}\left(\begin{array}{c} \varphi^+ \\ \varphi^0 \end{array} \right),$$

where the indices + and 0 indicate the electric charge (Q) of the components. The weak isospin (YW) of both components is 1.

Before symmetry breaking, the Higgs Lagrangian is:

$$\mathcal{L}_{\text{H}} = \varphi^\dagger \left(\partial^\mu - \frac{i}{2}\left(g'Y_{\text{W}} B^\mu + g\vec\tau\vec W^\mu \right) \right)\left(\partial_\mu + \frac{i}{2}\left(g'Y_{\text{W}} B_\mu + g\vec\tau\vec W_\mu \right) \right)\varphi - \frac{\lambda^2}{4}\left(\varphi^\dagger\varphi - v^2 \right)^2,$$

which can also be written as:

$$\mathcal{L}_{\text{H}} = \left| \left(\partial_\mu + \frac{i}{2}\left(g'Y_{\text{W}} B_\mu + g\vec\tau\vec W_\mu \right) \right)\varphi \right|^2 - \frac{\lambda^2}{4}\left(\varphi^\dagger\varphi - v^2 \right)^2.$$

4.5 Fundamental forces

Main article: Fundamental interaction

The Standard Model classified all four fundamental forces in nature. In the Standard Model, a force is described as an exchange of bosons between the objects affected, such as a photon for the electromagnetic force and a gluon for the strong interaction. Those particles are called force carriers.[26]

4.6 Tests and predictions

The Standard Model (SM) predicted the existence of the W and Z bosons, gluon, and the top and charm quarks before these particles were observed. Their predicted properties were experimentally confirmed with good precision. To give an idea of the success of the SM, the following table compares the measured masses of the W and Z bosons with the masses predicted by the SM:

The SM also makes several predictions about the decay of Z bosons, which have been experimentally confirmed by the Large Electron-Positron Collider at CERN.

In May 2012 BaBar Collaboration reported that their recently analyzed data may suggest possible flaws in the Standard Model of particle physics.[28][29] These data show that a particular type of particle decay called "B to D-star-tau-nu" happens more often than the Standard Model says it should. In this type of decay, a particle called the B-bar meson decays into a D meson, an antineutrino and a tau-lepton. While the level of certainty of the excess (3.4 sigma) is not enough to claim a break from the Standard Model, the results are a potential sign of something amiss and are likely to impact existing theories, including those attempting to deduce the properties of Higgs bosons.[30]

On December 13, 2012, physicists reported the constancy, over space and time, of a basic physical constant of nature that supports the *standard model of physics*. The scientists, studying methanol molecules in a distant galaxy, found the change ($\Delta\mu/\mu$) in the proton-to-electron mass ratio μ to be equal to "$(0.0 \pm 1.0) \times 10^{-7}$ at redshift z = 0.89" and consistent with "a null result".[31][32]

4.7 Challenges

See also: Physics beyond the Standard Model

Self-consistency of the Standard Model (currently formulated as a non-abelian gauge theory quantized through path-integrals) has not been mathematically proven. While regularized versions useful for approximate computations (for example lattice gauge theory) exist, it is not known whether they converge (in the sense of S-matrix elements) in the limit that the regulator is removed. A key question related to the consistency is the Yang–Mills existence and mass gap problem.

Experiments indicate that neutrinos have mass, which the classic Standard Model did not allow.[33] To accommodate this finding, the classic Standard Model can be modified to include neutrino mass.

If one insists on using only Standard Model particles, this can be achieved by adding a non-renormalizable interaction of leptons with the Higgs boson.[34] On a fundamental level, such an interaction emerges in the seesaw mechanism where heavy right-handed neutrinos are added to the theory. This is natural in the left-right symmetric extension of the Standard Model[35][36] and in certain grand unified theories.[37] As long as new physics appears below or around 10^{14} GeV, the neutrino masses can be of the right order of magnitude.

Theoretical and experimental research has attempted to extend the Standard Model into a Unified field theory or a Theory of everything, a complete theory explaining all physical phenomena including constants. Inadequacies of the Standard Model that motivate such research include:

- It does not attempt to explain gravitation, although a theoretical particle known as a graviton would help explain it, and unlike for the strong and electroweak interactions of the Standard Model, there is no known way of describing general relativity, the canonical theory of gravitation, consistently in terms of quantum field theory. The reason for this is, among other things, that quantum field theories of gravity generally break down before reaching the Planck scale. As a consequence, we have no reliable theory for the very early universe;

- Some consider it to be *ad hoc* and inelegant, requiring 19 numerical constants whose values are unrelated and arbitrary. Although the Standard Model, as it now stands, can explain why neutrinos have masses, the specifics of neutrino mass are still unclear. It is believed that explaining neutrino mass will require an additional 7 or 8 constants, which are also arbitrary parameters;

- The Higgs mechanism gives rise to the hierarchy problem if some new physics (coupled to the Higgs) is present at high energy scales. In these cases in order for the weak scale to be much smaller than the Planck scale, severe fine tuning of the parameters is required; there are, however, other scenarios that include quantum gravity in which such fine tuning can be avoided.[38] There are also issues of Quantum triviality, which suggests that it may not be possible to create a consistent quantum field theory involving elementary scalar particles.

- It should be modified so as to be consistent with the emerging "Standard Model of cosmology." In particular, the Standard Model cannot explain the observed amount of cold dark matter (CDM) and gives contributions to dark energy which are many orders of magnitude too large. It is also difficult to accommodate the observed predominance of matter over antimatter (matter/antimatter asymmetry). The isotropy and homogeneity of the visible universe over large distances seems to require a mechanism like cosmic inflation, which would also constitute an extension of the Standard Model.

- The existence of ultra-high-energy cosmic rays are difficult to explain under the Standard Model.

Currently, no proposed Theory of Everything has been widely accepted or verified.

4.8   See also

- Fundamental interaction:

  - Quantum electrodynamics

  - Strong interaction: Color charge, Quantum chromodynamics, Quark model

  - Weak interaction: Electroweak theory, Fermi theory of beta decay, Weak hypercharge, Weak isospin

- Gauge theory: Nontechnical introduction to gauge theory

- Generation

- Higgs mechanism: Higgs boson, Higgsless model

- J. C. Ward

- J. J. Sakurai Prize for Theoretical Particle Physics

- Lagrangian

- Open questions: BTeV experiment, CP violation, Neutrino masses, Quark matter, Quantum triviality

- Penguin diagram

- Quantum field theory

- Standard Model: Mathematical formulation of, Physics beyond the Standard Model

4.9   Notes and references

[1] Technically, there are nine such color–anticolor combinations. However, there is one color-symmetric combination that can be constructed out of a linear superposition of the nine combinations, reducing the count to eight.

4.10 References

[1] R. Oerter (2006). *The Theory of Almost Everything: The Standard Model, the Unsung Triumph of Modern Physics* (Kindle ed.). Penguin Group. p. 2. ISBN 0-13-236678-9.

[2] In fact, there are mathematical issues regarding quantum field theories still under debate (see e.g. Landau pole), but the predictions extracted from the Standard Model by current methods applicable to current experiments are all self-consistent. For a further discussion see e.g. Chapter 25 of R. Mann (2010). *An Introduction to Particle Physics and the Standard Model.* CRC Press. ISBN 978-1-4200-8298-2.

[3] Sean Carroll, Ph.D., Cal Tech, 2007, The Teaching Company, *Dark Matter, Dark Energy: The Dark Side of the Universe*, Guidebook Part 2 page 59, Accessed Oct. 7, 2013, "...Standard Model of Particle Physics: The modern theory of elementary particles and their interactions ... It does not, strictly speaking, include gravity, although it's often convenient to include gravitons among the known particles of nature..."

[4] S.L.Glashow(1961). "Partial-symmetries of weak interactions".*Nuclear Physics***22**(4): 579–588. Bibcode:1961NucPh..22... doi:10.1016/0029-5582(61)90469-2.

[5] S. Weinberg (1967). "A Model of Leptons". *Physical Review Letters* **19** (21): 1264–1266. Bibcode:1967PhRvL..19.1264W. doi:10.1103/PhysRevLett.19.1264.

[6] A. Salam (1968). N. Svartholm, ed. *Elementary Particle Physics: Relativistic Groups and Analyticity.* Eighth Nobel Symposium. Stockholm: Almquvist and Wiksell. p. 367.

[7] F. Englert, R. Brout (1964). "Broken Symmetry and the Mass of Gauge Vector Mesons". *Physical Review Letters* **13** (9): 321–323. Bibcode:1964PhRvL..13..321E. doi:10.1103/PhysRevLett.13.321.

[8] P.W. Higgs (1964). "Broken Symmetries and the Masses of Gauge Bosons". *Physical Review Letters* **13** (16): 508–509. Bibcode:1964PhRvL..13..508H. doi:10.1103/PhysRevLett.13.508.

[9] G.S. Guralnik, C.R. Hagen, T.W.B. Kibble (1964). "Global Conservation Laws and Massless Particles". *Physical Review Letters* **13** (20): 585–587. Bibcode:1964PhRvL..13..585G. doi:10.1103/PhysRevLett.13.585.

[10] F.J.Hasert;et al. (1973). "Search for elastic muon-neutrino electron scattering".*Physics Letters B***46**(1): 121. Bibcode:1973PhLB. doi:10.1016/0370-2693(73)90494-2.

[11] F.J. Hasert; et al. (1973). "Observation of neutrino-like interactions without muon or electron in the Gargamelle neutrino experiment". *Physics Letters B* **46** (1): 138. Bibcode:1973PhLB...46..138H. doi:10.1016/0370-2693(73)90499-1.

[12] F.J. Hasert; et al. (1974). "Observation of neutrino-like interactions without muon or electron in the Gargamelle neutrino experiment". *Nuclear Physics B* **73** (1): 1. Bibcode:1974NuPhB..73....1H. doi:10.1016/0550-3213(74)90038-8.

[13] D. Haidt (4 October 2004). "The discovery of the weak neutral currents". *CERN Courier.* Retrieved 8 May 2008.

[14] "Details can be worked out if the situation is simple enough for us to make an approximation, which is almost never, but often we can understand more or less what is happening." from *The Feynman Lectures on Physics*, Vol 1. pp. 2–7

[15] G.S. Guralnik (2009). "The History of the Guralnik, Hagen and Kibble development of the Theory of Spontaneous Symmetry Breaking and Gauge Particles". *International Journal of Modern Physics A* **24** (14): 2601–2627. arXiv:0907.3466. Bibcode:2009IJMPA..24.2601G. doi:10.1142/S0217751X09045431.

[16] B.W. Lee, C. Quigg, H.B. Thacker (1977). "Weak interactions at very high energies: The role of the Higgs-boson mass". *Physical Review D* **16** (5): 1519–1531. Bibcode:1977PhRvD..16.1519L. doi:10.1103/PhysRevD.16.1519.

[17] "Huge $10 billion collider resumes hunt for 'God particle'". CNN. 11 November 2009. Retrieved 2010-05-04.

[18] M. Strassler (10 July 2012). "Higgs Discovery: Is it a Higgs?". Retrieved 2013-08-06.

[19] "CERN experiments observe particle consistent with long-sought Higgs boson". CERN. 4 July 2012. Retrieved 2012-07-04.

[20] "Observation of a New Particle with a Mass of 125 GeV". CERN. 4 July 2012. Retrieved 2012-07-05.

[21] "ATLAS Experiment". ATLAS. 1 January 2006. Retrieved 2012-07-05.

[22] "Confirmed: CERN discovers new particle likely to be the Higgs boson". *YouTube*. Russia Today. 4 July 2012. Retrieved 2013-08-06.

[23] D. Overbye (4 July 2012). "A New Particle Could Be Physics' Holy Grail". *New York Times*. Retrieved 2012-07-04.

[24] "New results indicate that new particle is a Higgs boson". CERN. 14 March 2013. Retrieved 2013-08-06.

[25] S. Braibant, G. Giacomelli, M. Spurio (2009). *Particles and Fundamental Interactions: An Introduction to Particle Physics*. Springer. pp. 313–314. ISBN 978-94-007-2463-1.

[26] http://home.web.cern.ch/about/physics/standard-model Official CERN website

[27] http://www.pha.jhu.edu/~{}dfehling/particle.gif

[28] "BABAR Data in Tension with the Standard Model". SLAC. 31 May 2012. Retrieved 2013-08-06.

[29] BaBar Collaboration (2012). "Evidence for an excess of $B \rightarrow D^{(*)} \tau^- \nu\tau$ decays". *Physical Review Letters* **109** (10): 101802. arXiv:1205.5442. Bibcode:2012PhRvL.109j1802L. doi:10.1103/PhysRevLett.109.101802.

[30] "BaBar data hint at cracks in the Standard Model". *e! Science News*. 18 June 2012. Retrieved 2013-08-06.

[31] J. Bagdonaite; et al. (2012). "A Stringent Limit on a Drifting Proton-to-Electron Mass Ratio from Alcohol in the Early Universe". *Science* **339** (6115): 46. Bibcode:2013Sci...339...46B. doi:10.1126/science.1224898.

[32] C. Moskowitz (13 December 2012). "Phew! Universe's Constant Has Stayed Constant". Space.com. Retrieved 2012-12-14.

[33] "Particle chameleon caught in the act of changing". CERN. 31 May 2010. Retrieved 2012-07-05.

[34] S. Weinberg (1979). "Baryon and Lepton Nonconserving Processes". *Physical Review Letters* **43** (21): 1566. Bibcode:1979PhR. doi:10.1103/PhysRevLett.43.1566.

[35] P. Minkowski (1977). "$\mu \rightarrow e\gamma$ at a Rate of One Out of 10_9 Muon Decays?". doi:10.1016/0370-2693(77)90435-X.

[36] R. N. Mohapatra, G. Senjanovic (1980). "Neutrino Mass and Spontaneous Parity Nonconservation". *Physical Review Letters* **44** (14): 912–915. Bibcode:1980PhRvL..44..912M. doi:10.1103/PhysRevLett.44.912.

[37] M. Gell-Mann, P. Ramond and R. Slansky (1979). F. van Nieuwenhuizen and D. Z. Freedman, ed. *Supergravity*. North Holland. pp. 315–321. ISBN 0-444-85438-X.

[38] Salvio, Strumia (2014-03-17). "Agravity". *JHEP1406(2014)080*. arXiv:1403.4226. Bibcode:2014JHEP...06..080S.doi)080.

4.11   Further reading

- R. Oerter (2006). *The Theory of Almost Everything: The Standard Model, the Unsung Triumph of Modern Physics*. Plume.

- B.A. Schumm (2004). *Deep Down Things: The Breathtaking Beauty of Particle Physics*. Johns Hopkins University Press. ISBN 0-8018-7971-X.

- "The Standard Model of Particle Physics Interactive Graphic".

Introductory textbooks

- I. Aitchison, A. Hey (2003). *Gauge Theories in Particle Physics: A Practical Introduction*. Institute of Physics. ISBN 978-0-585-44550-2.

- W. Greiner, B. Müller (2000). *Gauge Theory of Weak Interactions*. Springer. ISBN 3-540-67672-4.

- G.D. Coughlan, J.E. Dodd, B.M. Gripaios (2006). *The Ideas of Particle Physics: An Introduction for Scientists*. Cambridge University Press.

- D.J. Griffiths (1987). *Introduction to Elementary Particles*. John Wiley & Sons. ISBN 0-471-60386-4.

- G.L. Kane (1987). *Modern Elementary Particle Physics*. Perseus Books. ISBN 0-201-11749-5.

Advanced textbooks

- T.P. Cheng, L.F. Li (2006). *Gauge theory of elementary particle physics*. Oxford University Press. ISBN 0-19-851961-3. Highlights the gauge theory aspects of the Standard Model.

- J.F. Donoghue, E. Golowich, B.R. Holstein (1994). *Dynamics of the Standard Model*. Cambridge University Press. ISBN 978-0-521-47652-2. Highlights dynamical and phenomenological aspects of the Standard Model.

- L. O'Raifeartaigh (1988). *Group structure of gauge theories*. Cambridge University Press. ISBN 0-521-34785-8.

- Nagashima Y. Elementary Particle Physics: Foundations of the Standard Model, Volume 2. (Wiley 2013) 920 рапуы

- Schwartz, M.D. Quantum Field Theory and the Standard Model (Cambridge University Press 2013) 952 pages

- Langacker P. The standard model and beyond. (CRC Press, 2010) 670 pages Highlights group-theoretical aspects of the Standard Model.

Journal articles

- E.S.Abers,B.W.Lee(1973). "Gauge theories".*Physics Reports***9**: 1–141. Bibcode:1973PhR.....9....1A.doi:10.1 1573(73)90027-6.

- M. Baak; et al. (2012). "The Electroweak Fit of the Standard Model after the Discovery of a New Boson at the LHC". *The European Physical Journal C* **72** (11). arXiv:1209.2716. Bibcode:2012EPJC...72.2205B. doi:10.1140/epjc/s10052-012-2205-9.

- Y. Hayato; et al. (1999). "Search for Proton Decay through $p \rightarrow \nu K^+$ in a Large Water Cherenkov Detector". *Physical Review Letters***83**(8): 1529. arXiv:hep-ex/9904020. Bibcode:1999PhRvL..83.1529H.doi:10.1103/Phy.

- S.F. Novaes (2000). "Standard Model: An Introduction". arXiv:hep-ph/0001283 [hep-ph].

- D.P. Roy (1999). "Basic Constituents of Matter and their Interactions — A Progress Report". arXiv:hep-ph/9912523 [hep-ph].

- F. Wilczek (2004). "The Universe Is A Strange Place". *Nuclear Physics B - Proceedings Supplements* **134**: 3. arXiv:astro-ph/0401347. Bibcode:2004NuPhS.134....3W. doi:10.1016/j.nuclphysbps.2004.08.001.

4.12 External links

- "The Standard Model explained in Detail by CERN's John Ellis" omega tau podcast.

- "LHC sees hint of lightweight Higgs boson" "New Scientist".

- "Standard Model may be found incomplete," *New Scientist*.

- "Observation of the Top Quark" at Fermilab.

- "The Standard Model Lagrangian." After electroweak symmetry breaking, with no explicit Higgs boson.

- "Standard Model Lagrangian" with explicit Higgs terms. PDF, PostScript, and LaTeX versions.

- "The particle adventure." Web tutorial.

- Nobes, Matthew (2002) "Introduction to the Standard Model of Particle Physics" on Kuro5hin: Part 1, Part 2, Part 3a, Part 3b.

- "The Standard Model" The Standard Model on the CERN web site explains how the basic building blocks of matter interact, governed by four fundamental forces.

Chapter 5

List of particles

This is a list of the different types of particles found or believed to exist in the whole of the universe. For individual lists of the different particles, see the list below.

5.1 Elementary particles

Main article: Elementary particle

Elementary particles are particles with no measurable internal structure; that is, they are not composed of other particles. They are the fundamental objects of quantum field theory. Many families and sub-families of elementary particles exist. Elementary particles are classified according to their spin. Fermions have half-integer spin while bosons have integer spin. All the particles of the Standard Model have been experimentally observed, recently including the Higgs boson.[1][2]

5.1.1 Fermions

Main article: Fermion

Fermions are one of the two fundamental classes of particles, the other being bosons. Fermion particles are described by Fermi–Dirac statistics and have quantum numbers described by the Pauli exclusion principle. They include the quarks and leptons, as well as any composite particles consisting of an odd number of these, such as all baryons and many atoms and nuclei.

Fermions have half-integer spin; for all known elementary fermions this is $\frac{1}{2}$. All known fermions, except neutrinos, are also Dirac fermions; that is, each known fermion has its own distinct antiparticle. It is not known whether the neutrino is a Dirac fermion or a Majorana fermion.[3] Fermions are the basic building blocks of all matter. They are classified according to whether they interact via the color force or not. In the Standard Model, there are 12 types of elementary fermions: six quarks and six leptons.

Quarks

Main article: Quark

Quarks are the fundamental constituents of hadrons and interact via the strong interaction. Quarks are the only known carriers of fractional charge, but because they combine in groups of three (baryons) or in groups of two with antiquarks (mesons), only integer charge is observed in nature. Their respective antiparticles are the antiquarks, which are identical except for the fact that they carry the opposite electric charge (for example the up quark carries charge $+\frac{2}{3}$, while the up antiquark carries charge $-\frac{2}{3}$), color charge, and baryon number. There are six flavors of quarks; the three positively charged quarks are called "up-type quarks" and the three negatively charged quarks are called "down-type quarks".

Leptons

Main article: Leptons

Leptons do not interact via the strong interaction. Their respective antiparticles are the antileptons which are identical, except for the fact that they carry the opposite electric charge and lepton number. The antiparticle of an electron is an antielectron, which is nearly always called a "positron" for historical reasons. There are six leptons in total; the three charged leptons are called "electron-like leptons", while the neutral leptons are called "neutrinos". Neutrinos are known to oscillate, so that neutrinos of definite flavor do not have definite mass, rather they exist in a superposition of mass eigenstates. The hypothetical heavy right-handed neutrino, called a "sterile neutrino", has been left off the list.

5.1.2   Bosons

Main article: Boson

Bosons are one of the two fundamental classes of particles, the other being fermions. Bosons are characterized by Bose–Einstein statistics and all have integer spins. Bosons may be either elementary, like photons and gluons, or composite, like mesons.

The fundamental forces of nature are mediated by gauge bosons, and mass is believed to be created by the Higgs field. According to the Standard Model the elementary bosons are:

The graviton is added to the list although it is not predicted by the Standard Model, but by other theories in the framework of quantum field theory. Furthermore, gravity is non-renormalizable. There are a total of eight independent gluons. The Higgs boson is postulated by the electroweak theory primarily to explain the origin of particle masses. In a process known as the "Higgs mechanism", the Higgs boson and the other gauge bosons in the Standard Model acquire mass via spontaneous symmetry breaking of the SU(2) gauge symmetry. The Minimal Supersymmetric Standard Model (MSSM) predicts several Higgs bosons. A new particle expected to be the Higgs boson was observed at the CERN/LHC on March 14, 2013, around the energy of 126.5GeV with an accuracy of close to five sigma (99.9999%, which is accepted as definitive). The Higgs mechanism giving mass to other particles has not been observed yet.

5.1.3   Hypothetical particles

Supersymmetric theories predict the existence of more particles, none of which have been confirmed experimentally as of 2014:

Note: just as the photon, Z boson and $W^{\pm}$ bosons are superpositions of the B^0, W^0, W^1, and W^2 fields – the photino, zino, and wino$^{\pm}$ are superpositions of the bino^0, wino^0, wino^1, and wino^2 by definition.

No matter if one uses the original gauginos or this superpositions as a basis, the only predicted physical particles are neutralinos and charginos as a superposition of them together with the Higgsinos.

Other theories predict the existence of additional bosons:

Mirror particles are predicted by theories that restore parity symmetry.

"Magnetic monopole" is a generic name for particles with non-zero magnetic charge. They are predicted by some GUTs.

"Tachyon" is a generic name for hypothetical particles that travel faster than the speed of light and have an imaginary rest mass.

Preons were suggested as subparticles of quarks and leptons, but modern collider experiments have all but ruled out their existence.

Kaluza–Klein towers of particles are predicted by some models of extra dimensions. The extra-dimensional momentum is manifested as extra mass in four-dimensional spacetime.

5.2   Composite particles

5.2.1   Hadrons

Main article: Hadron

Hadrons are defined as strongly interacting composite particles. Hadrons are either:

- Composite fermions, in which case they are called baryons.
- Composite bosons, in which case they are called mesons.

Quark models, first proposed in 1964 independently by Murray Gell-Mann and George Zweig (who called quarks "aces"), describe the known hadrons as composed of valence quarks and/or antiquarks, tightly bound by the color force, which is mediated by gluons. A "sea" of virtual quark-antiquark pairs is also present in each hadron.

Baryons

See also: List of baryons

Ordinary baryons (composite fermions) contain three valence quarks or three valence antiquarks each.

- Nucleons are the fermionic constituents of normal atomic nuclei:
    - Protons, composed of two up and one down quark (uud)
    - Neutrons, composed of two down and one up quark (ddu)
- Hyperons, such as the Λ, Σ, Ξ, and Ω particles, which contain one or more strange quarks, are short-lived and heavier than nucleons. Although not normally present in atomic nuclei, they can appear in short-lived hypernuclei.
- A number of charmed and bottom baryons have also been observed.

Some hints at the existence of exotic baryons have been found recently; however, negative results have also been reported. Their existence is uncertain.

- Pentaquarks consist of four valence quarks and one valence antiquark.

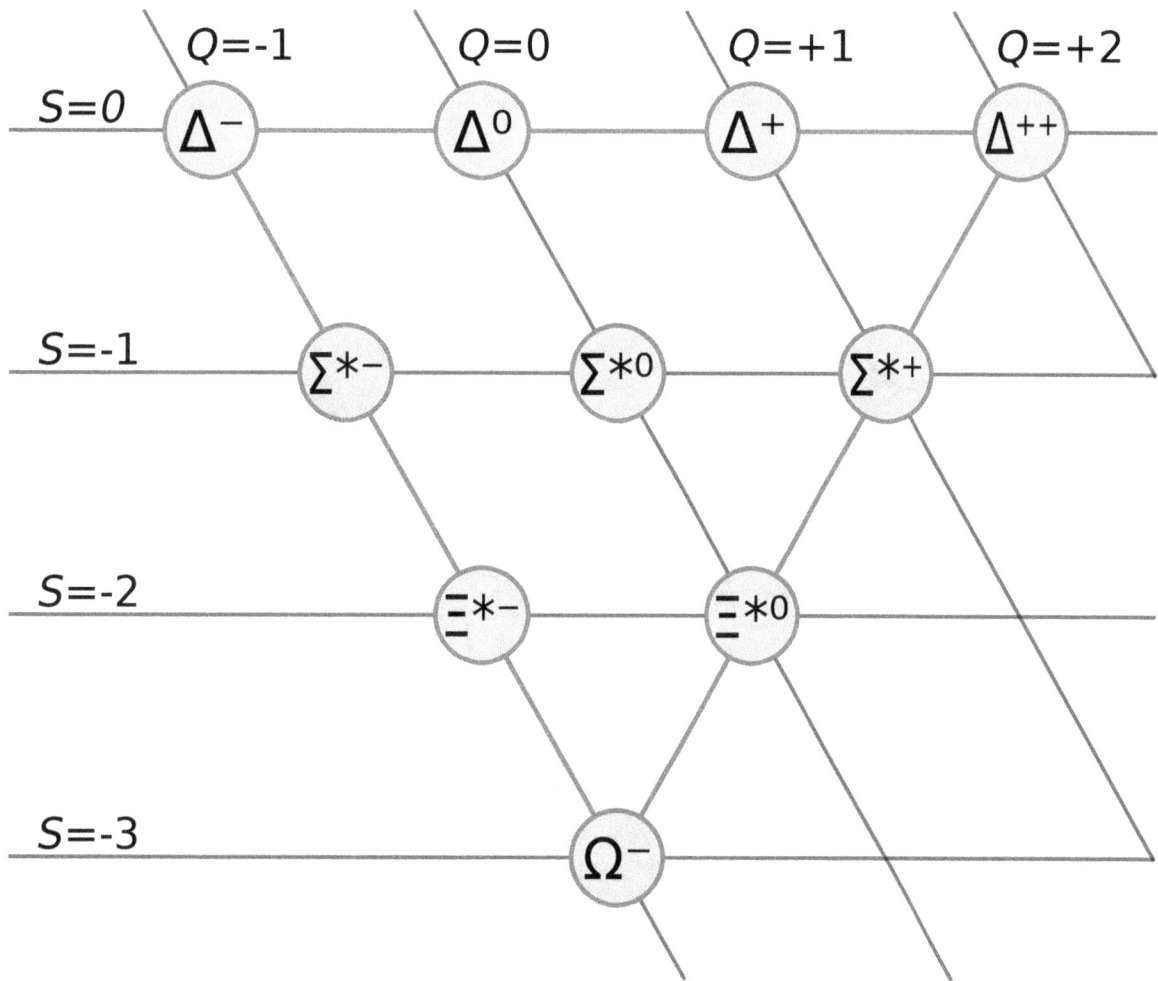

A combination of three u, d or s-quarks with a total spin of $^3\!/_2$ form the so-called "baryon decuplet".

Mesons

See also: List of mesons

Ordinary mesons are made up of a valence quark and a valence antiquark. Because mesons have spin of 0 or 1 and are not themselves elementary particles, they are "composite" bosons. Examples of mesons include the pion, kaon, and the J/ψ. In quantum hydrodynamic models, mesons mediate the residual strong force between nucleons.

At one time or another, positive signatures have been reported for all of the following exotic mesons but their existences have yet to be confirmed.

- A tetraquark consists of two valence quarks and two valence antiquarks;

- A glueball is a bound state of gluons with no valence quarks;

- Hybrid mesons consist of one or more valence quark-antiquark pairs and one or more real gluons.

5.2.2   Atomic nuclei

Atomic nuclei consist of protons and neutrons. Each type of nucleus contains a specific number of protons and a specific

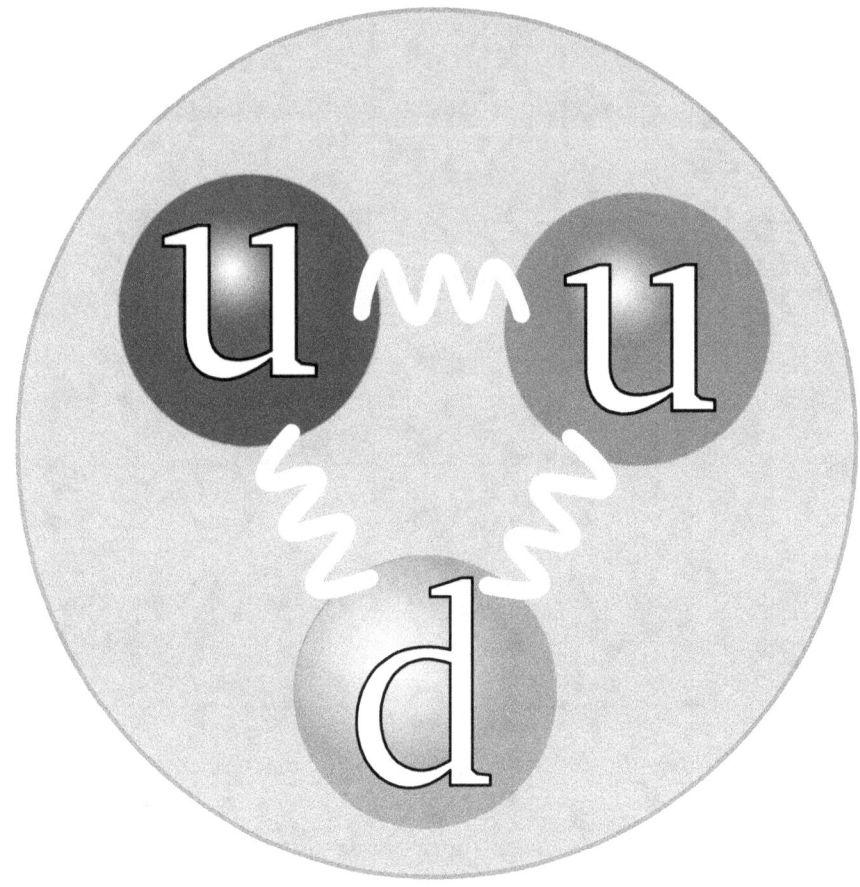

Proton quark structure: 2 up quarks and 1 down quark. The gluon tubes or flux tubes are now known to be Y shaped.

number of neutrons, and is called a "nuclide" or "isotope". Nuclear reactions can change one nuclide into another. See table of nuclides for a complete list of isotopes.

5.2.3 Atoms

Atoms are the smallest neutral particles into which matter can be divided by chemical reactions. An atom consists of a small, heavy nucleus surrounded by a relatively large, light cloud of electrons. Each type of atom corresponds to a specific chemical element. To date, 118 elements have been discovered, while only the elements 1-112,114, and 116 have received official names.

The atomic nucleus consists of protons and neutrons. Protons and neutrons are, in turn, made of quarks.

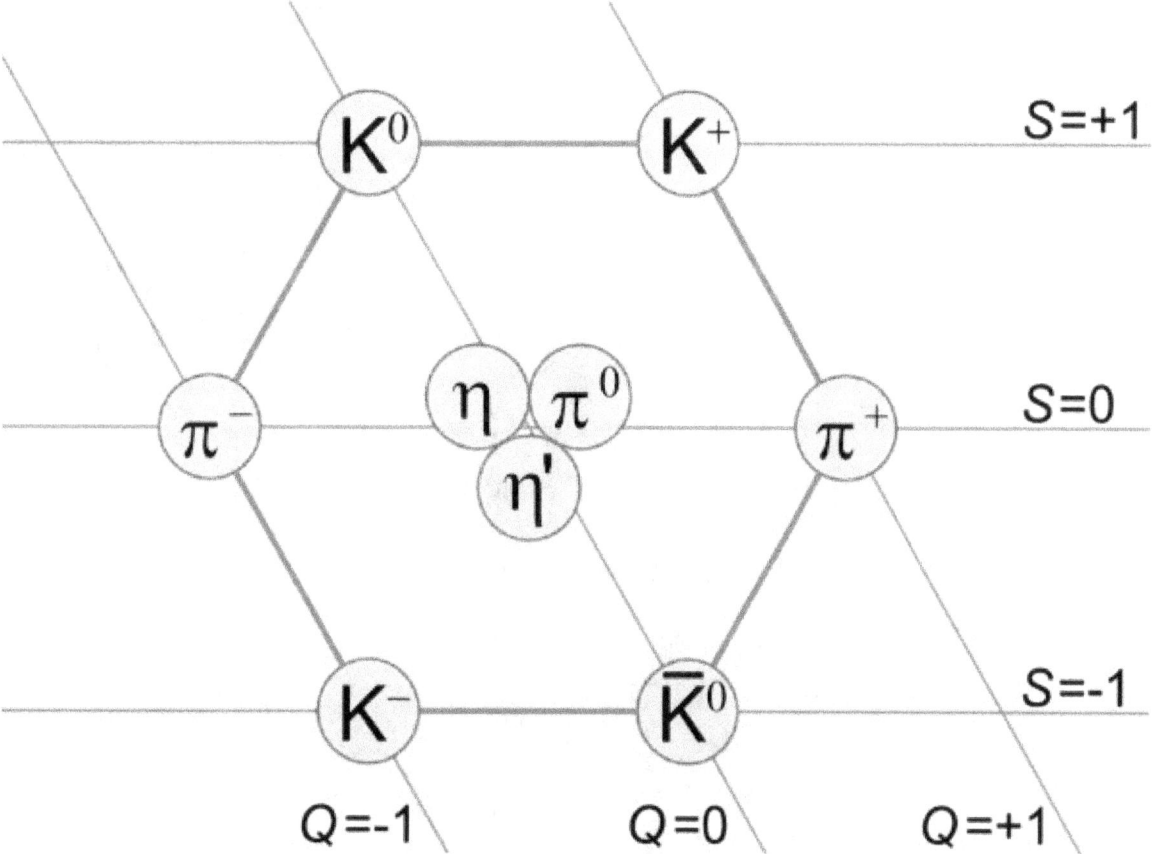

Mesons of spin 0 form a nonet

5.2.4 Molecules

Molecules are the smallest particles into which a non-elemental substance can be divided while maintaining the physical properties of the substance. Each type of molecule corresponds to a specific chemical compound. Molecules are a composite of two or more atoms. See list of compounds for a list of molecules.

5.3 Condensed matter

The field equations of condensed matter physics are remarkably similar to those of high energy particle physics. As a result, much of the theory of particle physics applies to condensed matter physics as well; in particular, there are a selection of field excitations, called quasi-particles, that can be created and explored. These include:

- Phonons are vibrational modes in a crystal lattice.

- Excitons are bound states of an electron and a hole.

- Plasmons are coherent excitations of a plasma.

- Polaritons are mixtures of photons with other quasi-particles.

- Polarons are moving, charged (quasi-) particles that are surrounded by ions in a material.

- Magnons are coherent excitations of electron spins in a material.

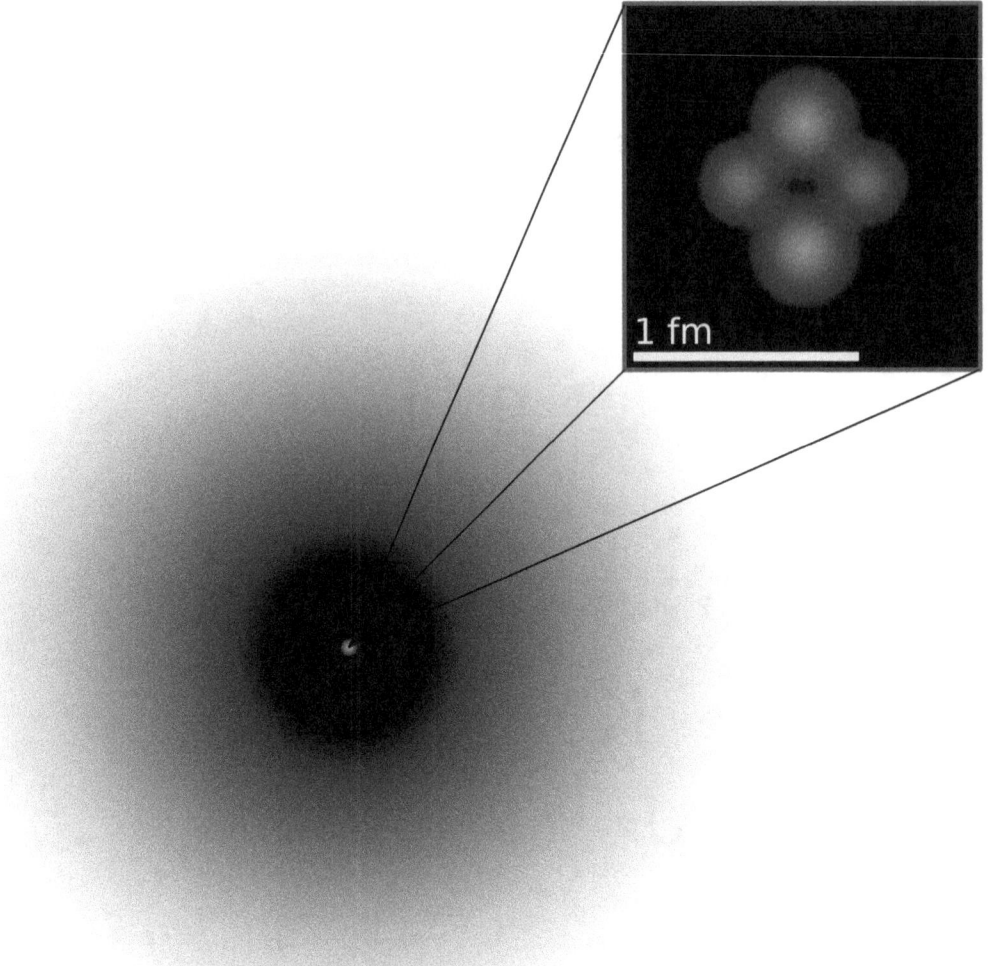

1 Å = 100,000 fm

A semi-accurate depiction of the helium atom. In the nucleus, the protons are in red and neutrons are in purple. In reality, the nucleus is also spherically symmetrical.

5.4 Other

- An anyon is a generalization of fermion and boson in two-dimensional systems like sheets of graphene that obeys braid statistics.

- A plekton is a theoretical kind of particle discussed as a generalization of the braid statistics of the anyon to dimension > 2.

- A WIMP (weakly interacting massive particle) is any one of a number of particles that might explain dark matter (such as the neutralino or the axion).

- The pomeron, used to explain the elastic scattering of hadrons and the location of Regge poles in Regge theory.

- The skyrmion, a topological solution of the pion field, used to model the low-energy properties of the nucleon, such as the axial vector current coupling and the mass.

- A genon is a particle existing in a closed timelike world line where spacetime is curled as in a Frank Tipler or Ronald Mallett time machine.

- A goldstone boson is a massless excitation of a field that has been spontaneously broken. The pions are quasi-goldstone bosons (quasi- because they are not exactly massless) of the broken chiral isospin symmetry of quantum chromodynamics.

- A goldstino is a goldstone fermion produced by the spontaneous breaking of supersymmetry.

- An instanton is a field configuration which is a local minimum of the Euclidean action. Instantons are used in nonperturbative calculations of tunneling rates.

- A dyon is a hypothetical particle with both electric and magnetic charges.

- A geon is an electromagnetic or gravitational wave which is held together in a confined region by the gravitational attraction of its own field energy.

- An inflaton is the generic name for an unidentified scalar particle responsible for the cosmic inflation.

- A spurion is the name given to a "particle" inserted mathematically into an isospin-violating decay in order to analyze it as though it conserved isospin.

- What is called "true muonium", a bound state of a muon and an antimuon, is a theoretical exotic atom which has never been observed.

5.5   Classification by speed

- A tardyon or bradyon travels slower than light and has a non-zero rest mass.

- A luxon travels at the speed of light and has no rest mass.

- A tachyon (mentioned above) is a hypothetical particle that travels faster than the speed of light and has an imaginary rest mass.

5.6   See also

- Acceleron

- List of baryons

- List of compounds for a list of molecules.

- List of fictional elements, materials, isotopes and atomic particles

- List of mesons

- Periodic table for an overview of atoms.

- Standard Model for the current theory of these particles.

- Table of nuclides

- Timeline of particle discoveries

5.7   References

[1] Observation of a new boson at a mass of 125 GeV with the CMS experiment at the LHC (2013). *arXiv:1207.7235*.

[2] Observation of a new particle in the search for the Standard Model Higgs boson with the ATLAS detector at the LHC (2012). *arXiv:1207.7214*.

[3] B. Kayser, *Two Questions About Neutrinos*, arXiv:1012.4469v1 [hep-ph] (2010).

[4] R. Maartens (2004). *Brane-World Gravity* (PDF). *Living Reviews in Relativity* **7**. p. 7. Also available in web format at http://www.livingreviews.org/lrr-2004-7.

- C. Amsler *et al.* (Particle Data Group) (2008). "Review of Particle Physics". *Physics Letters B* **667** (1–5): 1. Bibcode:2008PhLB..667....1P. doi:10.1016/j.physletb.2008.07.018. *(All information on this list, and more, can be found in the extensive, biannually-updated review by the Particle Data Group)*

Chapter 6

Fundamental interaction

Fundamental interactions, also known as **fundamental forces**, are the interactions in physical systems that don't appear to be reducible to more basic interactions. There are four conventionally accepted fundamental interactions—gravitational, electromagnetic, strong nuclear, and weak nuclear. Each one is understood as the dynamics of a *field*. The gravitational force is modeled as a continuous classical field. The other three are each modeled as discrete quantum fields, and exhibit a measurable unit or *elementary particle*.

Gravitation and electromagnetism act over a potentially infinite distance across the universe. They mediate macroscopic phenomena every day. The other two fields act over minuscule, subatomic distances. The strong nuclear interaction is responsible for the binding of atomic nuclei. The weak nuclear interaction also acts on the nucleus, mediating radioactive decay.

Theoretical physicists working beyond the Standard Model seek to quantize the gravitational field toward predictions that particle physicists can experimentally confirm, thus yielding acceptance to a theory of quantum gravity (QG). (Phenomena suitable to model as a fifth force—perhaps an added gravitational effect—remain widely disputed). Other theorists seek to unite the electroweak and strong fields within a Grand Unified Theory (GUT). While all four fundamental interactions are widely thought to align on a highly minuscule scale, particle accelerators cannot produce the massive energy levels required to experimentally probe at that Planck scale (which would experimentally confirm such theories). Yet some theories, such as the string theory, seek both QG and GUT within one framework, unifying all four fundamental interactions along with mass generation within a theory of everything (ToE).

6.1 General relativity

In his 1687 theory, Isaac Newton postulated space as an infinite and unalterable physical structure existing before, within, and around all objects while their states and relations unfold at a constant pace everywhere, thus absolute space and time. Inferring that all objects bearing mass approach at a constant rate, but collide by impact proportional to their masses, Newton inferred that matter exhibits an attractive force. His law of universal gravitation mathematically stated it to span the entire universe instantly (despite absolute time), or, if not actually a force, to be instant interaction among all objects (despite absolute space). As conventionally interpreted, Newton's theory of motion modeled a *central force* without a communicating medium.[2] Thus Newton's theory violated the first principle of mechanical philosophy, as stated by Descartes, *No action at a distance*. Conversely, during the 1820s, when explaining magnetism, Michael Faraday inferred a *field* filling space and transmitting that force. Faraday conjectured that ultimately, all forces unified into one.

In the early 1870s, James Clerk Maxwell unified electricity and magnetism as effects of an electromagnetic field whose third consequence was light, traveling at constant speed in a vacuum. The electromagnetic field theory contradicted predictions of Newton's theory of motion, unless physical states of the luminiferous aether—presumed to fill all space whether within matter or in a vacuum and to manifest the electromagnetic field—aligned all phenomena and thereby held valid the Newtonian principle relativity or invariance. Disfavoring hypotheses at unobservables, Albert Einstein discarded the aether, and aligned electrodynamics with relativity by denying absolute space and time, and stating relative space and time. The two phenomena altered in the vicinity of an object measured to be in motion—length contraction and time dilation for the object experienced to be in relative motion—Einstein's principle special relativity, published in 1905.

Special relativity was accepted as a theory, too. It rendered Newton's theory of motion apparently untenable, especially since Newtonian physics postulated an object's mass to be constant. A consequence of special relativity is mass being a variant form of energy, condensed into an object. By the equivalence principle, published by Einstein in 1907, gravitation is indistinguishable from acceleration, perhaps two phenomena sharing a mechanism. That year, Hermann Minkowski modeled special relativity to a unification of space and time, 4D spacetime. So stretching the three spatial dimensions onto the single dimension of time's arrow, Einstein arrived at general theory of relativity in 1915.[3] Einstein interpreted space as a substance, *Einstein-aether*, whose physical properties receive motion from an object and transmit it to other objects while modulating events' unfolding. Equivalent to energy, mass contracts space, which dilates time—events unfold more slowly—establishing local tension. The object relieves it in the likeness of a free fall at light speed along the pathway of least resistance, a straight line's equivalent on the curved surface of 4D spacetime, a pathway termed *worldline*.

Einstein abolished *action at a distance* by theorizing a gravitational field—4D spacetime—that waves while transmitting motion across the universe at light speed. All objects always travel at light speed in 4D spacetime. At zero relative speed, an object is observed to travel none through space, but age most rapidly. That is, an object at relative rest in 3D space exhibits its constant energy to an observer by exhibiting top speed along 1D time flow. Conversely, at highest relative speed, an object traverses 3D space at light speed, yet is ageless, none of its constant energy available to internal motion as flow along 1D time. Whereas Newtonian inertia is an idealized case of an object either keeping rest or holding constant velocity by its hypothetical existence in a universe otherwise devoid of matter, Einsteinian inertia is indistinguishable from an object experiencing no acceleration by existing in a gravitational field possibly full of matter distributed uniformly. Conversely, even massless energy manifests gravitation—which is acceleration—on local objects by "curving" the surface of 4D spacetime. Physicists renounced belief that motion must be mediated by a *force*.

6.2 Standard Model

Main article: Standard Model
See also: Lambda-CDM model

The electromagnetic, strong, and weak interactions associate with elementary particles, whose behaviors are modeled in quantum mechanics (QM). For predictive success with QM's probabilistic outcomes, particle physics conventionally models QM events across a field set to special relativity, altogether relativistic quantum field theory (QFT).[4] Force particles, called gauge bosons—*force carriers* or *messenger particles* of underlying fields—interact with matter particles, called fermions. Everyday matter is atoms, composed of three fermion types: up-quarks and down-quarks constituting, as well as electrons orbiting, the atom's nucleus. Atoms interact, form molecules, and manifest further properties through electromagnetic interactions among their electrons absorbing and emitting photons, the electromagnetic field's force carrier, which if unimpeded traverse potentially infinite distance. Electromagnetism's QFT is quantum electrodynamics (QED).

The electromagnetic interaction was modeled with the weak interaction, whose force carriers are W and Z bosons, traversing the minuscule distance, in electroweak theory (EWT). Electroweak interaction would operate at such high temperatures as soon after the presumed Big Bang, but, as the early universe cooled, split into electromagnetic and weak interactions. The strong interaction, whose force carrier is the gluon, traversing minuscule distance among quarks, is modeled in quantum chromodynamics (QCD). EWT, QCD, and the Higgs mechanism, whereby the Higgs field manifests Higgs bosons that interact with some quantum particles and thereby endow those particles with mass comprise particle physics' Standard Model (SM). Predictions are usually made using calculational approximation methods, although such perturbation theory is inadequate to model some experimental observations (for instance bound states and solitons). Still, physicists widely accept the Standard Model as science's most experimentally confirmed theory.

Beyond the Standard Model, some theorists work to unite the electroweak and strong interactions within a Grand Unified Theory (GUT). Some attempts at GUTs hypothesize "shadow" particles, such that every known matter particle associates with an undiscovered force particle, and vice versa, altogether supersymmetry (SUSY). Other theorists seek to quantize the gravitational field by the modeling behavior of its hypothetical force carrier, the graviton and achieve quantum gravity (QG). One approach to QG is loop quantum gravity (LQG). Still other theorists seek both QG and GUT within one framework, reducing all four fundamental interactions to a Theory of Everything (ToE). The most prevalent aim at a ToE is string theory, although to model matter particles, it added SUSY to force particles—and so, strictly speaking, became

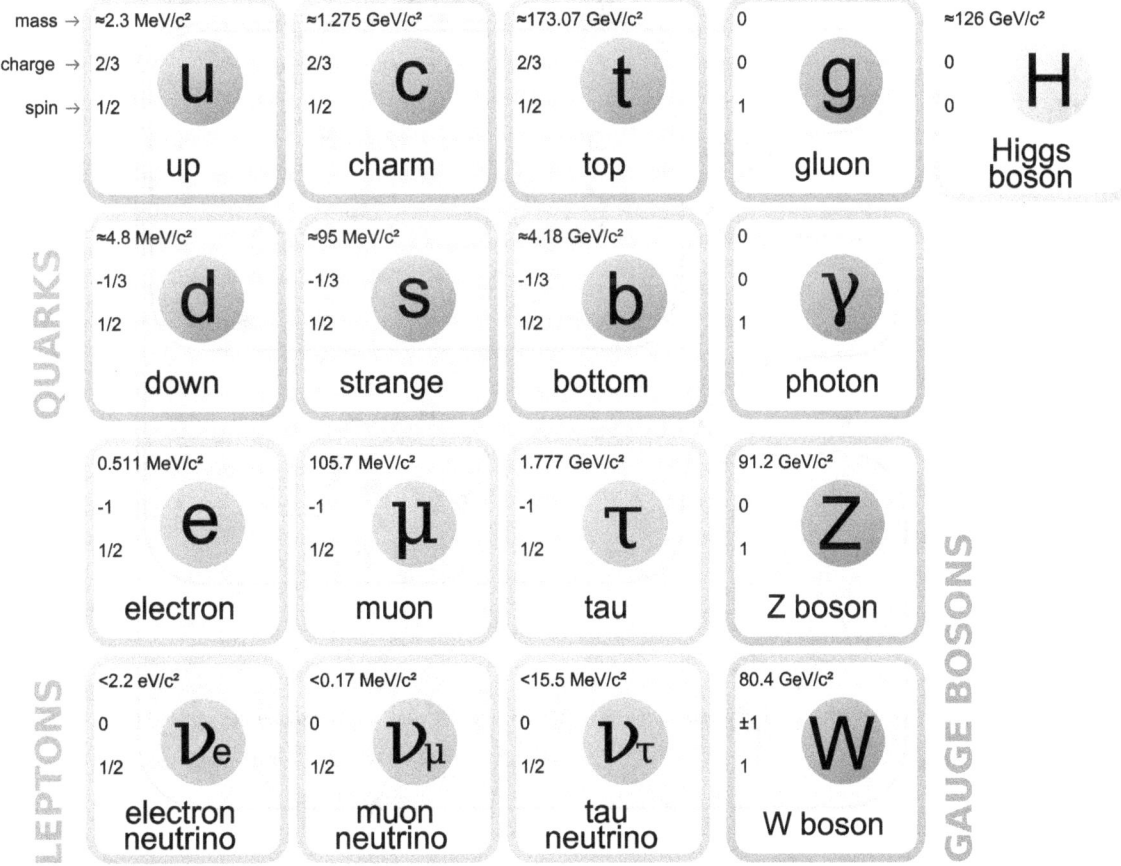

The Standard Model of elementary particles, with the fermions in the first three columns, the gauge bosons in the fourth column, and the Higgs boson in the fifth column

superstring theory. Multiple, seemingly disparate superstring theories were unified on a backbone, M theory. Theories beyond the Standard Model remain highly speculative, lacking great experimental support.

6.3   Overview of the fundamental interactions

In the conceptual model of fundamental interactions, matter consists of fermions, which carry properties called charges and spin $\pm\frac{1}{2}$ (intrinsic angular momentum $\pm\frac{\hbar}{2}$, where $\hbar$ is the reduced Planck constant). They attract or repel each other by exchanging bosons.

The interaction of any pair of fermions in perturbation theory can then be modeled thus:

Two fermions go in $\rightarrow$ *interaction* by boson exchange $\rightarrow$ Two changed fermions go out.

The exchange of bosons always carries energy and momentum between the fermions, thereby changing their speed and direction. The exchange may also transport a charge between the fermions, changing the charges of the fermions in the process (e.g., turn them from one type of fermion to another). Since bosons carry one unit of angular momentum, the fermion's spin direction will flip from $+\frac{1}{2}$ to $-\frac{1}{2}$ (or vice versa) during such an exchange (in units of the reduced Planck's constant).

Because an interaction results in fermions attracting and repelling each other, an older term for "interaction" is force.

According to the present understanding, there are four fundamental interactions or forces: gravitation, electromagnetism,

Elementary Particles

```
          Matter                              Force Carriers
            |                                      |
   ┌────────┼────────┐          ┌──────────┬───────┴────────┬──────────┐
 Quarks    ?     Leptons      Gluons   W & Z bosons     Photons     Gravitons
            |                    |          |               |            |
      Quark-Lepton            Strong      Weak      Electromagnetism  Gravity
      complementarity           |          |               |            |
         Hadrons              Quantum              Quantum         Quantum
            |              Chromodynamics       Electrodynamics    Gravity
     ┌──────┴──────┐             |                   |                |
  Mesons        Baryons          |        Electroweak Theory         |
                   |             └──────────────┬────────────────────┘
                 Nuclei              Grand Unified Theory
                   |                          |
                 Atoms                Theory of Everything
                   |
               Molecules
```

Composite Particles   **Forces**

An overview of the various families of elementary and composite particles, and the theories describing their interactions. Fermions are on the left, and Bosons are on the right.

the weak interaction, and the strong interaction. Their magnitude and behavior vary greatly, as described in the table below. Modern physics attempts to explain every observed physical phenomenon by these fundamental interactions. Moreover, reducing the number of different interaction types is seen as desirable. Two cases in point are the unification of:

- Electric and magnetic force into electromagnetism;

- The electromagnetic interaction and the weak interaction into the electroweak interaction; see below.

Both magnitude ("relative strength") and "range", as given in the table, are meaningful only within a rather complex theoretical framework. It should also be noted that the table below lists properties of a conceptual scheme that is still the subject of ongoing research.

The modern (perturbative) quantum mechanical view of the fundamental forces other than gravity is that particles of matter (fermions) do not directly interact with each other, but rather carry a charge, and exchange virtual particles (gauge bosons), which are the interaction carriers or force mediators. For example, photons mediate the interaction of electric charges, and gluons mediate the interaction of color charges.

6.4 The interactions

6.4.1 Gravitation

Main article: Gravitation

Gravitation is by far the weakest of the four interactions. The weakness of gravity can easily be demonstrated by suspending a pin using a simple magnet (such as a refrigerator magnet). The magnet is able to hold the pin against the gravitational pull of the entire Earth.

Yet gravitation is very important for macroscopic objects and over macroscopic distances for the following reasons. Gravitation:

- is the only interaction that acts on all particles having mass, energy and/or momentum;

- has an infinite range, like electromagnetism but unlike strong and weak interaction;

- cannot be absorbed, transformed, or shielded against;

- always attracts and never repels.

Even though electromagnetism is far stronger than gravitation, electrostatic attraction is not relevant for large celestial bodies, such as planets, stars, and galaxies, simply because such bodies contain equal numbers of protons and electrons and so have a net electric charge of zero. Nothing "cancels" gravity, since it is only attractive, unlike electric forces which can be attractive or repulsive. On the other hand, all objects having mass are subject to the gravitational force, which only attracts. Therefore, only gravitation matters on the large-scale structure of the universe.

The long range of gravitation makes it responsible for such large-scale phenomena as the structure of galaxies, black holes, and it retards the expansion of the universe. Gravitation also explains astronomical phenomena on more modest scales, such as planetary orbits, as well as everyday experience: objects fall; heavy objects act as if they were glued to the ground, and animals can only jump so high.

Gravitation was the first interaction to be described mathematically. In ancient times, Aristotle hypothesized that objects of different masses fall at different rates. During the Scientific Revolution, Galileo Galilei experimentally determined that this was not the case — neglecting the friction due to air resistance, and buoyancy forces if an atmosphere is present (e.g. the case of a dropped air-filled balloon vs a water-filled balloon) all objects accelerate toward the Earth at the same rate. Isaac Newton's law of Universal Gravitation (1687) was a good approximation of the behavior of gravitation. Our present-day understanding of gravitation stems from Albert Einstein's General Theory of Relativity of 1915, a more accurate (especially for cosmological masses and distances) description of gravitation in terms of the geometry of space-time.

Merging general relativity and quantum mechanics (or quantum field theory) into a more general theory of quantum gravity is an area of active research. It is hypothesized that gravitation is mediated by a massless spin-2 particle called the graviton.

Although general relativity has been experimentally confirmed (at least, in the weak field or Post-Newtonian case) on all but the smallest scales, there are rival theories of gravitation. Those taken seriously by the physics community all reduce to general relativity in some limit, and the focus of observational work is to establish limitations on what deviations from general relativity are possible.

Proposed extra dimensions could explain why the gravity force is so weak.[6]

6.4.2   Electroweak interaction

Main article: Electroweak interaction

Electromagnetism and weak interaction appear to be very different at everyday low energies. They can be modeled using two different theories. However, above unification energy, on the order of 100 GeV, they would merge into a single electroweak force.

Electroweak theory is very important for modern cosmology, particularly on how the universe evolved. This is because shortly after the Big Bang, the temperature was approximately above 10^{15} K. Electromagnetic force and weak force were merged into a combined electroweak force.

For contributions to the unification of the weak and electromagnetic interaction between elementary particles, Abdus Salam, Sheldon Glashow and Steven Weinberg were awarded the Nobel Prize in Physics in 1979.[7][8]

Electromagnetism

Main article: Electromagnetism

Electromagnetism is the force that acts between electrically charged particles. This phenomenon includes the electrostatic force acting between charged particles at rest, and the combined effect of electric and magnetic forces acting between charged particles moving relative to each other.

Electromagnetism is infinite-ranged like gravity, but vastly stronger, and therefore describes a number of macroscopic phenomena of everyday experience such as friction, rainbows, lightning, and all human-made devices using electric current, such as television, lasers, and computers. Electromagnetism fundamentally determines all macroscopic, and many atomic levels, properties of the chemical elements, including all chemical bonding.

In a four kilogram (~1 gallon) jug of water there are

$$4000 \text{ g } H_2O \cdot \frac{1 \text{ mol } H_2O}{18 \text{ g } H_2O} \cdot \frac{10 \text{ mol } e^-}{1 \text{ mol } H_2O} \cdot \frac{96{,}000 \text{ C}}{1 \text{ mol } e^-} = 2.1 \times 10^8 C$$

of total electron charge. Thus, if we place two such jugs a meter apart, the electrons in one of the jugs repel those in the other jug with a force of

$$\frac{1}{4\pi\varepsilon_0} \frac{(2.1 \times 10^8 C)^2}{(1m)^2} = 4.1 \times 10^{26} N.$$

This is larger than the planet Earth would weigh if weighed on another Earth. The atomic nuclei in one jug also repel those in the other with the same force. However, these repulsive forces are canceled by the attraction of the electrons in jug A with the nuclei in jug B and the attraction of the nuclei in jug A with the electrons in jug B, resulting in no net force. Electromagnetic forces are tremendously stronger than gravity but cancel out so that for large bodies gravity dominates.

Electrical and magnetic phenomena have been observed since ancient times, but it was only in the 19th century that it was discovered that electricity and magnetism are two aspects of the same fundamental interaction. By 1864, Maxwell's equations had rigorously quantified this unified interaction. Maxwell's theory, restated using vector calculus, is the classical theory of electromagnetism, suitable for most technological purposes.

The constant speed of light in a vacuum (customarily described with the letter "c") can be derived from Maxwell's equations, which are consistent with the theory of special relativity. Einstein's 1905 theory of special relativity, however, which flows from the observation that the speed of light is constant no matter how fast the observer is moving, showed that the theoretical result implied by Maxwell's equations has profound implications far beyond electromagnetism on the very nature of time and space.

In another work that departed from classical electro-magnetism, Einstein also explained the photoelectric effect by hypothesizing that light was transmitted in quanta, which we now call photons. Starting around 1927, Paul Dirac combined quantum mechanics with the relativistic theory of electromagnetism. Further work in the 1940s, by Richard Feynman, Freeman Dyson, Julian Schwinger, and Sin-Itiro Tomonaga, completed this theory, which is now called quantum electrodynamics, the revised theory of electromagnetism. Quantum electrodynamics and quantum mechanics provide a theoretical basis for electromagnetic behavior such as quantum tunneling, in which a certain percentage of electrically charged particles move in ways that would be impossible under the classical electromagnetic theory, that is necessary for everyday electronic devices such as transistors to function.

Weak interaction

Main article: Weak interaction

The *weak interaction* or *weak nuclear force* is responsible for some nuclear phenomena such as beta decay. Electromagnetism and the weak force are now understood to be two aspects of a unified electroweak interaction — this discovery was the first step toward the unified theory known as the Standard Model. In the theory of the electroweak interaction, the carriers of the weak force are the massive gauge bosons called the W and Z bosons. The weak interaction is the only known interaction which does not conserve parity; it is left-right asymmetric. The weak interaction even violates CP symmetry but does conserve CPT.

6.4.3 Strong interaction

Main article: Strong interaction

The *strong interaction*, or *strong nuclear force*, is the most complicated interaction, mainly because of the way it varies with distance. At distances greater than 10 femtometers, the strong force is practically unobservable. Moreover, it holds only inside the atomic nucleus.

After the nucleus was discovered in 1908, it was clear that a new force was needed to overcome the electrostatic repulsion, a manifestation of electromagnetism, of the positively charged protons. Otherwise, the nucleus could not exist. Moreover, the force had to be strong enough to squeeze the protons into a volume that is 10^{-15} of that of the entire atom. From the short range of this force, Hideki Yukawa predicted that it was associated with a massive particle, whose mass is approximately 100 MeV.

The 1947 discovery of the pion ushered in the modern era of particle physics. Hundreds of hadrons were discovered from the 1940s to 1960s, and an extremely complicated theory of hadrons as strongly interacting particles was developed. Most notably:

- The pions were understood to be oscillations of vacuum condensates;

- Jun John Sakurai proposed the rho and omega vector bosons to be force carrying particles for approximate symmetries of isospin and hypercharge;

- Geoffrey Chew, Edward K. Burdett and Steven Frautschi grouped the heavier hadrons into families that could be understood as vibrational and rotational excitations of strings.

While each of these approaches offered deep insights, no approach led directly to a fundamental theory.

Murray Gell-Mann along with George Zweig first proposed fractionally charged quarks in 1961. Throughout the 1960s, different authors considered theories similar to the modern fundamental theory of quantum chromodynamics (QCD) as simple models for the interactions of quarks. The first to hypothesize the gluons of QCD were Moo-Young Han and Yoichiro Nambu, who introduced the quark color charge and hypothesized that it might be associated with a force-carrying field. At that time, however, it was difficult to see how such a model could permanently confine quarks. Han and Nambu also assigned each quark color an integer electrical charge, so that the quarks were fractionally charged only on average, and they did not expect the quarks in their model to be permanently confined.

In 1971, Murray Gell-Mann and Harald Fritzsch proposed that the Han/Nambu color gauge field was the correct theory of the short-distance interactions of fractionally charged quarks. A little later, David Gross, Frank Wilczek, and David Politzer discovered that this theory had the property of asymptotic freedom, allowing them to make contact with experimental evidence. They concluded that QCD was the complete theory of the strong interactions, correct at all distance scales. The discovery of asymptotic freedom led most physicists to accept QCD since it became clear that even the long-distance properties of the strong interactions could be consistent with experiment if the quarks are permanently confined.

Assuming that quarks are confined, Mikhail Shifman, Arkady Vainshtein, and Valentine Zakharov were able to compute the properties of many low-lying hadrons directly from QCD, with only a few extra parameters to describe the vacuum. In 1980, Kenneth G. Wilson published computer calculations based on the first principles of QCD, establishing, to a level of confidence tantamount to certainty, that QCD will confine quarks. Since then, QCD has been the established theory of the strong interactions.

QCD is a theory of fractionally charged quarks interacting by means of 8 photon-like particles called gluons. The gluons interact with each other, not just with the quarks, and at long distances the lines of force collimate into strings. In this way, the mathematical theory of QCD not only explains how quarks interact over short distances but also the string-like behavior, discovered by Chew and Frautschi, which they manifest over longer distances.

6.4.4 Beyond the Standard Model

Main article: Physics beyond the Standard Model
See also: Elementary particle § Beyond the Standard Model

Numerous theoretical efforts have been made to systematize the existing four fundamental interactions on the model of electroweak unification.

Grand Unified Theories (GUTs) are proposals to show that all of the fundamental interactions, other than gravity, arise from a single interaction with symmetries that break down at low energy levels. GUTs predict relationships among constants of nature that are unrelated in the SM. GUTs also predict gauge coupling unification for the relative strengths of the electromagnetic, weak, and strong forces, a prediction verified at the Large Electron–Positron Collider in 1991 for supersymmetric theories.

Theories of everything, which integrate GUTs with a quantum gravity theory face a greater barrier, because no quantum gravity theories, which include string theory, loop quantum gravity, and twistor theory, have secured wide acceptance. Some theories look for a graviton to complete the Standard Model list of force-carrying particles, while others, like loop quantum gravity, emphasize the possibility that time-space itself may have a quantum aspect to it.

Some theories beyond the Standard Model include a hypothetical fifth force, and the search for such a force is an ongoing line of experimental research in physics. In supersymmetric theories, there are particles that acquire their masses only through supersymmetry breaking effects and these particles, known as moduli can mediate new forces. Another reason to look for new forces is the recent discovery that the expansion of the universe is accelerating (also known as dark energy), giving rise to a need to explain a nonzero cosmological constant, and possibly to other modifications of general relativity. Fifth forces have also been suggested to explain phenomena such as CP violations, dark matter, and dark flow.

6.5 See also

- Standard Model

  - Strong interaction
  - Electroweak interaction
  - Weak interaction
  - Gravity
    - Quantum gravity
    - String Theory
    - Theory of Everything

- Grand Unified Theory

  - Gauge coupling unification
  - Unified Field Theory

- Quintessence, a hypothesized fifth force.

- *People*: Isaac Newton, James Clerk Maxwell, Albert Einstein, Richard Feynman, Sheldon Glashow, Abdus Salam, Steven Weinberg, Gerardus 't Hooft, David Gross, Edward Witten, Howard Georgi.

6.6   References

[1] http://www.pha.jhu.edu/~{}dfehling/particle.gif

[2] Newton's absolute space was a medium, but not one transmitting gravitation.

[3] Special relativity holds for objects at vast speed but of negligible mass, for instance elementary particles. Yet by yielding gravitation, which is a manner of acceleration, notable mass breaks inertia—that is, constant speed and direction—and thereby violates special relativity. Special relativity could approximately predict a massive object's motion during barely an instant, however, and thus is a temporally limited case of general relativity.

[4] Meinard Kuhlmann, "Physicists debate whether the world is made of particles or fields—or something else entirely", *Scientific American*, 24 Jul 2013.

[5] Approximate. See Coupling constant for more exact strengths, depending on the particles and energies involved.

[6] CERN (20 January 2012). "Extra dimensions, gravitons, and tiny black holes".

[7] Bais, Sander (2005), *The Equations. Icons of knowledge*, ISBN 0-674-01967-9 p.84

[8] "The Nobel Prize in Physics 1979". The Nobel Foundation. Retrieved 2008-12-16.

Bibliography General:

- Davies, Paul (1986), *The Forces of Nature*, Cambridge Univ. Press 2nd ed.

- Feynman, Richard (1967), *The Character of Physical Law*, MIT Press, ISBN 0-262-56003-8

- Schumm, Bruce A. (2004), *Deep Down Things*, Johns Hopkins University Press While all interactions are discussed, discussion is especially thorough on the weak.

- Weinberg, Steven (1993), *The First Three Minutes: A Modern View of the Origin of the Universe*, Basic Books, ISBN 0-465-02437-8

- Weinberg, Steven (1994), *Dreams of a Final Theory*, Basic Books, ISBN 0-679-74408-8

Texts:

- Padmanabhan, T. (1998), *After The First Three Minutes: The Story of Our Universe*, Cambridge Univ. Press, ISBN 0-521-62972-1

- Perkins, Donald H. (2000), *Introduction to High Energy Physics*, Cambridge Univ. Press, ISBN 0-521-62196-8

- Riazuddin (December 29, 2009). "Non-standard interactions" (PDF). *NCP 5th Particle Physics Sypnoisis* (Islamabad: Riazuddin, Head of High-Energy Theory Group at National Center for Physics) **1** (1): 1–25. Retrieved March 19, 2011.

Chapter 7

Strong interaction

In particle physics, the **strong interaction** is the mechanism responsible for the strong nuclear force (also called the **strong force**, **nuclear strong force** or **colour force**), one of the four fundamental interactions of nature, the others being electromagnetism, the weak interaction and gravitation. Effective only at a distance of a femtometer, it is approximately 100 times stronger than electromagnetism, a million times stronger than the weak force interaction and 10^{38} times stronger than gravitation at that range.[1] It ensures the stability of ordinary matter, as it confines the quark elementary particles into hadron particles, such as the proton and neutron, the largest components of the mass of ordinary matter. Furthermore, most of the mass-energy of a common proton or neutron is in the form of the strong force field energy; the individual quarks provide only about 1% of the mass-energy of a proton.

The strong interaction is observable in two areas: on a larger scale (about 1 to 3 femtometers (fm)), it is the force that binds protons and neutrons (nucleons) together to form the nucleus of an atom. On the smaller scale (less than about 0.8 fm, the radius of a nucleon), it is the force (carried by gluons) that holds quarks together to form protons, neutrons, and other hadron particles. The strong force inherently has so high a strength that the energy of an object bound by the strong force (a hadron) is high enough to produce new massive particles. Thus, if hadrons are struck by high-energy particles, they give rise to new hadrons instead of emitting freely moving radiation (gluons). This property of the strong force is called colour confinement, and it prevents the free "emission" of the strong force: instead, in practice, jets of massive particles are observed.

In the context of binding protons and neutrons together to form atomic nuclei, the strong interaction is called the nuclear force (or *residual strong force*). In this case, it is the residuum of the strong interaction between the quarks that make up the protons and neutrons. As such, the residual strong interaction obeys a quite different distance-dependent behavior between nucleons, from when it is acting to bind quarks within nucleons. The binding energy that is partly released on the breakup of a nucleus is related to the residual strong force and is harnessed in nuclear power and fission-type nuclear weapons.[2][3]

The strong interaction is thought to be mediated by massless particles called gluons, that are exchanged between quarks, antiquarks, and other gluons. Gluons, in turn, are thought to interact with quarks and gluons as all carry a type of charge called colour charge. Colour charge is analogous to electromagnetic charge, but it comes in three types rather than one (+/- red, +/- green, +/- blue) that results in a different type of force, with different rules of behavior. These rules are detailed in the theory of quantum chromodynamics (QCD), which is the theory of quark-gluon interactions.

Just after the Big Bang, and during the electroweak epoch, the electroweak force separated from the strong force. Although it is expected that a Grand Unified Theory exists to describe this, no such theory has been successfully formulated, and the unification remains an unsolved problem in physics.

7.1 History

Before the 1970s, physicists were uncertain about the binding mechanism of the atomic nucleus. It was known that the nucleus was composed of protons and neutrons and that protons possessed positive electric charge, while neutrons

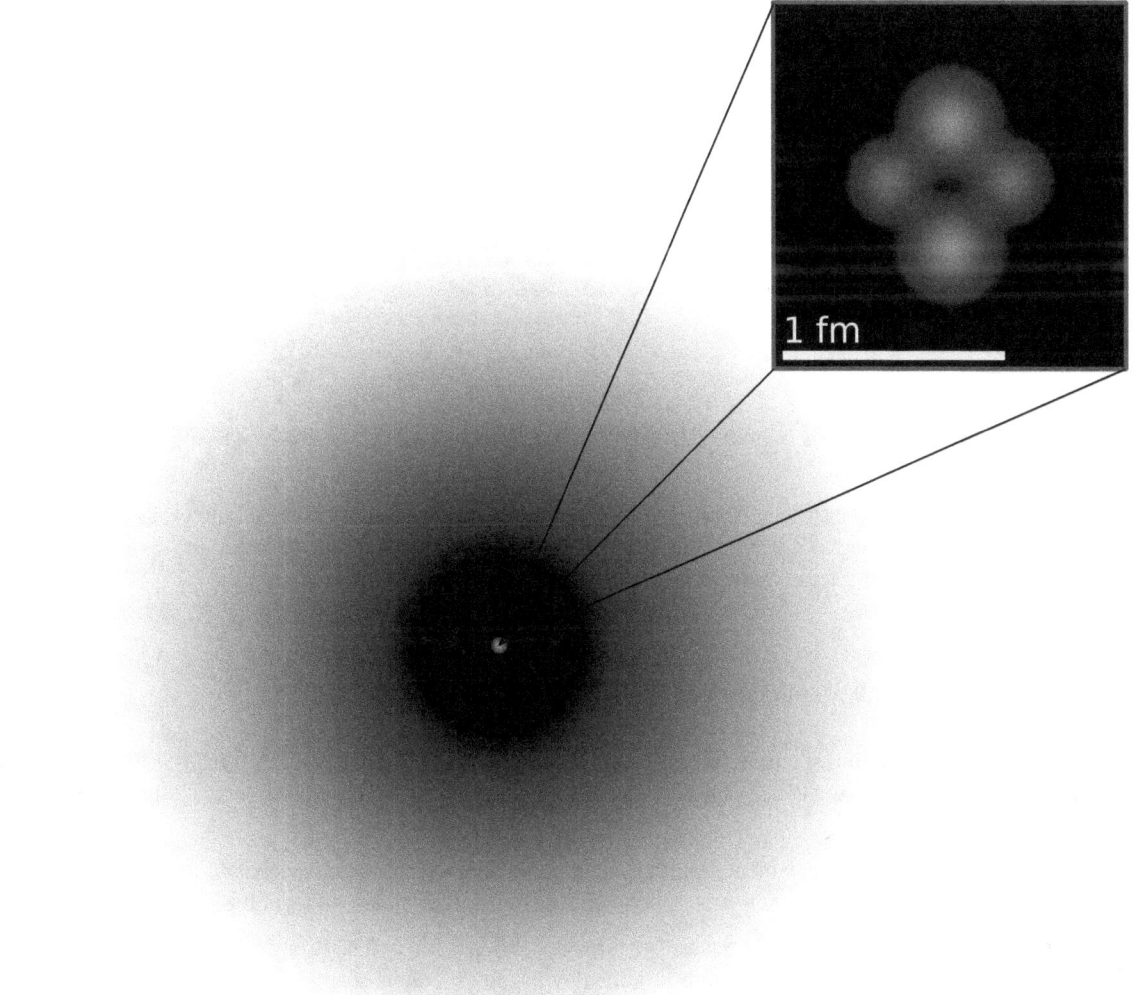

1 Å = 100,000 fm

The nucleus of a helium atom. The two protons have the same charge, but still stay together due to the residual nuclear force

were electrically neutral. However, these facts seemed to contradict one another. By physical understanding at that time, positive charges would repel one another and the nucleus should therefore fly apart. However, this was never observed. New physics was needed to explain this phenomenon.

A stronger attractive force was postulated to explain how the atomic nucleus was bound together despite the protons' mutual electromagnetic repulsion. This hypothesized force was called the *strong force*, which was believed to be a fundamental force that acted on the protons and neutrons that make up the nucleus.

It was later discovered that protons and neutrons were not fundamental particles, but were made up of constituent particles called quarks. The strong attraction between nucleons was the side-effect of a more fundamental force that bound the quarks together in the protons and neutrons. The theory of quantum chromodynamics explains that quarks carry what is called a colour charge, although it has no relation to visible colour.[4] Quarks with unlike colour charge attract one another as a result of the **strong interaction**, which is mediated by particles called gluons.

7.2 Details

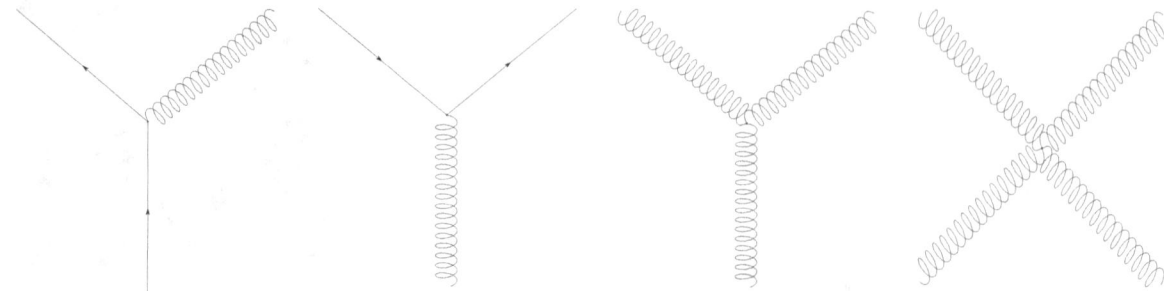

The fundamental couplings of the strong interaction, from left to right: gluon radiation, gluon splitting and gluon self-coupling.

The word *strong* is used since the strong interaction is the "strongest" of the four fundamental forces; its strength is around 10^2 times that of the electromagnetic force, some 10^6 times as great as that of the weak force, and about 10^{39} times that of gravitation, at a distance of a femtometer or less.

7.2.1 Behaviour of the strong force

The contemporary understanding of strong force is described by quantum chromodynamics (QCD), a part of the standard model of particle physics. Mathematically, QCD is a non-Abelian gauge theory based on a local (gauge) symmetry group called SU(3).

Quarks and gluons are the only fundamental particles that carry non-vanishing colour charge, and hence participate in strong interactions. The strong force itself acts directly only on elementary quark and gluon particles.

All quarks and gluons in QCD interact with each other through the strong force. The strength of interaction is parametrized by the strong coupling constant. This strength is modified by the gauge colour charge of the particle, a group theoretical property.

The strong force acts between quarks. Unlike all other forces (electromagnetic, weak, and gravitational), the strong force does not diminish in strength with increasing distance. After a limiting distance (about the size of a hadron) has been reached, it remains at a strength of about 10,000 newtons, no matter how much farther the distance between the quarks.[5] In QCD, this phenomenon is called colour confinement; it implies that only hadrons, not individual free quarks, can be observed. The explanation is that the amount of work done against a force of 10,000 newtons (about the weight of a one-metric ton mass on the surface of the Earth) is enough to create particle-antiparticle pairs within a very short distance of an interaction. In simple terms, the very energy applied to pull two quarks apart will create a pair of new quarks that will pair up with the original ones. The failure of all experiments that have searched for free quarks is considered to be evidence for this phenomenon.

The elementary quark and gluon particles affected are unobservable directly, but they instead emerge as jets of newly created hadrons, whenever energy is deposited into a quark-quark bond, as when a quark in a proton is struck by a very fast quark (in an impacting proton) during a particle accelerator experiment. However, quark–gluon plasmas have been observed.

Every quark in the universe does not attract every other quark in the above distance independent manner, since colour-confinement implies that the strong force acts without distance-diminishment only between pairs of single quarks, and that in collections of bound quarks (i.e., hadrons), the net colour-charge of the quarks cancels out, as seen from far away. Collections of quarks (hadrons) therefore appear (nearly) without colour-charge, and the strong force is therefore nearly absent between these hadrons (i.e., between baryons or mesons). However, the cancellation is not quite perfect. A small residual force remains (described below) known as the **residual strong force**. This residual force *does* diminish rapidly with distance, and is thus very short-range (effectively a few femtometers). It manifests as a force between the "colourless" hadrons, and is therefore sometimes known as the **strong nuclear force** or simply nuclear force.

7.2.2   Residual strong force

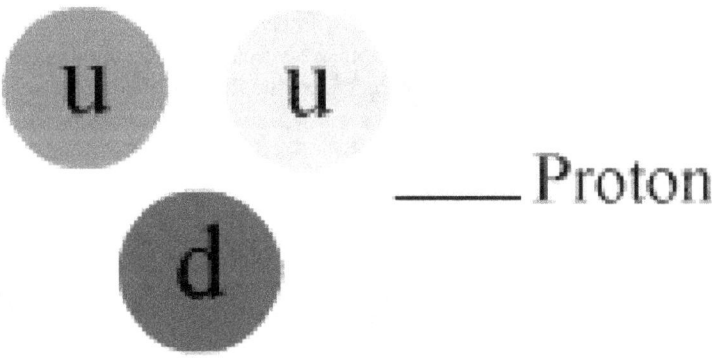

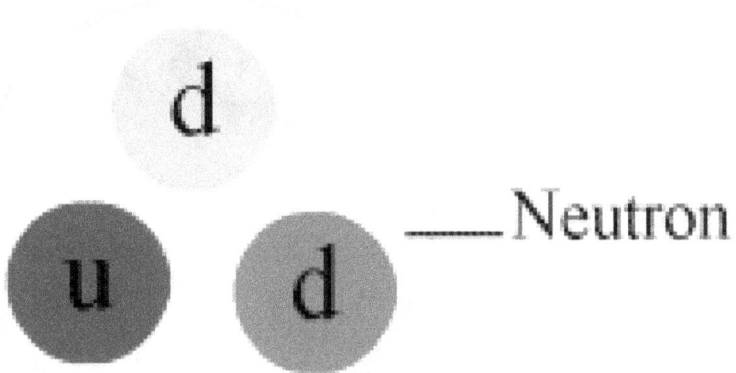

An animation of the nuclear force (or residual strong force) interaction between a proton and a neutron. The small coloured double circles are gluons, which can be seen binding the proton and neutron together. These gluons also hold the quark-antiquark combination called the pion together, and thus help transmit a residual part of the strong force even between colourless hadrons. Anticolours are shown as per this diagram. For a larger version, click here

The residual effect of the strong force is called the nuclear force. The nuclear force acts between hadrons, such as mesons or the nucleons in atomic nuclei. This "residual strong force", acting indirectly, transmits gluons that form part of the virtual pi and rho mesons, which, in turn, transmit the nuclear force between nucleons.

The residual strong force is thus a minor residuum of the strong force that binds quarks together into protons and neutrons.

This same force is much weaker *between* neutrons and protons, because it is mostly neutralized *within* them, in the same way that electromagnetic forces between neutral atoms (van der Waals forces) are much weaker than the electromagnetic forces that hold the atoms internally together.[6]

Unlike the strong force itself, the nuclear force, or residual strong force, *does* diminish in strength, and in fact diminishes rapidly with distance. The decrease is approximately as a negative exponential power of distance, though there is no simple expression known for this; see Yukawa potential. This fact, together with the less-rapid decrease of the disruptive electromagnetic force between protons with distance, causes the instability of larger atomic nuclei, such as all those with atomic numbers larger than 82 (the element lead).

7.3 See also

- Nuclear binding energy

- Colour charge

- Coupling constant

- Nuclear physics

- QCD matter

- Quantum field theory and Gauge theory

- Standard model of particle physics and Standard Model (mathematical formulation)

- Weak interaction, electromagnetism and gravity

- Intermolecular force

- Vortex

- Yukawa interaction

7.4 References

[1] Relative strength of interaction varies with distance. See for instance Matt Strassler's essay, "The strength of the known forces".

[2] on Binding energy: see Binding Energy, Mass Defect, Furry Elephant physics educational site, retr 2012 7 1

[3] on Binding energy: see Chapter 4 NUCLEAR PROCESSES, THE STRONG FORCE, M. Ragheb 1/27/2012, University of Illinois

[4] Feynman, R. P. (1985). *QED: The Strange Theory of Light and Matter*. Princeton University Press. p. 136. ISBN 0-691-08388-6. The idiot physicists, unable to come up with any wonderful Greek words anymore, call this type of polarization by the unfortunate name of 'colour,' which has nothing to do with colour in the normal sense.

[5] Fritzsch, op. cite, p. 164. The author states that the force between differently coloured quarks remains constant at any distance after they travel only a tiny distance from each other, and is equal to that need to raise one ton, which is 1000 kg x 9.8 m/s^2 = ~10,000 N.

[6] Fritzsch, H. (1983). *Quarks: The Stuff of Matter*. Basic Books. pp. 167–168. ISBN 978-0-465-06781-7.

7.5   Further reading

- Christman, J. R. (2001). "MISN-0-280: *The Strong Interaction*" (PDF). *Project PHYSNET*. External link in |work= (help)

- Griffiths, David (1987). *Introduction to Elementary Particles*. John Wiley & Sons. ISBN 0-471-60386-4.

- Halzen, F.; Martin, A. D. (1984). *Quarks and Leptons: An Introductory Course in Modern Particle Physics*. John Wiley & Sons. ISBN 0-471-88741-2.

- Kane, G. L. (1987). *Modern Elementary Particle Physics*. Perseus Books. ISBN 0-201-11749-5.

- Morris, R. (2003). *The Last Sorcerers: The Path from Alchemy to the Periodic Table*. Joseph Henry Press. ISBN 0-309-50593-3.

7.6   External links

- Strong force at *Encyclopædia Britannica*

Chapter 8

Free particle

In physics, a **free particle** is a particle that, in some sense, is not bound by an external force, or equivalently not in a region where its potential energy varies. In classical physics, this means the particle is present in a "field-free" space. In quantum mechanics, it means a region of uniform potential, usually set to zero in the region of interest since potential can be arbitrarily set to zero at any point (or surface in three dimensions) in space.

8.1 Classical free particle

The classical free particle is characterized simply by a fixed velocity **v**. The momentum is given by

$$\mathbf{p} = m\mathbf{v}$$

and the kinetic energy (equal to total energy) by

$$E = \frac{1}{2}mv^2$$

where m is the mass of the particle and **v** is the vector velocity of the particle.

8.2 Non-relativistic quantum free particle

8.2.1 Mathematical description

Main articles: Schrödinger equation and Matter wave

A free quantum particle is described by the Schrödinger equation:

$$-\frac{\hbar^2}{2m}\nabla^2\,\psi(\mathbf{r},t) = i\hbar\frac{\partial}{\partial t}\psi(\mathbf{r},t)$$

where ψ is the wavefunction of the particle at position **r** and time t. The solution for a particle with momentum **p** or wave vector **k**, at angular frequency ω or energy E, is given by the complex plane wave:

$$\psi(\mathbf{r},t) = Ae^{i(\mathbf{k}\cdot\mathbf{r}-\omega t)} = Ae^{i(\mathbf{p}\cdot\mathbf{r}-Et)/\hbar}$$

with amplitude *A*. As for *all* quantum particles free *or* bound, the Heisenberg uncertainty principles

$$\Delta p_x \Delta x \geq \frac{\hbar}{2}, \quad \Delta E \Delta t \geq \hbar$$

(similarly for the *y* and *z* directions), and the De Broglie relations:

$$\mathbf{p} = \hbar \mathbf{k}, \quad E = \hbar \omega$$

apply. Since the potential energy is (set to) zero, the total energy *E* is equal to the kinetic energy, which has the same form as in classical physics:

$$E = T \rightarrow \frac{\hbar^2 k^2}{2m} = \hbar \omega$$

8.2.2   Measurement and calculations

The integral of the probability density function

$$\rho(\mathbf{r}, t) = \psi^*(\mathbf{r}, t)\psi(\mathbf{r}, t) = |\psi(\mathbf{r}, t)|^2$$

where * denotes complex conjugate, over all space is the probability of finding the particle in all space, which must be unity if the particle exists:

$$\int_{\text{all space}} |\psi(\mathbf{r}, t)|^2 d^3\mathbf{r} = 1$$

This is the normalization condition for the wave function. The wavefunction is not normalizable for a plane wave, but is for a wavepacket.

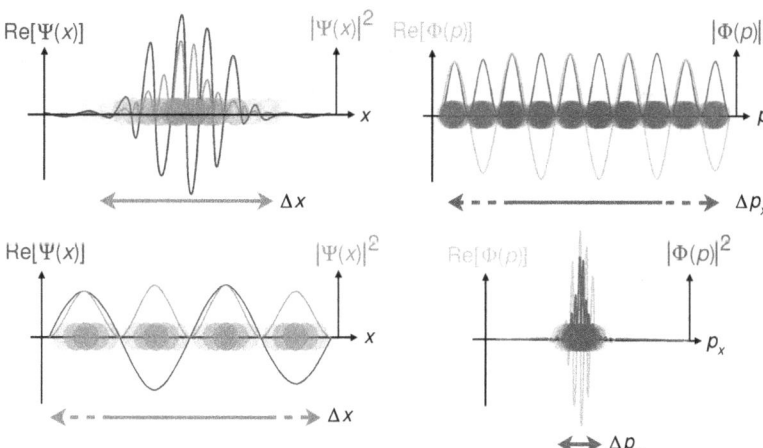

Increasing amounts of wavepacket localization, meaning the particle becomes more localized.

Perfect localization

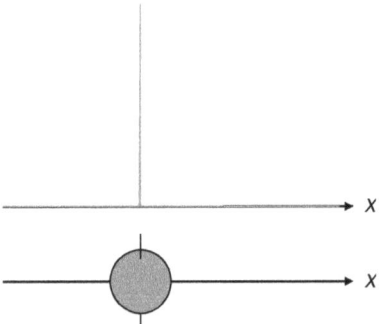

In the limit $\hbar \to 0$, the particle's position and momentum become known exactly.

Interpretation of wave function for one spin-0 particle in one dimension. The wavefunctions shown are continuous, finite, single-valued and normalized. The colour opacity (%) of the particles corresponds to the probability density (which can measure in %) of finding the particle at the points on the x-axis.

In this case, the free particle wavefunction may be represented by a superposition of free particle *momentum* eigenfunctions $\phi(\mathbf{k})$, the Fourier transform of the momentum space wavefunction:

$$\psi(\mathbf{r}, t) = \frac{1}{(\sqrt{2\pi\hbar})^3} \int_{\text{all p space}} A(\mathbf{p}) e^{i(\mathbf{p}\cdot\mathbf{r} - Et)/\hbar} d^3\mathbf{p} = \frac{1}{(\sqrt{2\pi})^3} \int_{\text{all k space}} A(\mathbf{k}) e^{i(\mathbf{k}\cdot\mathbf{r} - \omega t)} d^3\mathbf{k}$$

where the integral is over all $\mathbf{k}$-space, and $E = E(\mathbf{p}) = \frac{\mathbf{p}^2}{2m}$ and $\omega = \omega(\mathbf{k}) = \frac{\hbar \mathbf{k}^2}{2m}$ (to ensure that the wavepacket is a solution of the free particle Schrödinger equation). Note that here we abuse notation and denote $A(\mathbf{p}) = A(\hbar\mathbf{k})$ and $A(\mathbf{k})$ with the same symbol, when we should denote $\hat{A}(\mathbf{k}) = A(\hbar\mathbf{k})$, where A is the $\mathbf{p}$-space and $\hat{A}$ the $\mathbf{k}$-space function.

The expectation value of the momentum $\mathbf{p}$ for the complex plane wave is

$$\langle \mathbf{p} \rangle = \langle \psi \left| -i\hbar\nabla \right| \psi \rangle = \int_{\text{all space}} \psi^*(\mathbf{r}, t)(-i\hbar\nabla)\psi(\mathbf{r}, t) d^3\mathbf{r} = \hbar\mathbf{k}$$

and for the general wavepacket it is

$$\langle \mathbf{p} \rangle = \int_{\text{all space}} \psi^*(\mathbf{r}, t)(-i\hbar\nabla)\psi(\mathbf{r}, t) d^3\mathbf{r} = \int_{\text{all k space}} \hbar\mathbf{k} |A(\mathbf{k})|^2 d^3\mathbf{k}$$

The expectation value of the energy E is (for both plane wave and general wave packet; here one can observe the special status of time and hence energy in quantum mechanics as opposed to space and momentum)

$$\langle E \rangle = \left\langle \psi \left| i\hbar\frac{\partial}{\partial t} \right| \psi \right\rangle = \int_{\text{all space}} \psi^*(\mathbf{r}, t) \left(i\hbar\frac{\partial}{\partial t} \right) \psi(\mathbf{r}, t) d^3\mathbf{r} = \hbar\omega$$

For the plane wave, solving for $\mathbf{k}$ and ω and substituting into the constraint equation yields the familiar relationship between energy and momentum for non-relativistic massive particles

$$\langle E \rangle = \frac{\langle \mathbf{p} \rangle^2}{2m}.$$

In general, the identity holds in the form

$$\langle E \rangle = \frac{\langle p^2 \rangle}{2m}$$

where $p = |\mathbf{p}|$ is the magnitude of the momentum vector.

The group velocity of the plane wave is defined as

$$v_g = \frac{d\omega}{dk}$$

which turns out to be the classical velocity of the particle. The phase velocity of the plane wave is defined as

$$v_p = \frac{\omega}{k} = \frac{E}{p} = \frac{p}{2m} = \frac{v}{2}$$

8.3   Relativistic quantum free particle

Main article: Quantum field theory

There are a number of equations describing relativistic particles: see relativistic wave equations.

8.4   See also

- Particle in a box

- Finite square well

- Delta potential

- Wave packet

8.5   Sources

- *Quantum Mechanics Demystified*, D. McMahon, Mc Graw Hill (USA), 2006, ISBN(10-) 0-07-145546 9

- *Quantum Physics of Atoms, Molecules, Solids, Nuclei, and Particles (2nd Edition)*, R. Eisberg, R. Resnick, John Wiley & Sons, 1985, ISBN 978-0-471-87373-0

- *Quantum Mechanics*, E. Abers, Pearson Ed., Addison Wesley, Prentice Hall Inc, 2004, ISBN 978-0-13-146100-0

- *Elementary Quantum Mechanics*, N.F. Mott, Wykeham Science, Wykeham Press (Taylor & Francis Group), 1972, ISBN 0-85109-270-5

- *Stationary States*, A. Holden, College Physics Monographs (USA), Oxford University Press, 1971, ISBN 0-19-851121-3

- *Quantum mechanics*, E. Zaarur, Y. Peleg, R. Pnini, Schaum's Oulines, Mc Graw Hill (USA), 1998, ISBN (10-) 007-0540187

8.6 Further reading

- *The New Quantum Universe*, T.Hey, P.Walters, Cambridge University Press, 2009, ISBN 978-0-521-56457-1.

- *Quantum Field Theory*, D. McMahon, Mc Graw Hill (USA), 2008, ISBN 978-0-07-154382-8

- *Quantum mechanics*, E. Zaarur, Y. Peleg, R. Pnini, Schaum's Easy Oulines Crash Course, Mc Graw Hill (USA), 2006, ISBN (10-)007-145533-7 ISBN (13-)978-007-145533-6

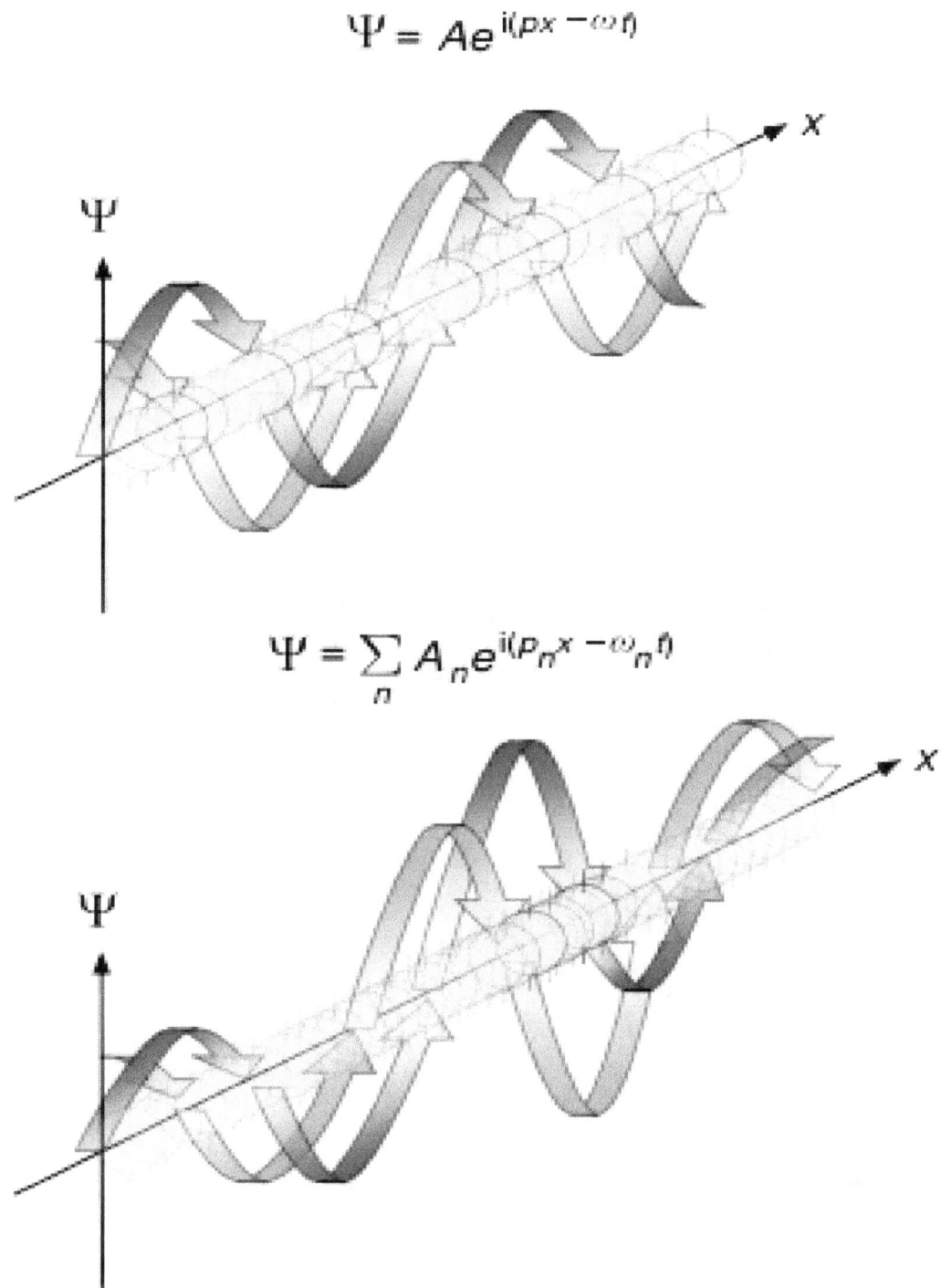

Propagation of de Broglie waves in 1d – real part of the complex amplitude is blue, imaginary part is green. The probability (shown as the colour opacity) of finding the particle at a given point x is spread out like a waveform, there is no definite position of the particle. As the amplitude increases above zero the curvature decreases, so the amplitude decreases again, and vice versa – the result is an alternating amplitude: a wave. Top: plane wave. Bottom: wave packet.

Chapter 9

Quark

This article is about the particle. For other uses, see Quark (disambiguation).

A **quark** (/ˈkwɔrk/ or /ˈkwɑrk/) is an elementary particle and a fundamental constituent of matter. Quarks combine to form composite particles called hadrons, the most stable of which are protons and neutrons, the components of atomic nuclei.[1] Due to a phenomenon known as *color confinement*, quarks are never directly observed or found in isolation; they can be found only within hadrons, such as baryons (of which protons and neutrons are examples), and mesons.[2][3] For this reason, much of what is known about quarks has been drawn from observations of the hadrons themselves.

Quarks have various intrinsic properties, including electric charge, mass, color charge and spin. Quarks are the only elementary particles in the Standard Model of particle physics to experience all four fundamental interactions, also known as *fundamental forces* (electromagnetism, gravitation, strong interaction, and weak interaction), as well as the only known particles whose electric charges are not integer multiples of the elementary charge.

There are six types of quarks, known as *flavors*: up, down, strange, charm, top, and bottom.[4] Up and down quarks have the lowest masses of all quarks. The heavier quarks rapidly change into up and down quarks through a process of particle decay: the transformation from a higher mass state to a lower mass state. Because of this, up and down quarks are generally stable and the most common in the universe, whereas strange, charm, bottom, and top quarks can only be produced in high energy collisions (such as those involving cosmic rays and in particle accelerators). For every quark flavor there is a corresponding type of antiparticle, known as an *antiquark*, that differs from the quark only in that some of its properties have equal magnitude but opposite sign.

The quark model was independently proposed by physicists Murray Gell-Mann and George Zweig in 1964.[5] Quarks were introduced as parts of an ordering scheme for hadrons, and there was little evidence for their physical existence until deep inelastic scattering experiments at the Stanford Linear Accelerator Center in 1968.[6][7] Accelerator experiments have provided evidence for all six flavors. The top quark was the last to be discovered at Fermilab in 1995.[5]

9.1 Classification

See also: Standard Model

The Standard Model is the theoretical framework describing all the currently known elementary particles. This model contains six flavors of quarks (q), named up (u), down (d), strange (s), charm (c), bottom (b), and top (t).[4] Antiparticles of quarks are called *antiquarks*, and are denoted by a bar over the symbol for the corresponding quark, such as u for an up antiquark. As with antimatter in general, antiquarks have the same mass, mean lifetime, and spin as their respective quarks, but the electric charge and other charges have the opposite sign.[8]

Quarks are spin-$\frac{1}{2}$ particles, implying that they are fermions according to the spin-statistics theorem. They are subject to the Pauli exclusion principle, which states that no two identical fermions can simultaneously occupy the same quantum state. This is in contrast to bosons (particles with integer spin), any number of which can be in the same state.[9] Unlike

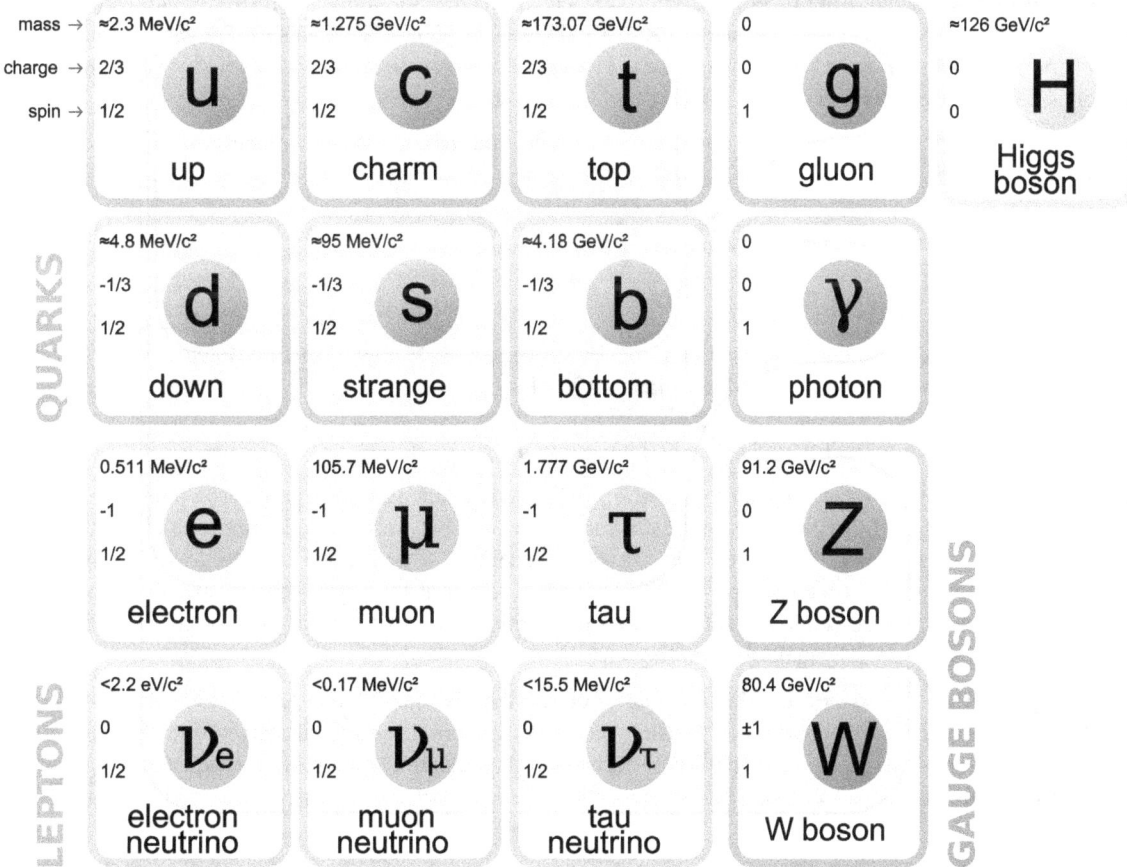

Six of the particles in the Standard Model are quarks (shown in purple). Each of the first three columns forms a generation of matter.

leptons, quarks possess color charge, which causes them to engage in the strong interaction. The resulting attraction between different quarks causes the formation of composite particles known as *hadrons* (see "Strong interaction and color charge" below).

The quarks which determine the quantum numbers of hadrons are called *valence quarks*; apart from these, any hadron may contain an indefinite number of virtual (or *sea*) quarks, antiquarks, and gluons which do not influence its quantum numbers.[10] There are two families of hadrons: baryons, with three valence quarks, and mesons, with a valence quark and an antiquark.[11] The most common baryons are the proton and the neutron, the building blocks of the atomic nucleus.[12] A great number of hadrons are known (see list of baryons and list of mesons), most of them differentiated by their quark content and the properties these constituent quarks confer. The existence of "exotic" hadrons with more valence quarks, such as tetraquarks (qqqq) and pentaquarks (qqqqq), has been conjectured[13] but not proven.[nb 1][13][14] However, on 13 July 2015, the LHCb collaboration at CERN reported results consistent with pentaquark states.[15]

Elementary fermions are grouped into three generations, each comprising two leptons and two quarks. The first generation includes up and down quarks, the second strange and charm quarks, and the third bottom and top quarks. All searches for a fourth generation of quarks and other elementary fermions have failed,[16] and there is strong indirect evidence that no more than three generations exist.[nb 2][17] Particles in higher generations generally have greater mass and less stability, causing them to decay into lower-generation particles by means of weak interactions. Only first-generation (up and down) quarks occur commonly in nature. Heavier quarks can only be created in high-energy collisions (such as in those involving cosmic rays), and decay quickly; however, they are thought to have been present during the first fractions of a second after the Big Bang, when the universe was in an extremely hot and dense phase (the quark epoch). Studies of heavier quarks are conducted in artificially created conditions, such as in particle accelerators.[18]

Having electric charge, mass, color charge, and flavor, quarks are the only known elementary particles that engage in all four fundamental interactions of contemporary physics: electromagnetism, gravitation, strong interaction, and weak interaction.[12] Gravitation is too weak to be relevant to individual particle interactions except at extremes of energy (Planck energy) and distance scales (Planck distance). However, since no successful quantum theory of gravity exists, gravitation is not described by the Standard Model.

See the table of properties below for a more complete overview of the six quark flavors' properties.

9.2 History

The quark model was independently proposed by physicists Murray Gell-Mann[19] (pictured) and George Zweig[20][21] in 1964.[5] The proposal came shortly after Gell-Mann's 1961 formulation of a particle classification system known as the *Eightfold Way*—or, in more technical terms, SU(3) flavor symmetry.[22] Physicist Yuval Ne'eman had independently developed a scheme similar to the Eightfold Way in the same year.[23][24]

At the time of the quark theory's inception, the "particle zoo" included, amongst other particles, a multitude of hadrons. Gell-Mann and Zweig posited that they were not elementary particles, but were instead composed of combinations of quarks and antiquarks. Their model involved three flavors of quarks, up, down, and strange, to which they ascribed properties such as spin and electric charge.[19][20][21] The initial reaction of the physics community to the proposal was mixed. There was particular contention about whether the quark was a physical entity or a mere abstraction used to explain concepts that were not fully understood at the time.[25]

In less than a year, extensions to the Gell-Mann–Zweig model were proposed. Sheldon Lee Glashow and James Bjorken predicted the existence of a fourth flavor of quark, which they called *charm*. The addition was proposed because it allowed for a better description of the weak interaction (the mechanism that allows quarks to decay), equalized the number of known quarks with the number of known leptons, and implied a mass formula that correctly reproduced the masses of the known mesons.[26]

In 1968, deep inelastic scattering experiments at the Stanford Linear Accelerator Center (SLAC) showed that the proton contained much smaller, point-like objects and was therefore not an elementary particle.[6][7][27] Physicists were reluctant to firmly identify these objects with quarks at the time, instead calling them "partons"—a term coined by Richard Feynman.[28][29][30] The objects that were observed at SLAC would later be identified as up and down quarks as the other flavors were discovered.[31] Nevertheless, "parton" remains in use as a collective term for the constituents of hadrons (quarks, antiquarks, and gluons).

The strange quark's existence was indirectly validated by SLAC's scattering experiments: not only was it a necessary component of Gell-Mann and Zweig's three-quark model, but it provided an explanation for the kaon (K) and pion (π) hadrons discovered in cosmic rays in 1947.[32]

In a 1970 paper, Glashow, John Iliopoulos and Luciano Maiani presented the so-called GIM mechanism to explain the experimental non-observation of flavor-changing neutral currents. This theoretical model required the existence of the as-yet undiscovered charm quark.[33][34] The number of supposed quark flavors grew to the current six in 1973, when Makoto Kobayashi and Toshihide Maskawa noted that the experimental observation of CP violation[nb 3][35] could be explained if there were another pair of quarks.

Charm quarks were produced almost simultaneously by two teams in November 1974 (see November Revolution)—one at SLAC under Burton Richter, and one at Brookhaven National Laboratory under Samuel Ting. The charm quarks were observed bound with charm antiquarks in mesons. The two parties had assigned the discovered meson two different symbols, J and ψ; thus, it became formally known as the J/ψ meson. The discovery finally convinced the physics community of the quark model's validity.[30]

In the following years a number of suggestions appeared for extending the quark model to six quarks. Of these, the 1975 paper by Haim Harari[36] was the first to coin the terms *top* and *bottom* for the additional quarks.[37]

In 1977, the bottom quark was observed by a team at Fermilab led by Leon Lederman.[38][39] This was a strong indicator of the top quark's existence: without the top quark, the bottom quark would have been without a partner. However, it was not until 1995 that the top quark was finally observed, also by the CDF[40] and DØ[41] teams at Fermilab.[5] It had a mass much larger than had been previously expected,[42] almost as large as that of a gold atom.[43]

Murray Gell-Mann at TED in 2007. Gell-Mann and George Zweig proposed the quark model in 1964.

9.3   Etymology

For some time, Gell-Mann was undecided on an actual spelling for the term he intended to coin, until he found the word *quark* in James Joyce's book *Finnegans Wake*:

> Three quarks for Muster Mark!
> Sure he has not got much of a bark

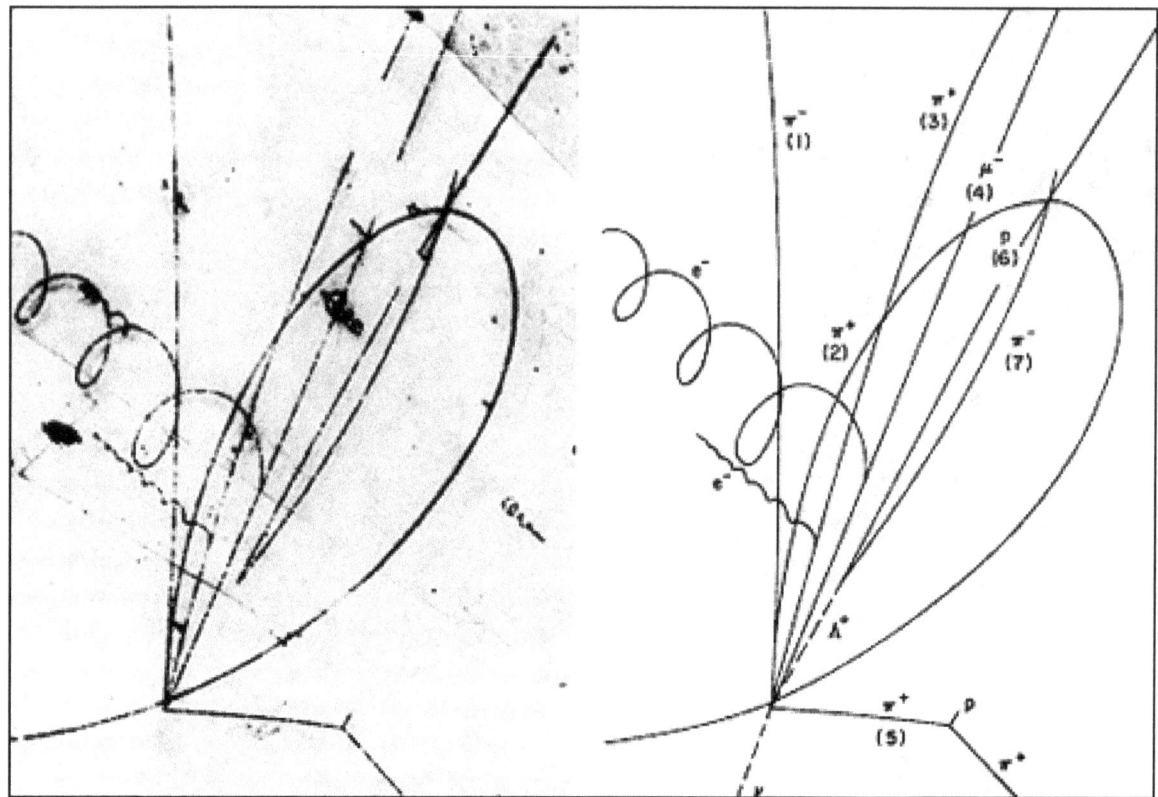

Photograph of the event that led to the discovery of the Σ++
c baryon, at the Brookhaven National Laboratory in 1974

And sure any he has it's all beside the mark.
— James Joyce, *Finnegans Wake*[44]

Gell-Mann went into further detail regarding the name of the quark in his book *The Quark and the Jaguar*:[45]

In 1963, when I assigned the name "quark" to the fundamental constituents of the nucleon, I had the sound first, without the spelling, which could have been "kwork". Then, in one of my occasional perusals of *Finnegans Wake*, by James Joyce, I came across the word "quark" in the phrase "Three quarks for Muster Mark". Since "quark" (meaning, for one thing, the cry of the gull) was clearly intended to rhyme with "Mark", as well as "bark" and other such words, I had to find an excuse to pronounce it as "kwork". But the book represents the dream of a publican named Humphrey Chimpden Earwicker. Words in the text are typically drawn from several sources at once, like the "portmanteau" words in "Through the Looking-Glass". From time to time, phrases occur in the book that are partially determined by calls for drinks at the bar. I argued, therefore, that perhaps one of the multiple sources of the cry "Three quarks for Muster Mark" might be "Three quarts for Mister Mark", in which case the pronunciation "kwork" would not be totally unjustified. In any case, the number three fitted perfectly the way quarks occur in nature.

Zweig preferred the name *ace* for the particle he had theorized, but Gell-Mann's terminology came to prominence once the quark model had been commonly accepted.[46]

The quark flavors were given their names for several reasons. The up and down quarks are named after the up and down components of isospin, which they carry.[47] Strange quarks were given their name because they were discovered to be components of the strange particles discovered in cosmic rays years before the quark model was proposed; these particles were deemed "strange" because they had unusually long lifetimes.[48] Glashow, who coproposed charm quark with Bjorken, is quoted as saying, "We called our construct the 'charmed quark', for we were fascinated and pleased by the symmetry it brought to the subnuclear world."[49] The names "bottom" and "top", coined by Harari, were chosen because they are "logical partners for up and down quarks".[36][37][48] In the past, bottom and top quarks were sometimes referred to as "beauty" and "truth" respectively, but these names have somewhat fallen out of use.[50] While "truth" never did catch on, accelerator complexes devoted to massive production of bottom quarks are sometimes called "beauty factories".[51]

9.4   Properties

9.4.1   Electric charge

See also: Electric charge

Quarks have fractional electric charge values – either $\frac{1}{3}$ or $\frac{2}{3}$ times the elementary charge (e), depending on flavor. Up, charm, and top quarks (collectively referred to as *up-type quarks*) have a charge of $+\frac{2}{3}$ e, while down, strange, and bottom quarks (*down-type quarks*) have $-\frac{1}{3}$ e. Antiquarks have the opposite charge to their corresponding quarks; up-type antiquarks have charges of $-\frac{2}{3}$ e and down-type antiquarks have charges of $+\frac{1}{3}$ e. Since the electric charge of a hadron is the sum of the charges of the constituent quarks, all hadrons have integer charges: the combination of three quarks (baryons), three antiquarks (antibaryons), or a quark and an antiquark (mesons) always results in integer charges.[52] For example, the hadron constituents of atomic nuclei, neutrons and protons, have charges of 0 e and +1 e respectively; the neutron is composed of two down quarks and one up quark, and the proton of two up quarks and one down quark.[12]

9.4.2   Spin

See also: Spin (physics)

Spin is an intrinsic property of elementary particles, and its direction is an important degree of freedom. It is sometimes visualized as the rotation of an object around its own axis (hence the name "spin"), though this notion is somewhat misguided at subatomic scales because elementary particles are believed to be point-like.[53]

Spin can be represented by a vector whose length is measured in units of the reduced Planck constant $\hbar$ (pronounced "h bar"). For quarks, a measurement of the spin vector component along any axis can only yield the values $+\hbar/2$ or $-\hbar/2$; for this reason quarks are classified as spin-$\frac{1}{2}$ particles.[54] The component of spin along a given axis – by convention the z axis – is often denoted by an up arrow ↑ for the value $+\frac{1}{2}$ and down arrow ↓ for the value $-\frac{1}{2}$, placed after the symbol for flavor. For example, an up quark with a spin of $+\frac{1}{2}$ along the z axis is denoted by u↑.[55]

9.4.3   Weak interaction

Main article: Weak interaction

A quark of one flavor can transform into a quark of another flavor only through the weak interaction, one of the four fundamental interactions in particle physics. By absorbing or emitting a W boson, any up-type quark (up, charm, and top quarks) can change into any down-type quark (down, strange, and bottom quarks) and vice versa. This flavor transformation mechanism causes the radioactive process of beta decay, in which a neutron (n) "splits" into a proton (p), an electron (e−) and an electron antineutrino (v

e) (see picture). This occurs when one of the down quarks in the neutron (udd) decays into an up quark by emitting a virtual W− boson, transforming the neutron into a proton (uud). The W− boson then decays into an electron and an electron antineutrino.[56]

Both beta decay and the inverse process of *inverse beta decay* are routinely used in medical applications such as positron emission tomography (PET) and in experiments involving neutrino detection.

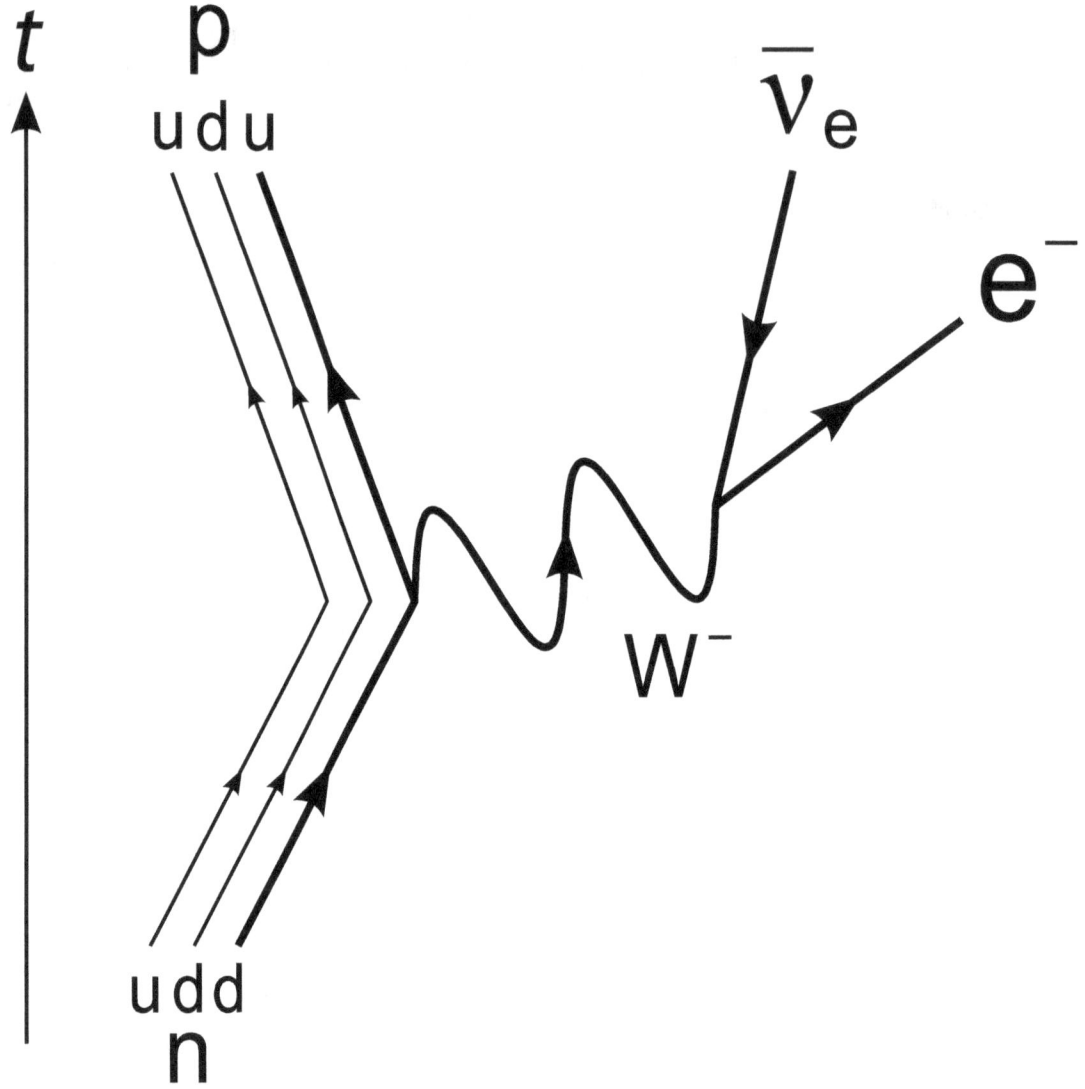

Feynman diagram of beta decay with time flowing upwards. The CKM matrix (discussed below) encodes the probability of this and other quark decays.

While the process of flavor transformation is the same for all quarks, each quark has a preference to transform into the quark of its own generation. The relative tendencies of all flavor transformations are described by a mathematical table, called the Cabibbo–Kobayashi–Maskawa matrix (CKM matrix). Enforcing unitarity, the approximate magnitudes of the entries of the CKM matrix are:[57]

$$
\begin{bmatrix} |V_{ud}| & |V_{us}| & |V_{ub}| \\ |V_{cd}| & |V_{cs}| & |V_{cb}| \\ |V_{td}| & |V_{ts}| & |V_{tb}| \end{bmatrix} \approx \begin{bmatrix} 0.974 & 0.225 & 0.003 \\ 0.225 & 0.973 & 0.041 \\ 0.009 & 0.040 & 0.999 \end{bmatrix},
$$

where Vij represents the tendency of a quark of flavor i to change into a quark of flavor j (or vice versa).[nb 4]

There exists an equivalent weak interaction matrix for leptons (right side of the W boson on the above beta decay diagram), called the Pontecorvo–Maki–Nakagawa–Sakata matrix (PMNS matrix).[58] Together, the CKM and PMNS matrices

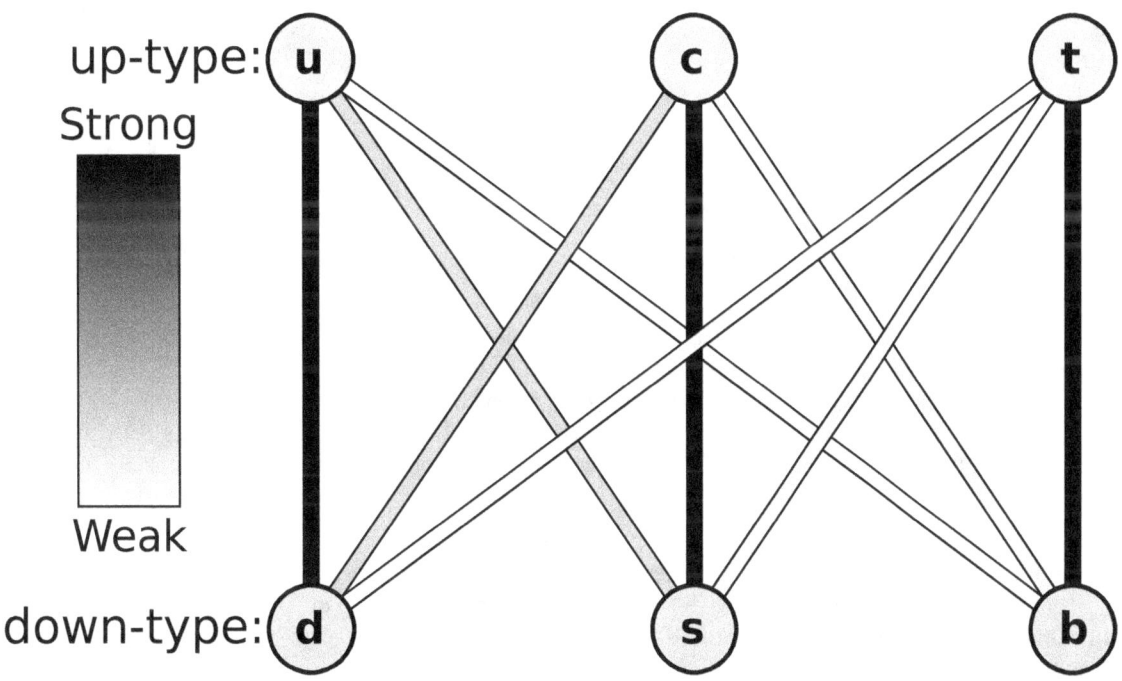

The strengths of the weak interactions between the six quarks. The "intensities" of the lines are determined by the elements of the CKM matrix.

describe all flavor transformations, but the links between the two are not yet clear.[59]

9.4.4 Strong interaction and color charge

See also: Color charge and Strong interaction

According to quantum chromodynamics (QCD), quarks possess a property called *color charge*. There are three types of color charge, arbitrarily labeled *blue*, *green*, and *red*.[nb 5] Each of them is complemented by an anticolor – *antiblue*, *antigreen*, and *antired*. Every quark carries a color, while every antiquark carries an anticolor.[60]

The system of attraction and repulsion between quarks charged with different combinations of the three colors is called strong interaction, which is mediated by force carrying particles known as *gluons*; this is discussed at length below. The theory that describes strong interactions is called quantum chromodynamics (QCD). A quark, which will have a single color value, can form a bound system with an antiquark carrying the corresponding anticolor. The result of two attracting quarks will be color neutrality: a quark with color charge ξ plus an antiquark with color charge $-\xi$ will result in a color charge of 0 (or "white" color) and the formation of a meson. This is analogous to the additive color model in basic optics. Similarly, the combination of three quarks, each with different color charges, or three antiquarks, each with anticolor charges, will result in the same "white" color charge and the formation of a baryon or antibaryon.[61]

In modern particle physics, gauge symmetries – a kind of symmetry group – relate interactions between particles (see gauge theories). Color SU(3) (commonly abbreviated to SU(3)$_c$) is the gauge symmetry that relates the color charge in quarks and is the defining symmetry for quantum chromodynamics.[62] Just as the laws of physics are independent of which directions in space are designated x, y, and z, and remain unchanged if the coordinate axes are rotated to a new orientation, the physics of quantum chromodynamics is independent of which directions in three-dimensional color space are identified as blue, red, and green. SU(3)$_c$ color transformations correspond to "rotations" in color space (which, mathematically speaking, is a complex space). Every quark flavor f, each with subtypes fB, fG, fR corresponding to the quark colors,[63] forms a triplet: a three-component quantum field which transforms under the fundamental representation of $SU(3)_c$.[64] The requirement that $SU(3)_c$ should be local – that is, that its transformations be allowed to vary with space and time – determines the properties of the strong interaction, in particular the existence of eight gluon types to act as its force carriers.[62][65]

9.4.5 Mass

See also: Invariant mass

Two terms are used in referring to a quark's mass: *current quark mass* refers to the mass of a quark by itself, while *constituent quark mass* refers to the current quark mass plus the mass of the gluon particle field surrounding the quark.[66] These masses typically have very different values. Most of a hadron's mass comes from the gluons that bind the constituent quarks together, rather than from the quarks themselves. While gluons are inherently massless, they possess energy – more specifically, quantum chromodynamics binding energy (QCBE) – and it is this that contributes so greatly to the overall mass of the hadron (see mass in special relativity). For example, a proton has a mass of approximately 938 MeV/c^2, of which the rest mass of its three valence quarks only contributes about 11 MeV/c^2; much of the remainder can be attributed to the gluons' QCBE.[67][68]

The Standard Model posits that elementary particles derive their masses from the Higgs mechanism, which is related to the Higgs boson. Physicists hope that further research into the reasons for the top quark's large mass of ~173 GeV/c^2, almost the mass of a gold atom,[67][69] might reveal more about the origin of the mass of quarks and other elementary particles.[70]

9.4.6 Table of properties

See also: Flavor (particle physics)

The following table summarizes the key properties of the six quarks. Flavor quantum numbers (isospin (I_3), charm (C), strangeness (S, not to be confused with spin), topness (T), and bottomness (B')) are assigned to certain quark flavors, and denote qualities of quark-based systems and hadrons. The baryon number (B) is $+\frac{1}{3}$ for all quarks, as baryons are made of three quarks. For antiquarks, the electric charge (Q) and all flavor quantum numbers (B, I_3, C, S, T, and B') are of opposite sign. Mass and total angular momentum (J; equal to spin for point particles) do not change sign for the antiquarks.

J = total angular momentum, B = baryon number, Q = electric charge, I_3 = isospin, C = charm, S = strangeness, T = topness, B' = bottomness.

* Notation such as 4190+180

−60 denotes measurement uncertainty. In the case of the top quark, the first uncertainty is statistical in nature, and the second is systematic.

9.5   Interacting quarks

See also: Color confinement and Gluon

As described by quantum chromodynamics, the strong interaction between quarks is mediated by gluons, massless vector gauge bosons. Each gluon carries one color charge and one anticolor charge. In the standard framework of particle interactions (part of a more general formulation known as perturbation theory), gluons are constantly exchanged between quarks through a virtual emission and absorption process. When a gluon is transferred between quarks, a color change occurs in both; for example, if a red quark emits a red–antigreen gluon, it becomes green, and if a green quark absorbs a red–antigreen gluon, it becomes red. Therefore, while each quark's color constantly changes, their strong interaction is preserved.[71][72][73]

Since gluons carry color charge, they themselves are able to emit and absorb other gluons. This causes *asymptotic freedom*: as quarks come closer to each other, the chromodynamic binding force between them weakens.[74] Conversely, as the distance between quarks increases, the binding force strengthens. The color field becomes stressed, much as an elastic band is stressed when stretched, and more gluons of appropriate color are spontaneously created to strengthen the field. Above a certain energy threshold, pairs of quarks and antiquarks are created. These pairs bind with the quarks being separated, causing new hadrons to form. This phenomenon is known as *color confinement*: quarks never appear in isolation.[75][76] This process of hadronization occurs before quarks, formed in a high energy collision, are able to interact in any other way. The only exception is the top quark, which may decay before it hadronizes.[77]

9.5.1   Sea quarks

Hadrons, along with the *valence quarks* (q
v) that contribute to their quantum numbers, contain virtual quark–antiquark (qq) pairs known as *sea quarks* (q
s). Sea quarks form when a gluon of the hadron's color field splits; this process also works in reverse in that the annihilation of two sea quarks produces a gluon. The result is a constant flux of gluon splits and creations colloquially known as "the sea".[78] Sea quarks are much less stable than their valence counterparts, and they typically annihilate each other within the interior of the hadron. Despite this, sea quarks can hadronize into baryonic or mesonic particles under certain circumstances.[79]

9.5.2   Other phases of quark matter

Main article: QCD matter

Under sufficiently extreme conditions, quarks may become deconfined and exist as free particles. In the course of asymptotic freedom, the strong interaction becomes weaker at higher temperatures. Eventually, color confinement would be lost and an extremely hot plasma of freely moving quarks and gluons would be formed. This theoretical phase of matter is called quark–gluon plasma.[82] The exact conditions needed to give rise to this state are unknown and have been the subject of a great deal of speculation and experimentation. A recent estimate puts the needed temperature at $(1.90\pm0.02)\times10^{12}$ kelvin.[83] While a state of entirely free quarks and gluons has never been achieved (despite numerous attempts by CERN in the 1980s and 1990s),[84] recent experiments at the Relativistic Heavy Ion Collider have yielded evidence for liquid-like quark matter exhibiting "nearly perfect" fluid motion.[85]

The quark–gluon plasma would be characterized by a great increase in the number of heavier quark pairs in relation to the number of up and down quark pairs. It is believed that in the period prior to 10^{-6} seconds after the Big Bang (the quark epoch), the universe was filled with quark–gluon plasma, as the temperature was too high for hadrons to be stable.[86]

Given sufficiently high baryon densities and relatively low temperatures – possibly comparable to those found in neutron stars – quark matter is expected to degenerate into a Fermi liquid of weakly interacting quarks. This liquid would be characterized by a condensation of colored quark Cooper pairs, thereby breaking the local $SU(3)_c$ symmetry. Because quark Cooper pairs harbor color charge, such a phase of quark matter would be color superconductive; that is, color charge would be able to pass through it with no resistance.[87]

9.6   See also

- Color–flavor locking

- Neutron magnetic moment

- Leptons

- Preons – Hypothetical particles which were once postulated to be subcomponents of quarks and leptons

- Quarkonium – Mesons made of a quark and antiquark of the same flavor

- Quark star – A hypothetical degenerate neutron star with extreme density

- Quark–lepton complementarity – Possible fundamental relation between quarks and leptons

9.7 Notes

[1] Several research groups claimed to have proven the existence of tetraquarks and pentaquarks in the early 2000s. While the status of tetraquarks is still under debate, all known pentaquark candidates have previously been established as non-existent.

[2] The main evidence is based on the resonance width of the Z0 boson, which constrains the 4th generation neutrino to have a mass greater than ~45 GeV/c^2. This would be highly contrasting with the other three generations' neutrinos, whose masses cannot exceed 2 MeV/c^2.

[3] CP violation is a phenomenon which causes weak interactions to behave differently when left and right are swapped (P symmetry) and particles are replaced with their corresponding antiparticles (C symmetry).

[4] The actual probability of decay of one quark to another is a complicated function of (amongst other variables) the decaying quark's mass, the masses of the decay products, and the corresponding element of the CKM matrix. This probability is directly proportional (but not equal) to the magnitude squared ($|Vij|^2$) of the corresponding CKM entry.

[5] Despite its name, color charge is not related to the color spectrum of visible light.

9.8 References

[1] "Quark (subatomic particle)". *Encyclopædia Britannica*. Retrieved 2008-06-29.

[2] R. Nave. "Confinement of Quarks". *HyperPhysics*. Georgia State University, Department of Physics and Astronomy. Retrieved 2008-06-29.

[3] R. Nave. "Bag Model of Quark Confinement". *HyperPhysics*. Georgia State University, Department of Physics and Astronomy. Retrieved 2008-06-29.

[4] R. Nave. "Quarks". *HyperPhysics*. Georgia State University, Department of Physics and Astronomy. Retrieved 2008-06-29.

[5] B. Carithers, P. Grannis (1995). "Discovery of the Top Quark" (PDF). *Beam Line* (SLAC) **25** (3): 4–16. Retrieved 2008-09-23.

[6] E.D. Bloom; et al. (1969). "High-Energy Inelastic *e–p* Scattering at 6° and 10°". *Physical Review Letters* **23** (16): 930–934. Bibcode:1969PhRvL..23..930B. doi:10.1103/PhysRevLett.23.930.

[7] M. Breidenbach; et al. (1969). "Observed Behavior of Highly Inelastic Electron–Proton Scattering". *Physical Review Letters* **23** (16): 935–939. Bibcode:1969PhRvL..23..935B. doi:10.1103/PhysRevLett.23.935.

[8] S.S.M. Wong (1998). *Introductory Nuclear Physics* (2nd ed.). Wiley Interscience. p. 30. ISBN 0-471-23973-9.

[9] K.A. Peacock (2008). *The Quantum Revolution*. Greenwood Publishing Group. p. 125. ISBN 0-313-33448-X.

[10] B. Povh, C. Scholz, K. Rith, F. Zetsche (2008). *Particles and Nuclei*. Springer. p. 98. ISBN 3-540-79367-4.

[11] Section 6.1. in P.C.W. Davies (1979). *The Forces of Nature*. Cambridge University Press. ISBN 0-521-22523-X.

[12] M. Munowitz (2005). *Knowing*. Oxford University Press. p. 35. ISBN 0-19-516737-6.

[13] W.-M. Yao (Particle Data Group); et al. (2006). "Review of Particle Physics: Pentaquark Update" (PDF). *Journal of Physics G* **33** (1): 1–1232. arXiv:astro-ph/0601168. Bibcode:2006JPhG...33....1Y. doi:10.1088/0954-3899/33/1/001.

[14] C. Amsler (Particle Data Group); et al. (2008). "Review of Particle Physics: Pentaquarks" (PDF). *Physics Letters B* **667** (1): 1–1340. Bibcode:2008PhLB..667....1P. doi:10.1016/j.physletb.2008.07.018.
C. Amsler (Particle Data Group); et al. (2008). "Review of Particle Physics: New Charmonium-Like States" (PDF). *Physics Letters B* **667** (1): 1–1340. Bibcode:2008PhLB..667....1P. doi:10.1016/j.physletb.2008.07.018.
E.V. Shuryak (2004). *The QCD Vacuum, Hadrons and Superdense Matter*. World Scientific. p. 59. ISBN 981-238-574-6.

[15] R. Aaij et al. (LHCb collaboration) (2015). "Observation of J/ψp resonances consistent with pentaquark states in Λ0
      b→J/ψK−
      p decays". *Physical Review Letters* **115** (7). doi:10.1103/PhysRevLett.115.072001.

[16] C. Amsler (Particle Data Group); et al. (2008). "Review of Particle Physics: b′ (4th Generation) Quarks, Searches for" (PDF).
      Physics Letters B **667** (1): 1–1340. Bibcode:2008PhLB..667....1P. doi:10.1016/j.physletb.2008.07.018.
      C. Amsler (Particle Data Group); et al. (2008). "Review of Particle Physics: t′ (4th Generation) Quarks, Searches for" (PDF).
      Physics Letters B **667** (1): 1–1340. Bibcode:2008PhLB..667....1P. doi:10.1016/j.physletb.2008.07.018.

[17] D. Decamp; Deschizeaux, B.; Lees, J.-P.; Minard, M.-N.; Crespo, J.M.; Delfino, M.; Fernandez, E.; Martinez, M.; et al. (1989).
      "Determination of the number of light neutrino species". *Physics Letters B* **231** (4): 519. Bibcode:1989PhLB..231..519D.
      doi:10.1016/0370-2693(89)90704-1.
      A. Fisher (1991). "Searching for the Beginning of Time: Cosmic Connection". *Popular Science* **238** (4): 70.
      J.D. Barrow (1997) [1994]. "The Singularity and Other Problems". *The Origin of the Universe* (Reprint ed.). Basic Books.
      ISBN 978-0-465-05314-8.

[18] D.H. Perkins (2003). *Particle Astrophysics*. Oxford University Press. p. 4. ISBN 0-19-850952-9.

[19] M.Gell-Mann(1964). "A Schematic Model of Baryons and Mesons".*Physics Letters***8**(3): 214–215. Bibcode:1964PhL.
      doi:10.1016/S0031-9163(64)92001-3.

[20] G. Zweig (1964). "An SU(3) Model for Strong Interaction Symmetry and its Breaking" (PDF). *CERN Report No.8182/TH.401*.

[21] G. Zweig (1964). "An SU(3) Model for Strong Interaction Symmetry and its Breaking: II" (PDF). *CERN Report No.8419/TH.412*.

[22] M. Gell-Mann (2000) [1964]. "The Eightfold Way: A theory of strong interaction symmetry". In M. Gell-Mann, Y. Ne'eman.
      The Eightfold Way. Westview Press. p. 11. ISBN 0-7382-0299-1.
      Original: M. Gell-Mann (1961). "The Eightfold Way: A theory of strong interaction symmetry". *Synchrotron Laboratory
      Report CTSL-20* (California Institute of Technology).

[23] Y. Ne'eman (2000) [1964]. "Derivation of strong interactions from gauge invariance". In M. Gell-Mann, Y. Ne'eman. *The
      Eightfold Way*. Westview Press. ISBN 0-7382-0299-1.
      Original Y.Ne'eman(1961). "Derivation of strong interactions from gauge invariancN.doi:10.1016/0029-5582(61)90134-1.

[24] R.C. Olby, G.N. Cantor (1996). *Companion to the History of Modern Science*. Taylor & Francis. p. 673. ISBN 0-415-14578-3.

[25] A. Pickering (1984). *Constructing Quarks*. University of Chicago Press. pp. 114–125. ISBN 0-226-66799-5.

[26] B.J.Bjorken,S.L.Glashow;Glashow(1964). "Elementary Particles and SU(4)".*Physics Letters***11**(3): 255–257. Bibcode:1964PB.
      doi:10.1016/0031-9163(64)90433-0.

[27] J.I. Friedman. "The Road to the Nobel Prize". Hue University. Retrieved 2008-09-29.

[28] R.P.Feynman(1969). "Very High-Energy Collisions of Hadrons".*Physical Review Letters***23**(24): 1415–1417. Bibcode:1F.
      doi:10.1103/PhysRevLett.23.1415.

[29] S. Kretzer; et al. (2004). "CTEQ6 Parton Distributions with Heavy Quark Mass Effects". *Physical Review D* **69** (11): 114005.
      arXiv:hep-ph/0307022. Bibcode:2004PhRvD..69k4005K. doi:10.1103/PhysRevD.69.114005.

[30] D.J. Griffiths (1987). *Introduction to Elementary Particles*. John Wiley & Sons. p. 42. ISBN 0-471-60386-4.

[31] M.E. Peskin, D.V. Schroeder (1995). *An introduction to quantum field theory*. Addison–Wesley. p. 556. ISBN 0-201-50397-2.

[32] V.V. Ezhela (1996). *Particle physics*. Springer. p. 2. ISBN 1-56396-642-5.

[33] S.L. Glashow, J. Iliopoulos, L. Maiani; Iliopoulos; Maiani (1970). "Weak Interactions with Lepton–Hadron Symmetry".
      Physical Review D **2** (7): 1285–1292. Bibcode:1970PhRvD...2.1285G. doi:10.1103/PhysRevD.2.1285.

[34] D.J. Griffiths (1987). *Introduction to Elementary Particles*. John Wiley & Sons. p. 44. ISBN 0-471-60386-4.

[35] M. Kobayashi, T. Maskawa; Maskawa (1973). "CP-Violation in the Renormalizable Theory of Weak Interaction". *Progress of
      Theoretical Physics* **49** (2): 652–657. Bibcode:1973PThPh..49..652K. doi:10.1143/PTP.49.652.

[36] H. Harari (1975). "A new quark model for hadrons". *Physics Letters B* **57B** (3): 265. Bibcode:1975PhLB...57..265H. doi:10.1016/0370-2693(75)90072-6.

[37] K.W. Staley (2004). *The Evidence for the Top Quark*. Cambridge University Press. pp. 31–33. ISBN 978-0-521-82710-2.

[38] S.W. Herb; et al. (1977). "Observation of a Dimuon Resonance at 9.5 GeV in 400-GeV Proton-Nucleus Collisions". *Physical Review Letters* **39** (5): 252. Bibcode:1977PhRvL..39..252H. doi:10.1103/PhysRevLett.39.252.

[39] M. Bartusiak (1994). *A Positron named Priscilla*. National Academies Press. p. 245. ISBN 0-309-04893-1.

[40] F. Abe (CDF Collaboration); et al. (1995). "Observation of Top Quark Production in pp Collisions with the Collider Detector at Fermilab". *Physical Review Letters* **74** (14): 2626–2631. Bibcode:1995PhRvL..74.2626A. doi:10.1103/PhysRevLett.74.2626. PMID 10057978.

[41] S. Abachi (DØ Collaboration); et al. (1995). "Search for High Mass Top Quark Production in pp Collisions at $\sqrt{s}$ = 1.8 TeV". *Physical Review Letters* **74** (13): 2422–2426. Bibcode:1995PhRvL..74.2422A. doi:10.1103/PhysRevLett.74.2422.

[42] K.W. Staley (2004). *The Evidence for the Top Quark*. Cambridge University Press. p. 144. ISBN 0-521-82710-8.

[43] "New Precision Measurement of Top Quark Mass". Brookhaven National Laboratory News. 2004. Retrieved 2013-11-03.

[44] J. Joyce (1982) [1939]. *Finnegans Wake*. Penguin Books. p. 383. ISBN 0-14-006286-6.

[45] M. Gell-Mann (1995). *The Quark and the Jaguar: Adventures in the Simple and the Complex*. Henry Holt and Co. p. 180. ISBN 978-0-8050-7253-2.

[46] J. Gleick (1992). *Genius: Richard Feynman and modern physics*. Little Brown and Company. p. 390. ISBN 0-316-90316-7.

[47] J.J. Sakurai (1994). S.F Tuan, ed. *Modern Quantum Mechanics* (Revised ed.). Addison–Wesley. p. 376. ISBN 0-201-53929-2.

[48] D.H. Perkins (2000). *Introduction to high energy physics*. Cambridge University Press. p. 8. ISBN 0-521-62196-8.

[49] M. Riordan (1987). *The Hunting of the Quark: A True Story of Modern Physics*. Simon & Schuster. p. 210. ISBN 978-0-671-50466-3.

[50] F. Close (2006). *The New Cosmic Onion*. CRC Press. p. 133. ISBN 1-58488-798-2.

[51] J.T. Volk; et al. (1987). "Letter of Intent for a Tevatron Beauty Factory" (PDF). Fermilab Proposal #783.

[52] G. Fraser (2006). *The New Physics for the Twenty-First Century*. Cambridge University Press. p. 91. ISBN 0-521-81600-9.

[53] "The Standard Model of Particle Physics". BBC. 2002. Retrieved 2009-04-19.

[54] F. Close (2006). *The New Cosmic Onion*. CRC Press. pp. 80–90. ISBN 1-58488-798-2.

[55] D. Lincoln (2004). *Understanding the Universe*. World Scientific. p. 116. ISBN 981-238-705-6.

[56] "Weak Interactions". *Virtual Visitor Center*. Stanford Linear Accelerator Center. 2008. Retrieved 2008-09-28.

[57] K. Nakamura; et al. (2010). "Review of Particles Physics: The CKM Quark-Mixing Matrix" (PDF). *J. Phys. G* **37** (75021): 150.

[58] Z. Maki, M. Nakagawa, S. Sakata (1962). "Remarks on the Unified Model of Elementary Particles". *Progress of Theoretical Physics* **28** (5): 870. Bibcode:1962PThPh..28..870M. doi:10.1143/PTP.28.870.

[59] B.C. Chauhan, M. Picariello, J. Pulido, E. Torrente-Lujan (2007). "Quark–lepton complementarity, neutrino and standard model data predict θPMNS
13 = 9°+1°
−2°".*European Physical Journal*C**50**(3): 573–578. arXiv:hep-ph/0605032. Bibcode:2007EPJC...50..573C.doi:-007-0212-z.

[60] R. Nave. "The Color Force". *HyperPhysics*. Georgia State University, Department of Physics and Astronomy. Retrieved 2009-04-26.

[61] B.A. Schumm (2004). *Deep Down Things*. Johns Hopkins University Press. pp. 131–132. ISBN 0-8018-7971-X. OCLC 55229065.

[62] Part III of M.E. Peskin, D.V. Schroeder (1995). *An Introduction to Quantum Field Theory*. Addison–Wesley. ISBN 0-201-50397-2.

[63] V. Icke (1995). *The force of symmetry*. Cambridge University Press. p. 216. ISBN 0-521-45591-X.

[64] M.Y. Han (2004). *A story of light*. World Scientific. p. 78. ISBN 981-256-034-3.

[65] C. Sutton. "Quantum chromodynamics (physics)". *Encyclopædia Britannica Online*. Retrieved 2009-05-12.

[66] A. Watson (2004). *The Quantum Quark*. Cambridge University Press. pp. 285–286. ISBN 0-521-82907-0.

[67] K.A. Olive *et al.* (Particle Data Group), Chin. Phys. **C38**, 090001 (2014) (URL: http://pdg.lbl.gov)

[68] W. Weise, A.M. Green (1984). *Quarks and Nuclei*. World Scientific. pp. 65–66. ISBN 9971-966-61-1.

[69] D. McMahon (2008). *Quantum Field Theory Demystified*. McGraw–Hill. p. 17. ISBN 0-07-154382-1.

[70] S.G. Roth (2007). *Precision electroweak physics at electron–positron colliders*. Springer. p. VI. ISBN 3-540-35164-7.

[71] R.P. Feynman (1985). *QED: The Strange Theory of Light and Matter* (1st ed.). Princeton University Press. pp. 136–137. ISBN 0-691-08388-6.

[72] M. Veltman (2003). *Facts and Mysteries in Elementary Particle Physics*. World Scientific. pp. 45–47. ISBN 981-238-149-X.

[73] F. Wilczek, B. Devine (2006). *Fantastic Realities*. World Scientific. p. 85. ISBN 981-256-649-X.

[74] F. Wilczek, B. Devine (2006). *Fantastic Realities*. World Scientific. pp. 400ff. ISBN 981-256-649-X.

[75] M. Veltman (2003). *Facts and Mysteries in Elementary Particle Physics*. World Scientific. pp. 295–297. ISBN 981-238-149-X.

[76] T. Yulsman (2002). *Origin*. CRC Press. p. 55. ISBN 0-7503-0765-X.

[77] F. Garberson (2008). "Top Quark Mass and Cross Section Results from the Tevatron". arXiv:0808.0273 [hep-ex].

[78] J. Steinberger (2005). *Learning about Particles*. Springer. p. 130. ISBN 3-540-21329-5.

[79] C.-Y. Wong (1994). *Introduction to High-energy Heavy-ion Collisions*. World Scientific. p. 149. ISBN 981-02-0263-6.

[80] S.B. Rüester, V. Werth, M. Buballa, I.A. Shovkovy, D.H. Rischke; Werth; Buballa; Shovkovy; Rischke (2005). "The phase diagram of neutral quark matter: Self-consistent treatment of quark masses". *Physical Review D* **72** (3): 034003. arXiv:hep-ph/0503184. Bibcode:2005PhRvD..72c4004R. doi:10.1103/PhysRevD.72.034004.

[81] M.G. Alford, K. Rajagopal, T. Schaefer, A. Schmitt; Schmitt; Rajagopal; Schäfer (2008). "Color superconductivity in dense quark matter".*Reviews of Modern Physics***80**(4): 1455–1515. arXiv:0709.4635. Bibcode:2008RvMP...80.1455A.doi:1455.

[82] S.Mrowczynski(1998). "Quark–Gluon Plasma".*Acta Physica Polonica B***29**: 3711. arXiv:nucl-th/9905005. Bibcode:1998.

[83] Z. Fodor, S.D. Katz; Katz (2004). "Critical point of QCD at finite T and μ, lattice results for physical quark masses". *Journal of High Energy Physics* **2004** (4): 50. arXiv:hep-lat/0402006. Bibcode:2004JHEP...04..050F. doi:10.1088/1126-6708/2004/04/050.

[84] U. Heinz, M. Jacob (2000). "Evidence for a New State of Matter: An Assessment of the Results from the CERN Lead Beam Programme". arXiv:nucl-th/0002042.

[85] "RHIC Scientists Serve Up "Perfect" Liquid". Brookhaven National Laboratory News. 2005. Retrieved 2009-05-22.

[86] T. Yulsman (2002). *Origins: The Quest for Our Cosmic Roots*. CRC Press. p. 75. ISBN 0-7503-0765-X.

[87] A. Sedrakian, J.W. Clark, M.G. Alford (2007). *Pairing in fermionic systems*. World Scientific. pp. 2–3. ISBN 981-256-907-3.

9.9 Further reading

- A. Ali, G. Kramer; Kramer (2011). "JETS and QCD: A historical review of the discovery of the quark and gluon jets and its impact on QCD". *European Physical Journal H* **36** (2): 245. arXiv:1012.2288. Bibcode:2011EPJH...3 .doi:10.1140/epjh/e2011-10047-1.

- D.J. Griffiths (2008). *Introduction to Elementary Particles* (2nd ed.). Wiley–VCH. ISBN 3-527-40601-8.

- I.S. Hughes (1985). *Elementary particles* (2nd ed.). Cambridge University Press. ISBN 0-521-26092-2.

- R. Oerter (2005). *The Theory of Almost Everything: The Standard Model, the Unsung Triumph of Modern Physics.* Pi Press. ISBN 0-13-236678-9.

- A. Pickering (1984). *Constructing Quarks: A Sociological History of Particle Physics.* The University of Chicago Press. ISBN 0-226-66799-5.

- B. Povh (1995). *Particles and Nuclei: An Introduction to the Physical Concepts.* Springer–Verlag. ISBN 0-387-59439-6.

- M. Riordan (1987). *The Hunting of the Quark: A true story of modern physics.* Simon & Schuster. ISBN 0-671-64884-5.

- B.A. Schumm (2004). *Deep Down Things: The Breathtaking Beauty of Particle Physics.* Johns Hopkins University Press. ISBN 0-8018-7971-X.

9.10 External links

- 1969 Physics Nobel Prize lecture by Murray Gell-Mann

- 1976 Physics Nobel Prize lecture by Burton Richter

- 1976 Physics Nobel Prize lecture by Samuel C.C. Ting

- 2008 Physics Nobel Prize lecture by Makoto Kobayashi

- 2008 Physics Nobel Prize lecture by Toshihide Maskawa

- The Top Quark And The Higgs Particle by T.A. Heppenheimer – A description of CERN's experiment to count the families of quarks.

- Bowley, Roger; Copeland, Ed. "Quarks". *Sixty Symbols*. Brady Haran for the University of Nottingham.

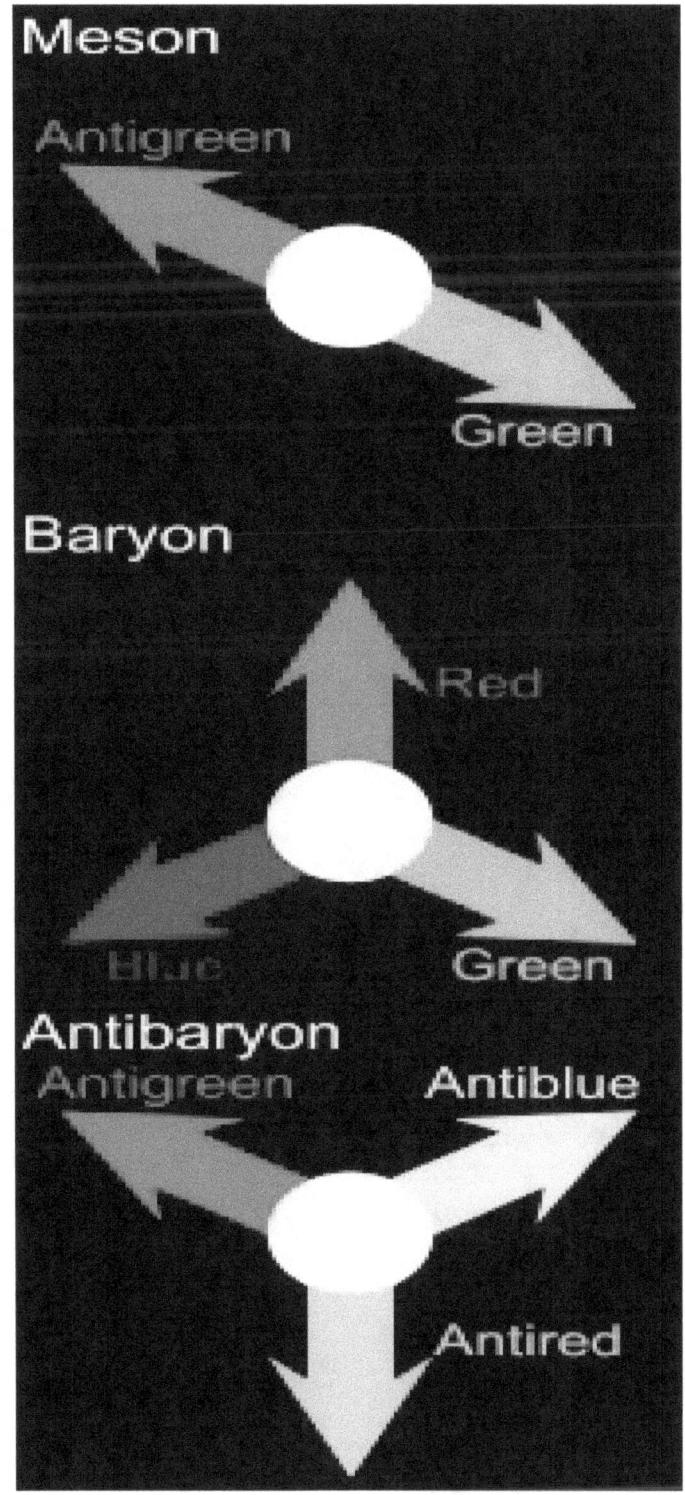

All types of hadrons have zero total color charge. (three examples shown)

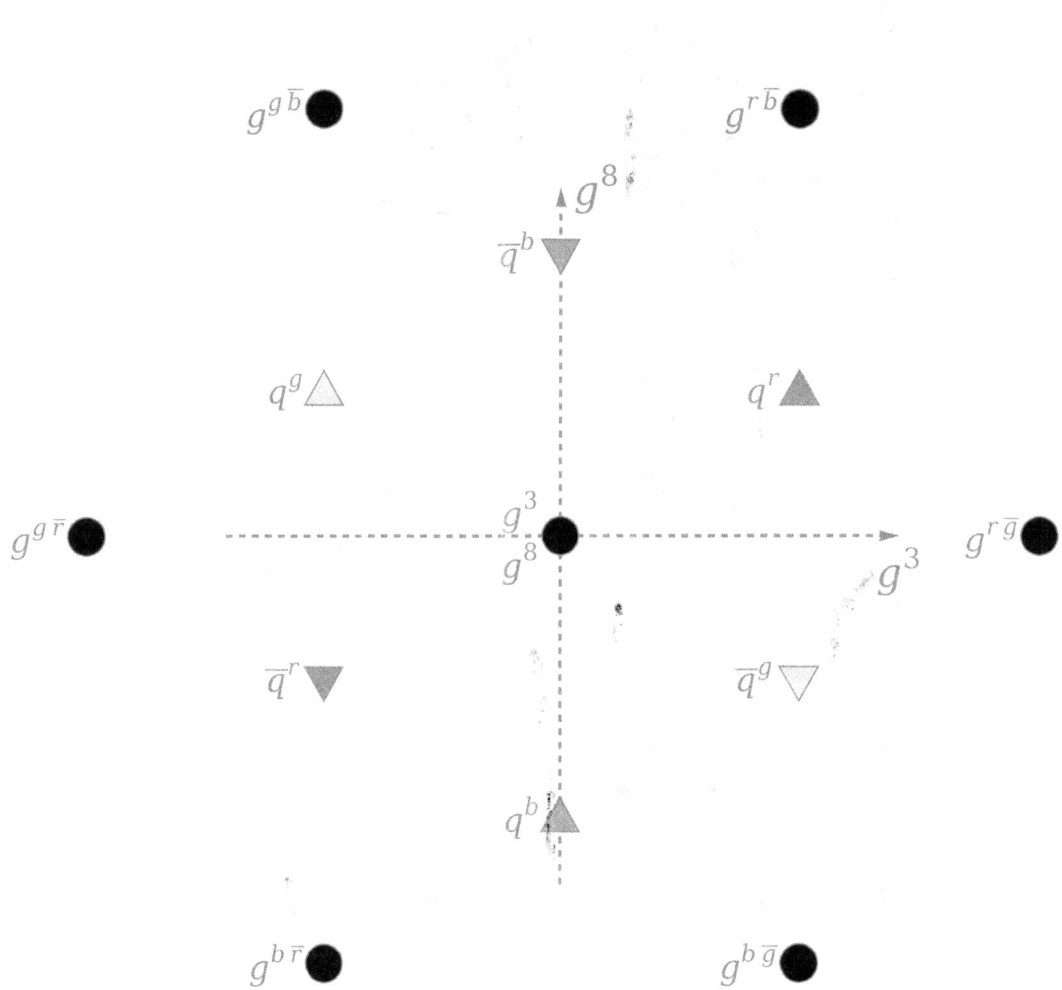

The pattern of strong charges for the three colors of quark, three antiquarks, and eight gluons (with two of zero charge overlapping).

Current quark masses for all six flavors in comparison, as balls of proportional volumes. Proton and electron (red) are shown in bottom left corner for scale

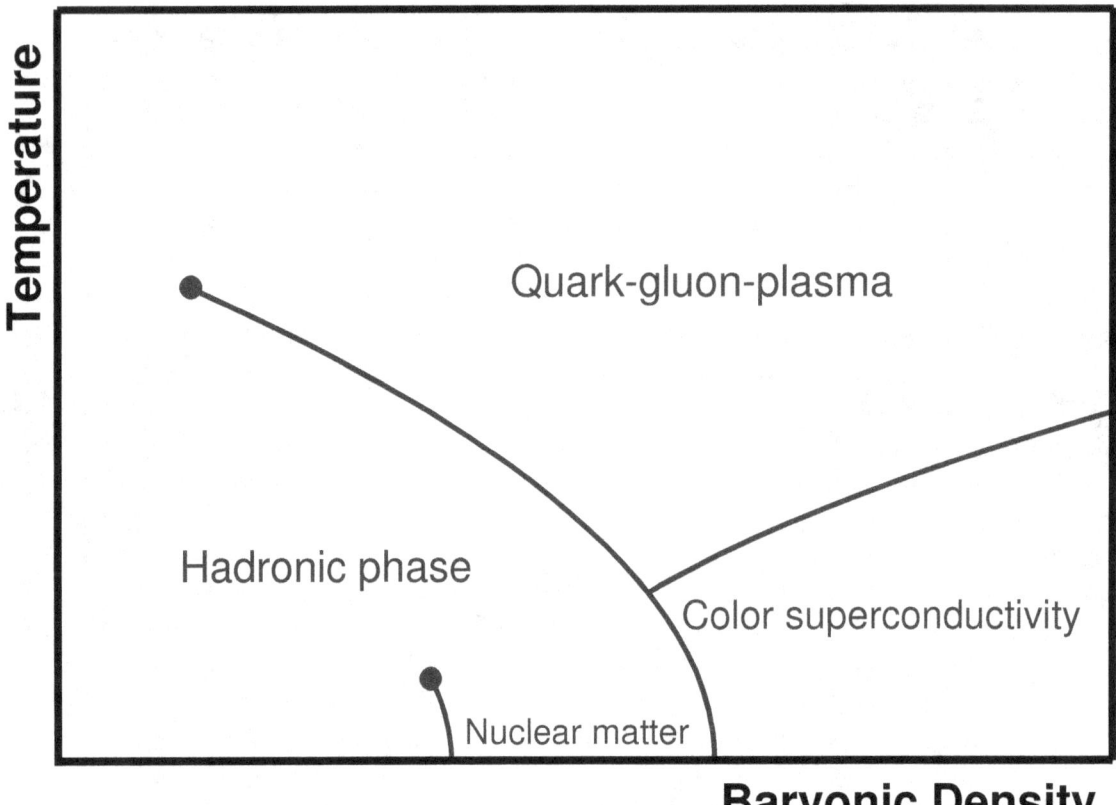

A qualitative rendering of the phase diagram of quark matter. The precise details of the diagram are the subject of ongoing research.[80][81]

Chapter 10

Quark model

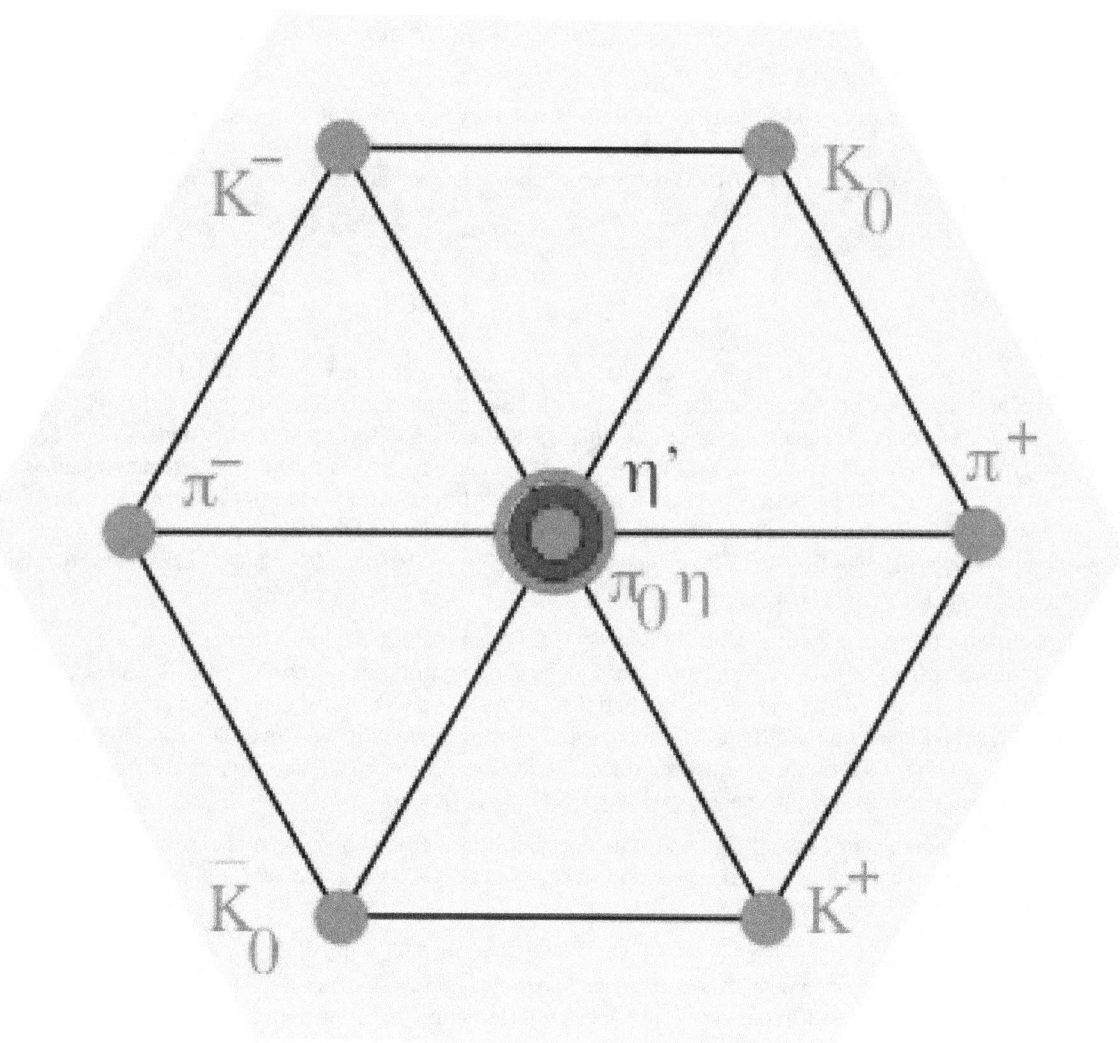

Figure 1: *The pseudoscalar meson nonet. Members of the octet are shown in green, the singlet in magenta. The name of the* Eightfold Way *derives from this classification.*

In particle physics, the **quark model** is a classification scheme for hadrons in terms of their valence quarks—the quarks and antiquarks which give rise to the quantum numbers of the hadrons. The quark model underlies "flavor SU(3)", or the *Eightfold Way*, the successful classification scheme organizing the large number of lighter hadrons that were being discovered starting in the 1950s and continuing through the 1960s. It received experimental verification beginning in the late 1960s and is a valid effective classification of them to date. The quark model was independently proposed by physicists Murray Gell-Mann,[1] and George Zweig[2][3] (also see [4]) in 1964. Today, the model has essentially been absorbed as a component of the established quantum field theory of strong and electroweak particle interactions, dubbed the Standard Model.

Hadrons are not really "elementary", and can be regarded as bound states of their "valence quarks" and antiquarks, which give rise to the quantum numbers of the hadrons. These quantum numbers are labels identifying the hadrons, and are of two kinds. One set comes from the Poincaré symmetry—J^{PC}, where J, P and C stand for the total angular momentum, P-symmetry, and C-symmetry, respectively.

The remaining are flavor quantum numbers such as the isospin, strangeness, charm, and so on. The strong interactions binding the quarks together are insensitive to these quantum numbers, so variation of them leads to systematic mass and coupling relationships among the hadrons in the same flavor multiplet.

All quarks are assigned a baryon number of ⅓. Up, charm and top quarks have an electric charge of +⅔, while the down, strange, and bottom quarks have an electric charge of −⅓. Antiquarks have the opposite quantum numbers. Quarks are spin-½ particles, and thus fermions. Each quark or antiquark obeys the Gell-Mann–Nishijima formula individually, so any additive assembly of them will as well.

Mesons are made of a valence quark–antiquark pair (thus have a baryon number of 0), while baryons are made of three quarks (thus have a baryon number of 1). This article discusses the quark model for the up, down, and strange flavors of quark (which form an approximate flavor SU(3) symmetry). There are generalizations to larger number of flavors.

10.1 History

Developing classification schemes for hadrons became a timely question after new experimental techniques uncovered so many of them, that it became clear that they could not all be elementary. These discoveries led Wolfgang Pauli to exclaim "Had I foreseen that, I would have gone into botany," and Enrico Fermi to advise his student Leon Lederman: "Young man, if I could remember the names of these particles, I would have been a botanist." These new schemes earned Nobel prizes for experimental particle physicists, including Luis Alvarez, who was at the forefront of many of these developments. Constructing hadrons as bound states of fewer constituents would thus organize the "zoo" at hand. Several early proposals, such as the ones by Enrico Fermi and Chen-Ning Yang (1949), and by Shoichi Sakata (1956), ended up satisfactorily covering the mesons, but failed with baryons, and so were unable to explain all the data.

The Gell-Mann–Nishijima formula, developed by Murray Gell-Mann and Kazuhiko Nishijima, led to the Eightfold way classification, invented by Gell-Mann, with important independent contributions from Yuval Ne'eman, in 1961. The hadrons were organized into SU(3) representation multiplets, octets and decuplets, of roughly the same mass, due to the strong interactions; and smaller mass differences linked to the flavor quantum numbers, invisible to the strong interactions. The Gell-Mann–Okubo mass formula systematized the quantification of these small mass differences among members of a hadronic multiplet, controlled by the explicit symmetry breaking of SU(3).

The spin-³⁄₂ Ω− baryon, a member of the ground-state decuplet, was a crucial prediction of that classification. After it was discovered in an experiment at Brookhaven National Laboratory, Gell-Mann received a Nobel prize in physics for his work on the Eightfold Way, in 1969.

Finally, in 1964, Gell-Mann, and, independently, George Zweig, discerned what the Eightfold Way picture encodes. They posited elementary fermionic constituents, unobserved, and possibly unobservable in a free form, underlying and elegantly encoding the Eightfold Way classification, in an economical, tight structure, resulting in further simplicity. Hadronic mass differences were now linked to the different masses of the constituent quarks.

It would take about a decade for the unexpected nature—and physical reality—of these quarks to be appreciated more fully (See Quarks). Counter-intuitively, they cannot ever be observed in isolation (color confinement), but instead always combine with other quarks to form full hadrons, which then furnish ample indirect information on the trapped quarks themselves. Conversely, the quarks serve in the definition of Quantum chromodynamics, the fundamental theory fully describing the strong interactions; and the Eightfold Way is now understood to be a consequence of the flavor symmetry structure of the lightest three of them. To date, no Nobel prize has been awarded to Gell-Mann and Zweig for this discovery.

10.2   Mesons

See also: Meson and List of mesons

The Eightfold Way classification is named after the following fact. If we take three flavors of quarks, then the quarks

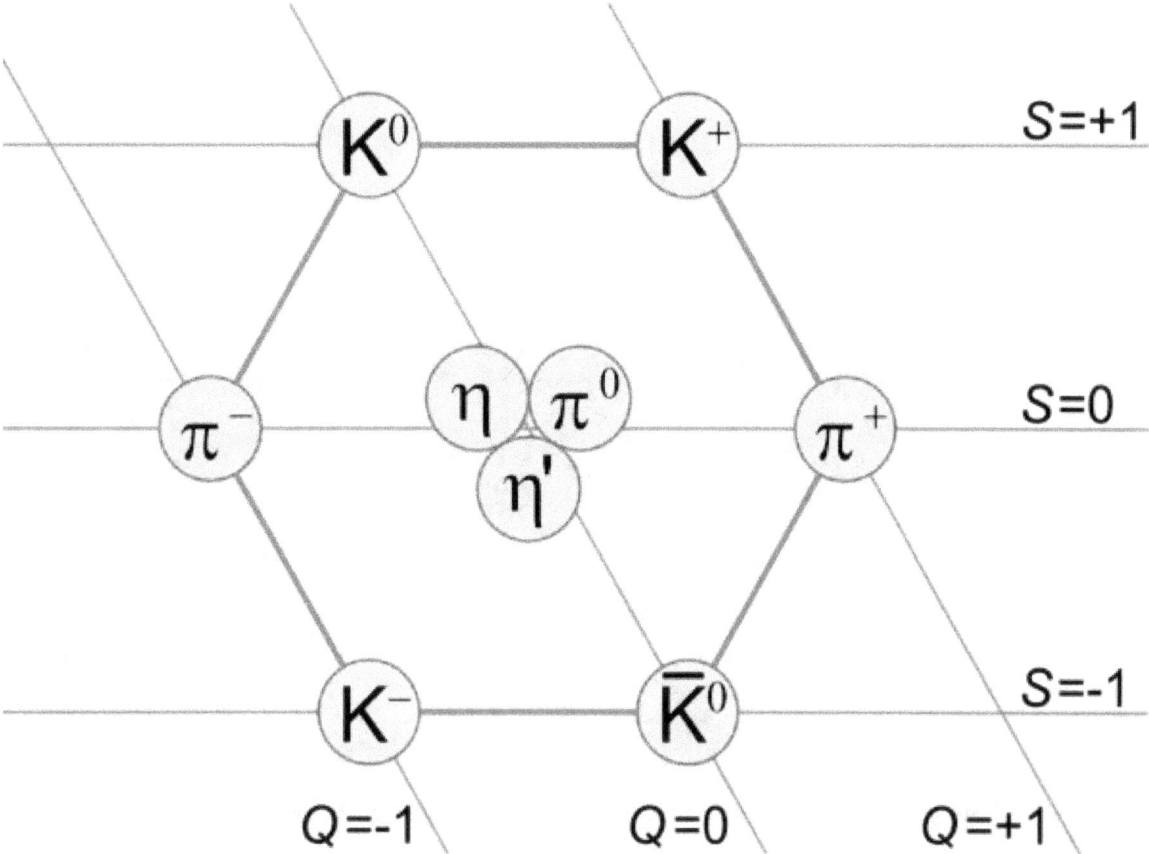

Figure 2: *Pseudoscalar mesons of spin 0 form a nonet*

lie in the fundamental representation, **3** (called the triplet) of flavor SU(3). The antiquarks lie in the complex conjugate representation $\bar{\mathbf{3}}$. The nine states (nonet) made out of a pair can be decomposed into the trivial representation, **1** (called the singlet), and the adjoint representation, **8** (called the octet). The notation for this decomposition is

$$\mathbf{3} \otimes \bar{\mathbf{3}} = \mathbf{8} \oplus \mathbf{1}$$

Figure 1 shows the application of this decomposition to the mesons. If the flavor symmetry were exact (as in the limit that only the strong interactions operate, but the electroweak interactions are notionally switched off), then all nine mesons would have the same mass. However, the physical content of the full theory includes consideration of the symmetry breaking induced by the quark mass differences, and considerations of mixing between various multiplets (such as the octet and the singlet).

N.B. Nevertheless, the mass splitting between the η and the η′ is larger than the quark model can accommodate, and this "η–η′ puzzle" has its origin in topological peculiarities of the strong interaction vacuum, such as instanton configurations.

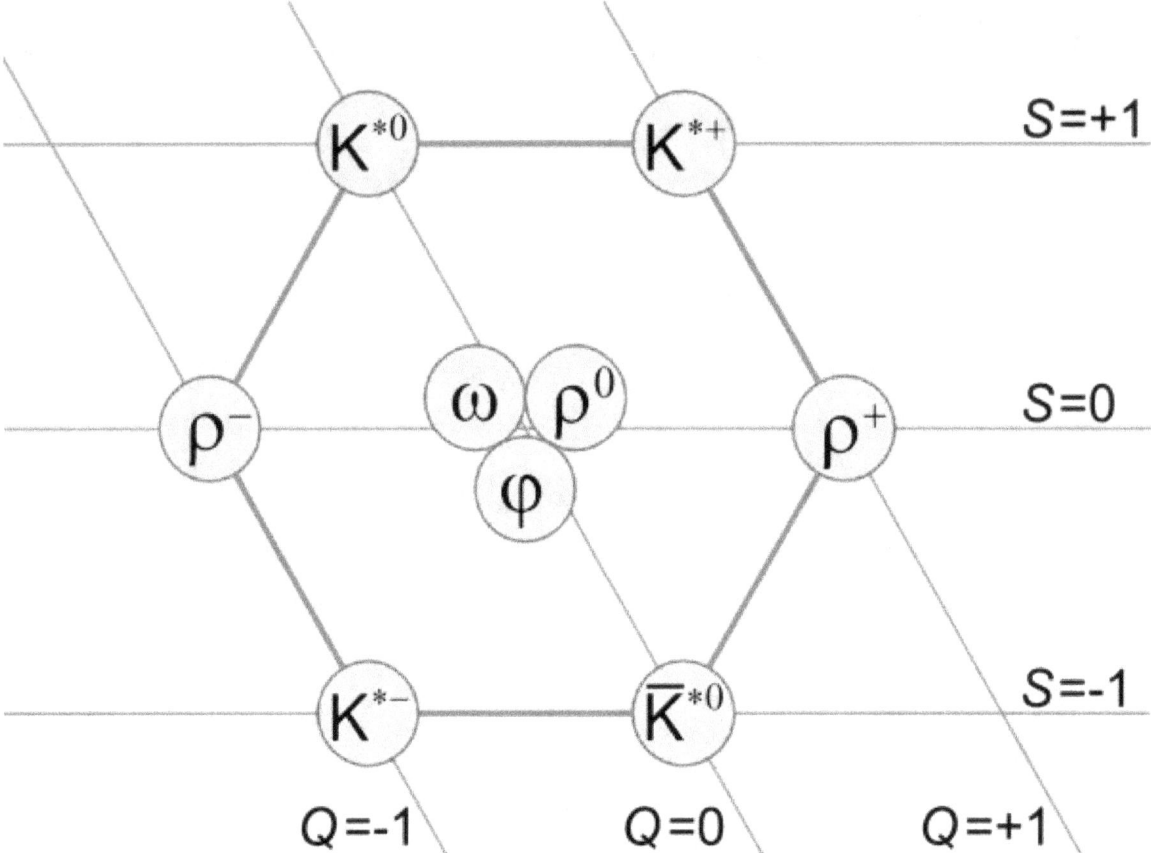

Figure 3: *Mesons of spin 1 form a nonet*

Mesons are hadrons with zero baryon number. If the quark–antiquark pair are in an orbital angular momentum L state, and have spin S, then

- $|L - S| \leq J \leq L + S$, where $S = 0$ or 1,

- $P = (-1)^{L+1}$, where the 1 in the exponent arises from the intrinsic parity of the quark–antiquark pair.

- $C = (-1)^{L+S}$ for mesons which have no flavor. Flavored mesons have indefinite value of C.

- For isospin $I = 1$ and 0 states, one can define a new multiplicative quantum number called the *G-parity* such that $G = (-1)^{I+L+S}$.

If $P = (-1)^J$, then it follows that $S = 1$, thus $PC = 1$. States with these quantum numbers are called *natural parity states*; while all other quantum numbers are thus called *exotic* (for example the state $J^{PC} = 0^{--}$).

10.3 Baryons

Main article: Baryon
See also: List of baryons
 Since quarks are fermions, the spin-statistics theorem implies that the wavefunction of a baryon must be antisymmetric under exchange of any two quarks. This antisymmetric wavefunction is obtained by making it fully antisymmetric in color, discussed below, and symmetric in flavor, spin and space put together. With three flavors, the decomposition in flavor is

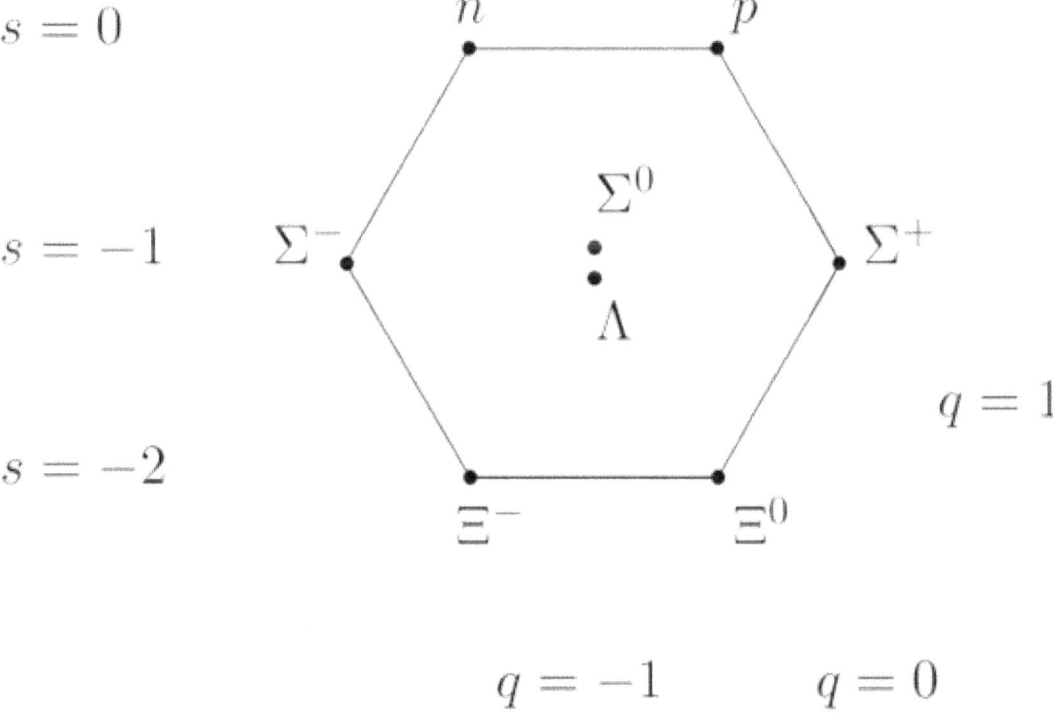

Figure 4. *The S = $^1\!/_2$ ground state baryon octet*

$3 \otimes 3 \otimes 3 = 10_S \oplus 8_M \oplus 8_M \oplus 1_A$

The decuplet is symmetric in flavor, the singlet antisymmetric and the two octets have mixed symmetry. The space and spin parts of the states are thereby fixed once the orbital angular momentum is given.

It is sometimes useful to think of the basis states of quarks as the six states of three flavors and two spins per flavor. This approximate symmetry is called spin-flavor SU(6). In terms of this, the decomposition is

$$6 \otimes 6 \otimes 6 = 56_S \oplus 70_M \oplus 70_M \oplus 20_A \ .$$

The 56 states with symmetric combination of spin and flavour decompose under flavor SU(3) into

$$56 = 10^{\frac{3}{2}} \oplus 8^{\frac{1}{2}} \ ,$$

where the superscript denotes the spin, S, of the baryon. Since these states are symmetric in spin and flavor, they should also be symmetric in space—a condition that is easily satisfied by making the orbital angular momentum $L = 0$. These are the ground state baryons.

The $S = {}^1\!/_2$ octet baryons are the two nucleons (p+, n0), the three Sigmas (Σ+, Σ0, Σ−), the two Xis (Ξ0, Ξ−), and the Lambda (Λ0). The $S = {}^3\!/_2$ decuplet baryons are the four Deltas (Δ++, Δ+, Δ0, Δ−), three Sigmas (Σ∗+, Σ∗0, Σ∗−), two Xis (Ξ∗0, Ξ∗−), and the Omega (Ω−).

Mixing of baryons, mass splittings within and between multiplets, and magnetic moments are some of the other questions that the model predicts successfully.

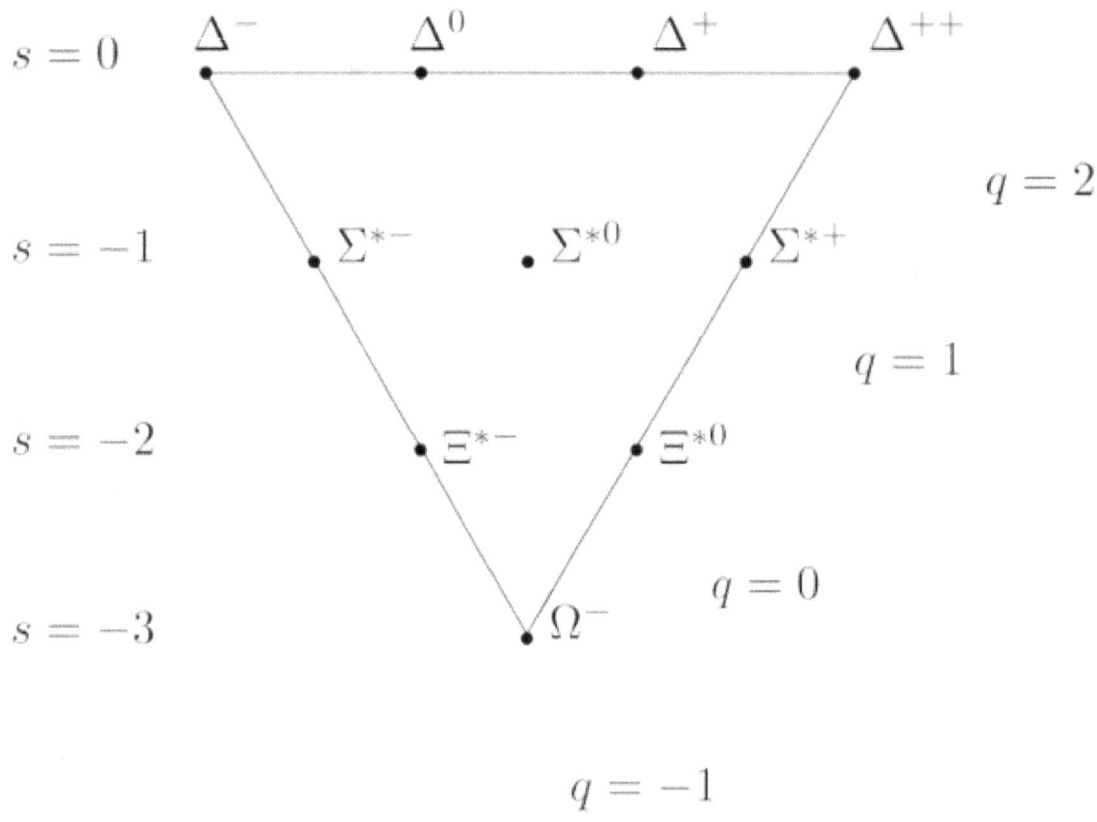

Figure 5. *The* S = $^3\!/_2$ *baryon decuplet*

10.3.1 The discovery of color

Main article: Color charge

Color quantum numbers are the characteristic charges of the strong force, and are completely uninvolved in electroweak interactions. They were discovered as a consequence of the quark model classification, when it was appreciated that the spin $S = ^3\!/_2$ baryon, the Δ++, required three up quarks with parallel spins and vanishing orbital angular momentum. Therefore, it could not have an antisymmetric wave function, (due to the Pauli exclusion principle), *unless there were a hidden quantum number.* Oscar Greenberg noted this problem in 1964, suggesting that quarks should be para-fermions.[5]

Instead, six months later, Moo-Young Han and Yoichiro Nambu suggested the existence of three triplets of quarks to solve this problem, but flavor and color intertwined in that model--- they did not commute.[6]

The modern concept of color completely commuting with all other charges and providing the strong force charge was articulated in 1973, by William Bardeen, Harald Fritzsch, and Murray Gell-Mann.[7][8]

10.4   States outside the quark model

While the quark model is derivable from the theory of quantum chromodynamics, the structure of hadrons is more complicated than this model allows. The full quantum mechanical wave function of any hadron must include virtual quark pairs as well as virtual gluons, and allows for a variety of mixings. There may be hadrons which lie outside the quark model. Among these are the *glueballs* (which contain only valence gluons), *hybrids* (which contain valence quarks as well as gluons) and "exotic hadrons" (such as tetraquarks or pentaquarks).

10.5   See also

- Subatomic particles

- Hadrons, baryons, mesons and quarks

- Exotic hadrons: exotic mesons and exotic baryons

- Quantum chromodynamics, flavor, the QCD vacuum

10.6   References and external links

[1] M. Gell-Mann (1964). "A Schematic Model of Baryons and Mesons". *Physics Letters* **8** (3): 214–215. Bibcode:1964PhL......8..214G. doi:10.1016/S0031-9163(64)92001-3.

[2] G. Zweig (1964). "An SU(3) Model for Strong Interaction Symmetry and its Breaking" (PDF). *CERN Report No.8182/TH.401.*

[3] G. Zweig (1964). "An SU(3) Model for Strong Interaction Symmetry and its Breaking: II" (PDF). *CERN Report No.8419/TH.412.*

[4] Petermann, A. (1965). "Propriétés de l'étrangeté et une formule de masse pour les mésons vectoriels". *Nuclear Physics* **63** (2): 349. doi:10.1016/0029-5582(65)90348-2. which gingerly touched upon the central ideas, without quantitative substantiation;

[5] O.W. Greenberg (1964). "Spin and Unitary-Spin Independence in a Paraquark Model of Baryons and Mesons". *Physical Review Letters* **13** (20): 598. Bibcode:1964PhRvL..13..598G. doi:10.1103/PhysRevLett.13.598.

[6] M.Y. Han, Y. Nambu (1965). "Three-Triplet Model with Double SU(3) Symmetry". *Physical Review* **139** (4B): B1006. Bibcode:1965PhRv..139.1006H. doi:10.1103/PhysRev.139.B1006.

[7] W. Bardeen, H. Fritzsch, M. Gell-Mann (1973). "Light cone current algebra, π^0 decay, and $e^+ e^-$ annihilation". In R. Gatto. *Scale and conformal symmetry in hadron physics.* John Wiley & Sons. p. 139. arXiv:hep-ph/0211388. ISBN 0-471-29292-3.

[8] Fritzsch, H.; Gell-Mann, M.; Leutwyler, H. (1973). "Advantages of the color octet gluon picture". *Physics Letters B* **47** (4): 365. doi:10.1016/0370-2693(73)90625-4.

- S. Eidelman *et al.* Particle Data Group (2004). "Review of Particle Physics" (PDF). *Physics Letters B* **592**: 1. arXiv:astro-ph/0406663. Bibcode:2004PhLB..592....1P. doi:10.1016/j.physletb.2004.06.001.

- Lichtenberg, D B (1970). *Unitary Symmetry and Elementary Particles.* Academic Press. ISBN 978-1483242729.

- Thomson, M A (2011), Lecture notes

- J.J.J. Kokkedee (1969). *The quark model.* W. A. Benjamin. ASIN B001RAVDIA.

Chapter 11

Electric charge

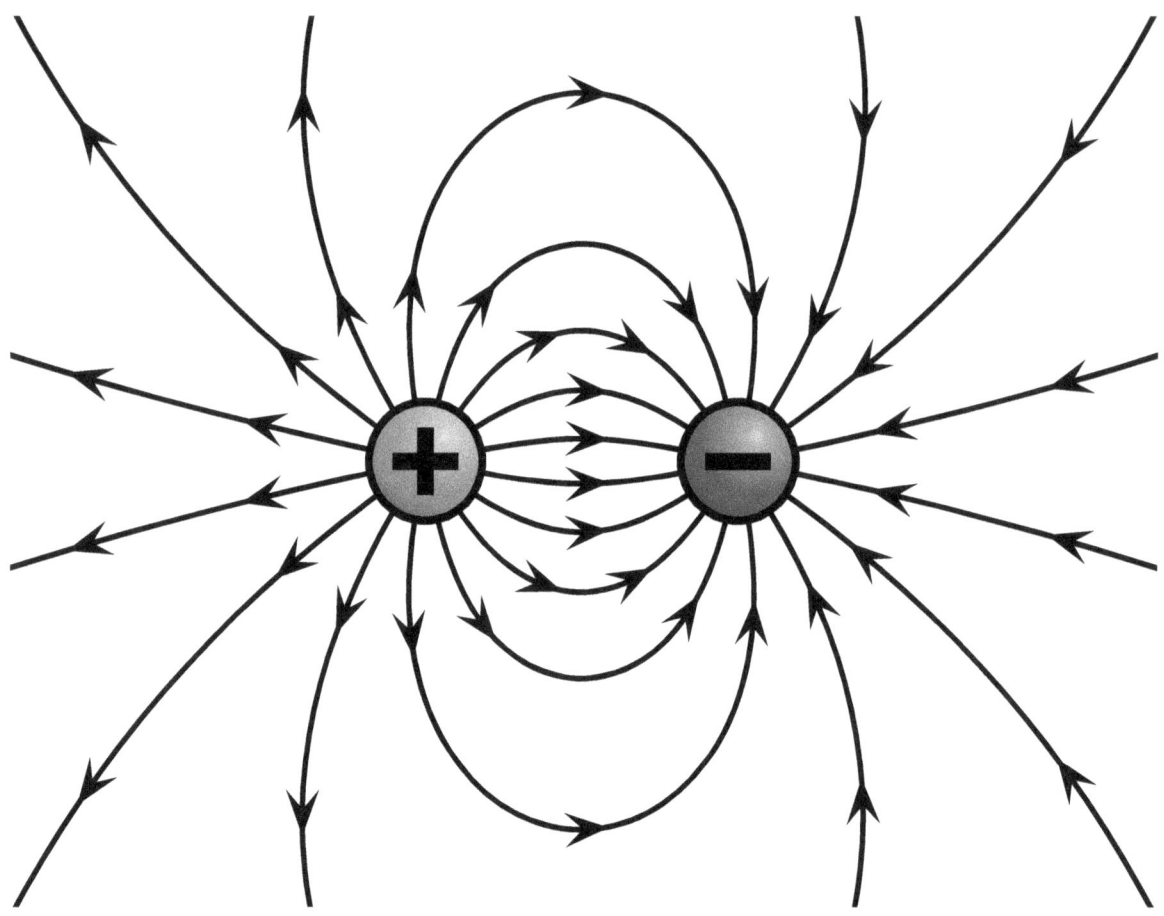

Electric field of a positive and a negative point charge.

Electric charge is the physical property of matter that causes it to experience a force when placed in an electromagnetic field. There are two types of electric charges: positive and negative. Positively charged substances are repelled from other positively charged substances, but attracted to negatively charged substances; negatively charged substances are repelled from negative and attracted to positive. An object is negatively charged if it has an excess of electrons, and is otherwise positively charged or uncharged. The SI derived unit of electric charge is the coulomb (C), although in electrical engineering it is also common to use the ampere-hour (Ah), and in chemistry it is common to use the elementary charge (e) as a unit. The symbol Q is often used to denote charge. The early knowledge of how charged substances interact is now called classical electrodynamics, and is still very accurate if quantum effects do not need to be considered.

The *electric charge* is a fundamental conserved property of some subatomic particles, which determines their electromagnetic interaction. Electrically charged matter is influenced by, and produces, electromagnetic fields. The interaction between a moving charge and an electromagnetic field is the source of the electromagnetic force, which is one of the four fundamental forces (See also: magnetic field).

Twentieth-century experiments demonstrated that electric charge is *quantized*; that is, it comes in integer multiples of individual small units called the elementary charge, e, approximately equal to 1.602×10^{-19} coulombs (except for particles called quarks, which have charges that are integer multiples of $e/3$). The proton has a charge of $+e$, and the electron has a charge of $-e$. The study of charged particles, and how their interactions are mediated by photons, is called quantum electrodynamics.

11.1　Overview

Charge is the fundamental property of forms of matter that exhibit electrostatic attraction or repulsion in the presence of other matter. Electric charge is a characteristic property of many subatomic particles. The charges of free-standing particles are integer multiples of the elementary charge e; we say that electric charge is *quantized*. Michael Faraday, in his electrolysis experiments, was the first to note the discrete nature of electric charge. Robert Millikan's oil-drop experiment demonstrated this fact directly, and measured the elementary charge.

By convention, the charge of an electron is -1, while that of a proton is $+1$. Charged particles whose charges have the same sign repel one another, and particles whose charges have different signs attract. Coulomb's law quantifies the electrostatic force between two particles by asserting that the force is proportional to the product of their charges, and inversely proportional to the square of the distance between them.

The charge of an antiparticle equals that of the corresponding particle, but with opposite sign. Quarks have fractional charges of either $-\frac{1}{3}$ or $+\frac{2}{3}$, but free-standing quarks have never been observed (the theoretical reason for this fact is asymptotic freedom).

The electric charge of a macroscopic object is the sum of the electric charges of the particles that make it up. This charge is often small, because matter is made of atoms, and atoms typically have equal numbers of protons and electrons, in which case their charges cancel out, yielding a net charge of zero, thus making the atom neutral.

An *ion* is an atom (or group of atoms) that has lost one or more electrons, giving it a net positive charge (cation), or that has gained one or more electrons, giving it a net negative charge (anion). *Monatomic ions* are formed from single atoms, while *polyatomic ions* are formed from two or more atoms that have been bonded together, in each case yielding an ion with a positive or negative net charge.

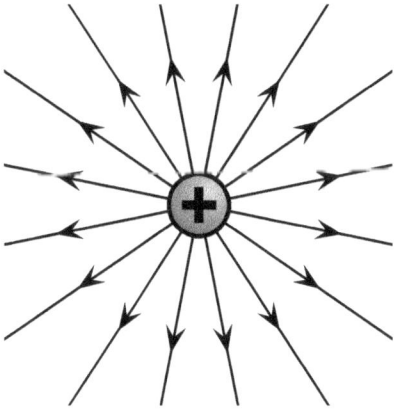

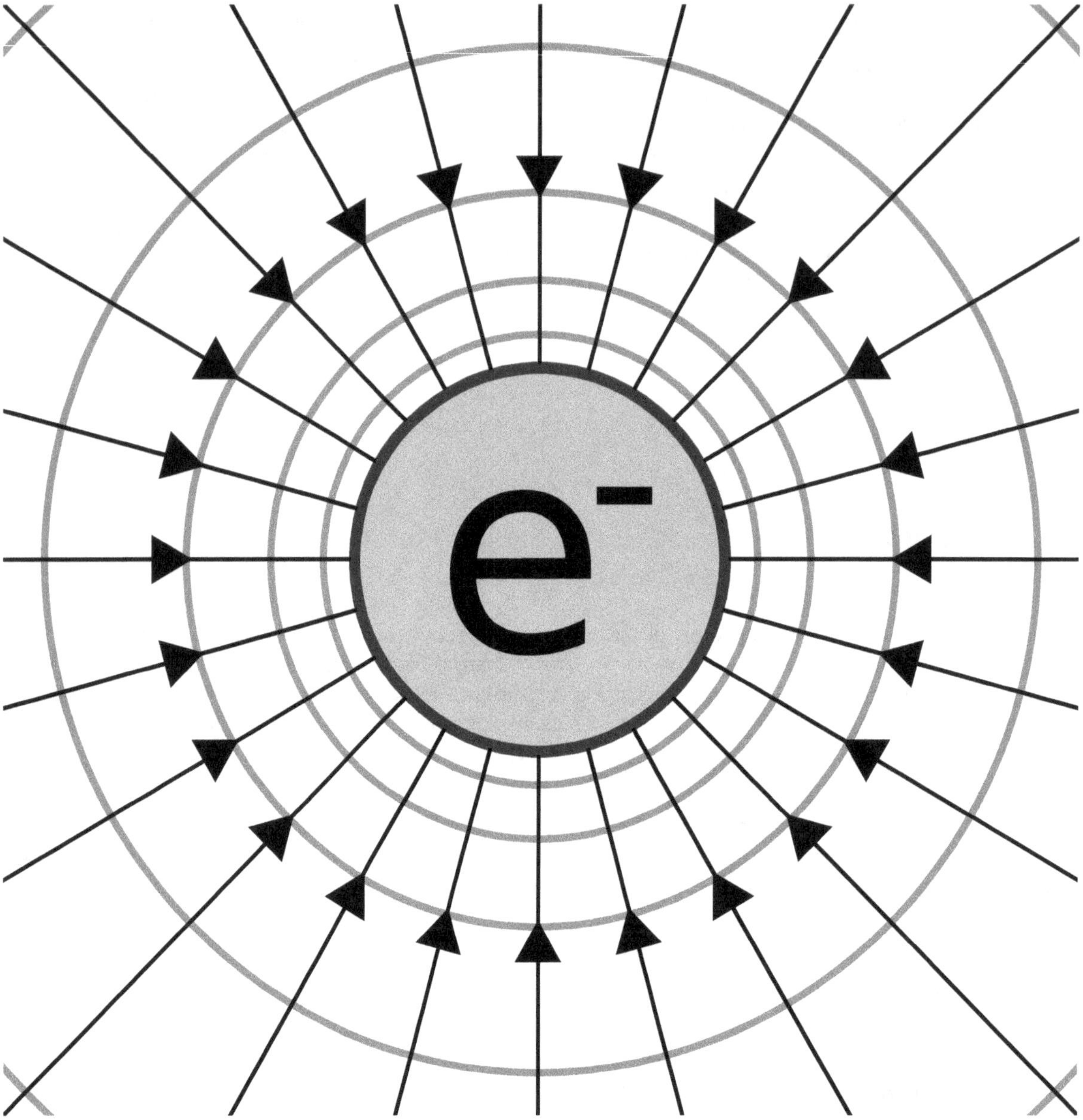

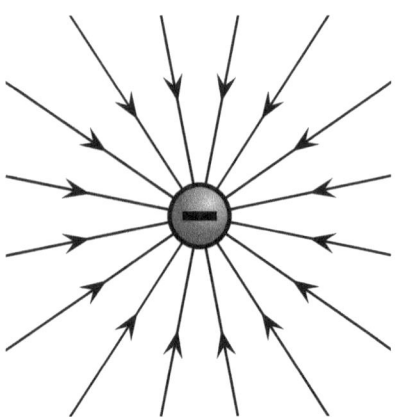

Diagram showing field lines and equipotentials around an electron, a negatively charged particle. In an electrically neutral atom, the number of electrons is equal to the number of protons (which are positively charged), resulting in a net zero overall charge

Electric field induced by a positive electric charge (left) and a field induced by a negative electric charge (right).

During formation of macroscopic objects, constituent atoms and ions usually combine to form structures composed of neutral *ionic compounds* electrically bound to neutral atoms. Thus macroscopic objects tend toward being neutral overall, but macroscopic objects are rarely perfectly net neutral.

Sometimes macroscopic objects contain ions distributed throughout the material, rigidly bound in place, giving an overall net positive or negative charge to the object. Also, macroscopic objects made of conductive elements, can more or less easily (depending on the element) take on or give off electrons, and then maintain a net negative or positive charge indefinitely. When the net electric charge of an object is non-zero and motionless, the phenomenon is known as static electricity. This can easily be produced by rubbing two dissimilar materials together, such as rubbing amber with fur or glass with silk. In this way non-conductive materials can be charged to a significant degree, either positively or negatively. Charge taken from one material is moved to the other material, leaving an opposite charge of the same magnitude behind. The law of *conservation of charge* always applies, giving the object from which a negative charge has been taken a positive charge of the same magnitude, and vice versa.

Even when an object's net charge is zero, charge can be distributed non-uniformly in the object (e.g., due to an external electromagnetic field, or bound polar molecules). In such cases the object is said to be polarized. The charge due to polarization is known as bound charge, while charge on an object produced by electrons gained or lost from outside the object is called *free charge*. The motion of electrons in conductive metals in a specific direction is known as electric current.

11.2   Units

The SI unit of quantity of electric charge is the coulomb, which is equivalent to about 6.242×10^{18} e (e is the charge of a proton). Hence, the charge of an electron is approximately -1.602×10^{-19} C. The coulomb is defined as the quantity of charge that has passed through the cross section of an electrical conductor carrying one ampere within one second. The symbol Q is often used to denote a quantity of electricity or charge. The quantity of electric charge can be directly measured with an electrometer, or indirectly measured with a ballistic galvanometer.

After finding the quantized character of charge, in 1891 George Stoney proposed the unit 'electron' for this fundamental unit of electrical charge. This was before the discovery of the particle by J.J. Thomson in 1897. The unit is today treated as nameless, referred to as "elementary charge", "fundamental unit of charge", or simply as "e". A measure of charge should be a multiple of the elementary charge e, even if at large scales charge seems to behave as a real quantity. In some contexts it is meaningful to speak of fractions of a charge; for example in the charging of a capacitor, or in the fractional quantum Hall effect.

In systems of units other than SI such as cgs, electric charge is expressed as combination of only three fundamental quantities such as length, mass and time and not four as in SI where electric charge is a combination of length, mass, time and electric current.

11.3   History

As reported by the ancient Greek mathematician Thales of Miletus around 600 BC, charge (or *electricity*) could be accumulated by rubbing fur on various substances, such as amber. The Greeks noted that the charged amber buttons could attract light objects such as hair. They also noted that if they rubbed the amber for long enough, they could even get an electric spark to jump. This property derives from the triboelectric effect.

In 1600, the English scientist William Gilbert returned to the subject in *De Magnete*, and coined the New Latin word *electricus* from ηλεκτρον (*elektron*), the Greek word for *amber*, which soon gave rise to the English words "electric" and "electricity." He was followed in 1660 by Otto von Guericke, who invented what was probably the first electrostatic generator. Other European pioneers were Robert Boyle, who in 1675 stated that electric attraction and repulsion can act across a vacuum; Stephen Gray, who in 1729 classified materials as conductors and insulators; and C. F. du Fay, who

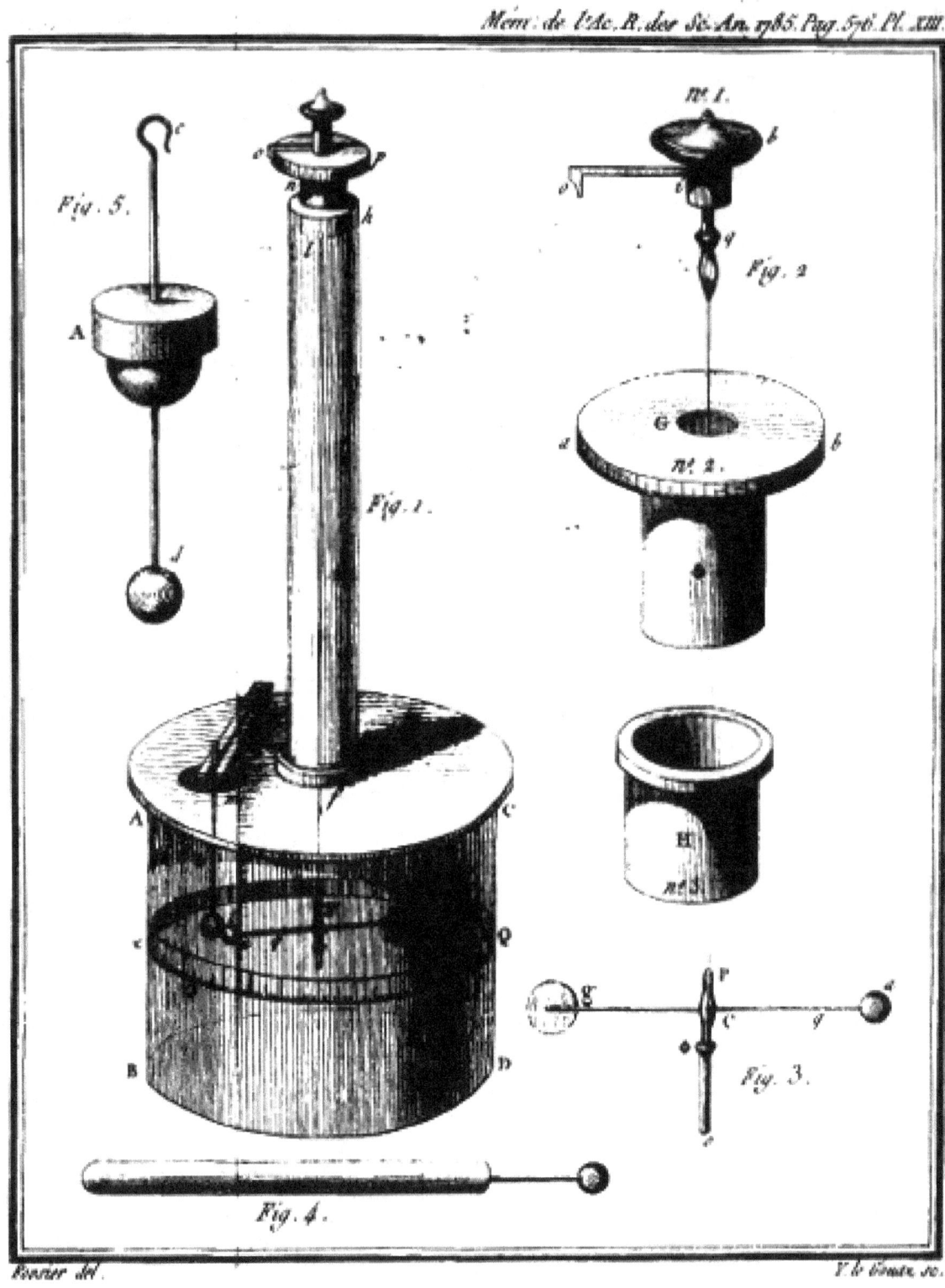

Mem. de l'Ac.R.der Sc.An.1785.Pag.576.Pl.XIII.

Coulomb's torsion balance

proposed in 1733[1] that electricity comes in two varieties that cancel each other, and expressed this in terms of a two-fluid theory. When glass was rubbed with silk, du Fay said that the glass was charged with *vitreous electricity*, and, when amber was rubbed with fur, the amber was said to be charged with *resinous electricity*. In 1839, Michael Faraday showed that the apparent division between static electricity, current electricity, and bioelectricity was incorrect, and all were a consequence of the behavior of a single kind of electricity appearing in opposite polarities. It is arbitrary which polarity is called positive and which is called negative. Positive charge can be defined as the charge left on a glass rod after being rubbed with silk.[2]

One of the foremost experts on electricity in the 18th century was Benjamin Franklin, who argued in favour of a one-fluid theory of electricity. Franklin imagined electricity as being a type of invisible fluid present in all matter; for example, he believed that it was the glass in a Leyden jar that held the accumulated charge. He posited that rubbing insulating surfaces together caused this fluid to change location, and that a flow of this fluid constitutes an electric current. He also posited that when matter contained too little of the fluid it was "negatively" charged, and when it had an excess it was "positively" charged. For a reason that was not recorded, he identified the term "positive" with vitreous electricity and "negative" with resinous electricity. William Watson arrived at the same explanation at about the same time.

11.4   Static electricity and electric current

Static electricity and electric current are two separate phenomena. They both involve electric charge, and may occur simultaneously in the same object. Static electricity refers to the electric charge of an object and the related electrostatic discharge when two objects are brought together that are not at equilibrium. An electrostatic discharge creates a change in the charge of each of the two objects. In contrast, electric current is the flow of electric charge through an object, which produces no net loss or gain of electric charge.

11.4.1   Electrification by friction

Further information: triboelectric effect

When a piece of glass and a piece of resin—neither of which exhibit any electrical properties—are rubbed together and left with the rubbed surfaces in contact, they still exhibit no electrical properties. When separated, they attract each other.

A second piece of glass rubbed with a second piece of resin, then separated and suspended near the former pieces of glass and resin causes these phenomena:

- The two pieces of glass repel each other.
- Each piece of glass attracts each piece of resin.
- The two pieces of resin repel each other.

This attraction and repulsion is an *electrical phenomena,* and the bodies that exhibit them are said to be *electrified,* or *electrically charged.* Bodies may be electrified in many other ways, as well as by friction. The electrical properties of the two pieces of glass are similar to each other but opposite to those of the two pieces of resin: The glass attracts what the resin repels and repels what the resin attracts.

If a body electrified in any manner whatsoever behaves as the glass does, that is, if it repels the glass and attracts the resin, the body is said to be 'vitreously' electrified, and if it attracts the glass and repels the resin it is said to be 'resinously' electrified. All electrified bodies are found to be either vitreously or resinously electrified.

It is the established convention of the scientific community to define the vitreous electrification as positive, and the resinous electrification as negative. The exactly opposite properties of the two kinds of electrification justify our indicating them by opposite signs, but the application of the positive sign to one rather than to the other kind must be considered as a matter of arbitrary convention, just as it is a matter of convention in mathematical diagram to reckon positive distances towards the right hand.

No force, either of attraction or of repulsion, can be observed between an electrified body and a body not electrified.[3]

Actually, all bodies are electrified, but may appear not to be so by the relative similar charge of neighboring objects in the environment. An object further electrified + or − creates an equivalent or opposite charge by default in neighboring objects, until those charges can equalize. The effects of attraction can be observed in high-voltage experiments, while lower voltage effects are merely weaker and therefore less obvious. The attraction and repulsion forces are codified by Coulomb's Law (attraction falls off at the square of the distance, which has a corollary for acceleration in a gravitational field, suggesting that gravitation may be merely electrostatic phenomenon between relatively weak charges in terms of scale). See also the Casimir effect.

It is now known that the Franklin/Watson model was fundamentally correct. There is only one kind of electrical charge, and only one variable is required to keep track of the amount of charge.[4] On the other hand, just knowing the charge is not a complete description of the situation. Matter is composed of several kinds of electrically charged particles, and these particles have many properties, not just charge.

The most common charge carriers are the positively charged proton and the negatively charged electron. The movement of any of these charged particles constitutes an electric current. In many situations, it suffices to speak of the *conventional current* without regard to whether it is carried by positive charges moving in the direction of the conventional current or by negative charges moving in the opposite direction. This macroscopic viewpoint is an approximation that simplifies electromagnetic concepts and calculations.

At the opposite extreme, if one looks at the microscopic situation, one sees there are many ways of carrying an electric current, including: a flow of electrons; a flow of electron "holes" that act like positive particles; and both negative and positive particles (ions or other charged particles) flowing in opposite directions in an electrolytic solution or a plasma.

Beware that, in the common and important case of metallic wires, the direction of the conventional current is opposite to the drift velocity of the actual charge carriers, i.e., the electrons. This is a source of confusion for beginners.

11.5 Properties

Aside from the properties described in articles about electromagnetism, charge is a relativistic invariant. This means that any particle that has charge Q, no matter how fast it goes, always has charge Q. This property has been experimentally verified by showing that the charge of *one* helium nucleus (two protons and two neutrons bound together in a nucleus and moving around at high speeds) is the same as *two* deuterium nuclei (one proton and one neutron bound together, but moving much more slowly than they would if they were in a helium nucleus).

11.6 Conservation of electric charge

Main article: Charge conservation

The total electric charge of an isolated system remains constant regardless of changes within the system itself. This law is inherent to all processes known to physics and can be derived in a local form from gauge invariance of the wave function. The conservation of charge results in the charge-current continuity equation. More generally, the net change in charge density ϱ within a volume of integration V is equal to the area integral over the current density $\mathbf{J}$ through the closed surface $S = \partial V$, which is in turn equal to the net current I:

$$-\tfrac{d}{dt} \int_V \rho \, dV = \oiint_{\partial V} \mathbf{J} \cdot d\mathbf{S} = \int J dS \cos\theta = I.$$

Thus, the conservation of electric charge, as expressed by the continuity equation, gives the result:

$$I = \frac{dQ}{dt}.$$

The charge transferred between times t_i and t_f is obtained by integrating both sides:

$$Q = \int_{t_i}^{t_f} I \, \mathrm{d}t$$

where I is the net outward current through a closed surface and Q is the electric charge contained within the volume defined by the surface.

11.7   See also

- Quantity of electricity

- SI electromagnetism units

11.8   References

[1] Two Kinds of Electrical Fluid: Vitreous and Resinous – 1733

[2] Electromagnetic Fields (2nd Edition), Roald K. Wangsness, Wiley, 1986. ISBN 0-471-81186-6 (intermediate level textbook)

[3] James Clerk Maxwell *A Treatise on Electricity and Magnetism*, pp. 32-33, Dover Publications Inc., 1954 ASIN: B000HFDK0K, 3rd ed. of 1891

[4] One Kind of Charge

11.9   External links

- How fast does a charge decay?

- Science Aid: Electrostatic charge Easy-to-understand page on electrostatic charge.

- History of the electrical units.

Chapter 12

Mass

This article is about the scientific concept. For the substance of which all physical objects consist, see Matter. For other uses, see Mass (disambiguation).

In physics, **mass** is a property of a physical body which determines the strength of its mutual gravitational attraction to other bodies, its resistance to being accelerated by a force, and in the theory of relativity gives the mass–energy content of a system. The SI unit of mass is the kilogram (kg).

Mass is not the same as weight, even though we often calculate an object's mass by measuring its weight with a spring scale instead of comparing it to known masses. An object on the Moon would weigh less than it would on Earth because of the lower gravity, but it would still have the same mass.

For everyday objects and energies well-described by Newtonian physics, mass describes the amount of matter in an object. However, at very high speeds or for subatomic particles, special relativity shows that energy is an additional source of mass. Thus, any stationary body having mass has an equivalent amount of energy, and all forms of energy resist acceleration by a force and have gravitational attraction.

There are several distinct phenomena which can be used to measure mass. Although some theorists have speculated some of these phenomena could be independent of each other,[1] current experiments have found no difference among any of the ways used to measure mass:

- *Inertial mass* measures an object's resistance to being accelerated by a force (represented by the relationship $F = ma$).

- *Active gravitational mass* measures the gravitational force exerted by an object.

- *Passive gravitational mass* measures the gravitational force experienced by an object in a known gravitational field.

- *Mass–energy* measures the total amount of energy contained within a body, using $E = mc^2$.

The mass of an object determines its acceleration in the presence of an applied force. This phenomenon is called inertia. According to Newton's second law of motion, if a body of fixed mass m is subjected to a single force F, its acceleration a is given by F/m. A body's mass also determines the degree to which it generates or is affected by a gravitational field. If a first body of mass m_A is placed at a distance r (center of mass to center of mass) from a second body of mass m_B, each body experiences an attractive force $F_g = Gm_Am_B/r^2$, where $G = 6.67{\times}10^{-11}$ N kg^{-2} m^2 is the "universal gravitational constant". This is sometimes referred to as gravitational mass.[note 1] Repeated experiments since the 17th century have demonstrated that inertial and gravitational mass are identical; since 1915, this observation has been entailed *a priori* in the equivalence principle of general relativity.

12.1   Units of mass

Further information: Orders of magnitude (mass)
The standard International System of Units (SI) unit of mass is the kilogram (kg). The kilogram is 1000 grams (g), first

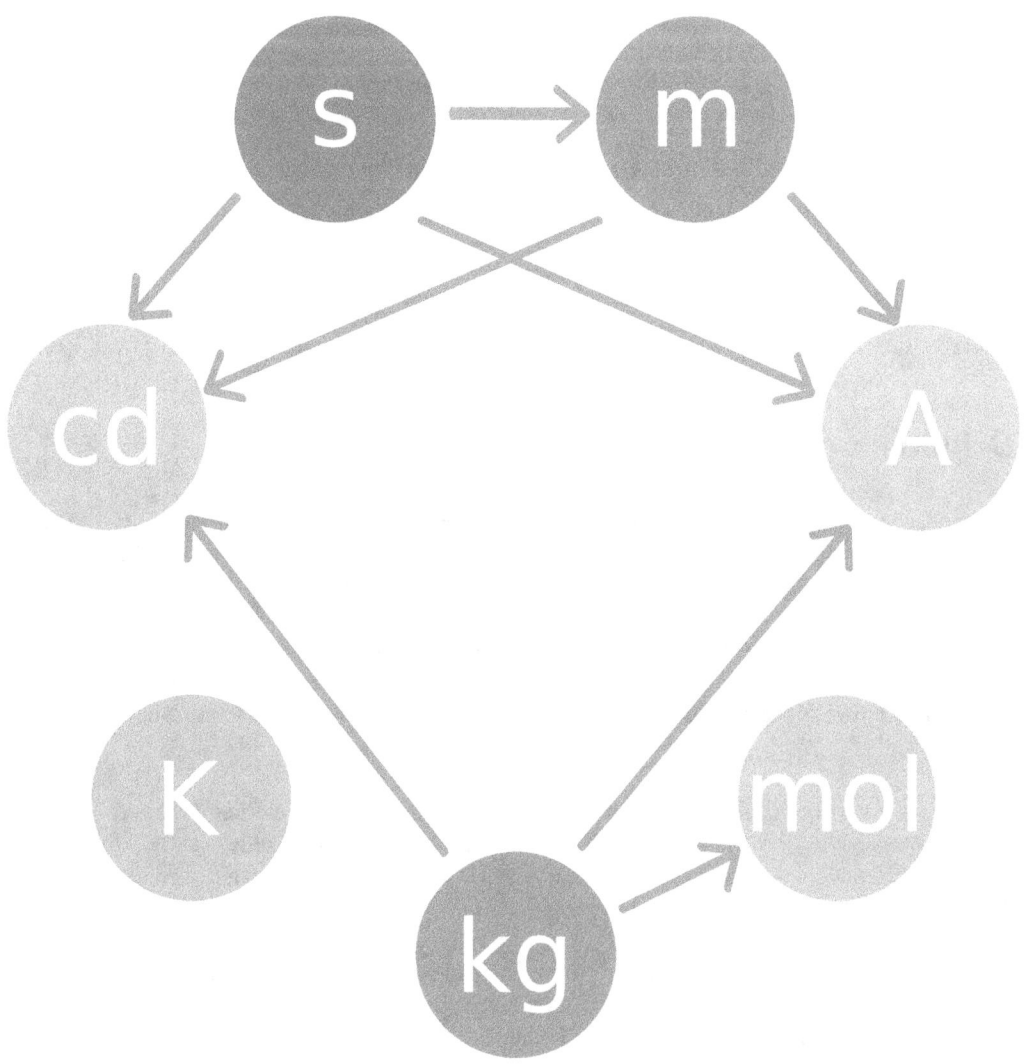

The kilogram is one of the seven SI base units and one of three which is defined ad hoc *(i.e. without reference to another base unit).*

defined in 1795 as one cubic decimeter of water at the melting point of ice. Then in 1889, the kilogram was redefined as the mass of the international prototype kilogram, and as such is independent of the meter, or the properties of water. As of January 2013, there are several proposals for redefining the kilogram yet again, including a proposal for defining it in terms of the Planck constant.[2]

Other units are accepted for use in SI:

- the tonne (t) (or "metric ton") is equal to 1000 kg.

- the electronvolt (eV) is a unit of energy, but because of the mass–energy equivalence it can easily be converted to a unit of mass, and is often used like one. In this context, the mass has units of eV/c^2. The electronvolt is common in particle physics.

- the atomic mass unit (u) is 1/12 of the mass of a carbon-12 atom, approximately 1.66×10^{-27} kg.[note 2] The atomic mass unit is convenient for expressing the masses of atoms and molecules.

Outside the SI system, other units include:

- the slug (sl) is an Imperial unit of mass (about 14.6 kg) similar to the kilogram.

- the pound (lb) is a unit of both mass and force, used mainly in the United States (about 0.45 kg or 4.5 N). In scientific contexts where pound (force) and pound (mass) need to be distinguished, SI units are usually used instead.

- the Planck mass (m_P) is the maximum mass of point particles (about 2.18×10^{-8} kg). It is used in particle physics.

- the solar mass ($M\odot$) is defined as the mass of the sun. It is primarily used in astronomy to compare large masses such as stars or galaxies ($\approx 1.99 \times 10^{30}$ kg).

- the mass of a very small particle may be identified with its inverse Compton wavelength (1 cm$^{-1} \approx 3.52 \times 10^{-41}$ kg).

- the mass of a very large star or black hole may be identified with its Schwarzschild radius (1 cm $\approx 6.73 \times 10^{24}$ kg).

12.2 Definitions of mass

In physical science, one may distinguish conceptually between at least seven different aspects of *mass*, or seven physical notions that involve the concept of *mass*:[3] Every experiment to date has shown these seven values to be proportional, and in some cases equal, and this proportionality gives rise to the abstract concept of mass.

- The amount of matter in certain types of samples can be exactly determined through electrodeposition or other precise processes. The mass of an exact sample is determined in part by the number and type of atoms or molecules it contains, and in part by the energy involved in binding it together (which contributes a negative "missing mass," or mass deficit).

- Inertial mass is a measure of an object's resistance to changing its state of motion when a force is applied. It is determined by applying a force to an object and measuring the acceleration that results from that force. An object with small inertial mass will accelerate more than an object with large inertial mass when acted upon by the same force. One says the body of greater mass has greater inertia.

- Active gravitational mass [note 3] is a measure of the strength of an object's gravitational flux (gravitational flux is equal to the surface integral of gravitational field over an enclosing surface). Gravitational field can be measured by allowing a small 'test object' to freely fall and measuring its free-fall acceleration. For example, an object in free-fall near the Moon will experience less gravitational field, and hence accelerate more slowly than the same object would if it were in free-fall near the Earth. The gravitational field near the Moon is weaker because the Moon has less active gravitational mass.

- Passive gravitational mass is a measure of the strength of an object's interaction with a gravitational field. Passive gravitational mass is determined by dividing an object's weight by its free-fall acceleration. Two objects within the same gravitational field will experience the same acceleration; however, the object with a smaller passive gravitational mass will experience a smaller force (less weight) than the object with a larger passive gravitational mass.

- Energy also has mass according to the principle of mass–energy equivalence. This equivalence is exemplified in a large number of physical processes including pair production, nuclear fusion, and the gravitational bending of light. Pair production and nuclear fusion are processes through which measurable amounts of mass and energy are converted into each other. In the gravitational bending of light, photons of pure energy are shown to exhibit a behavior similar to passive gravitational mass.

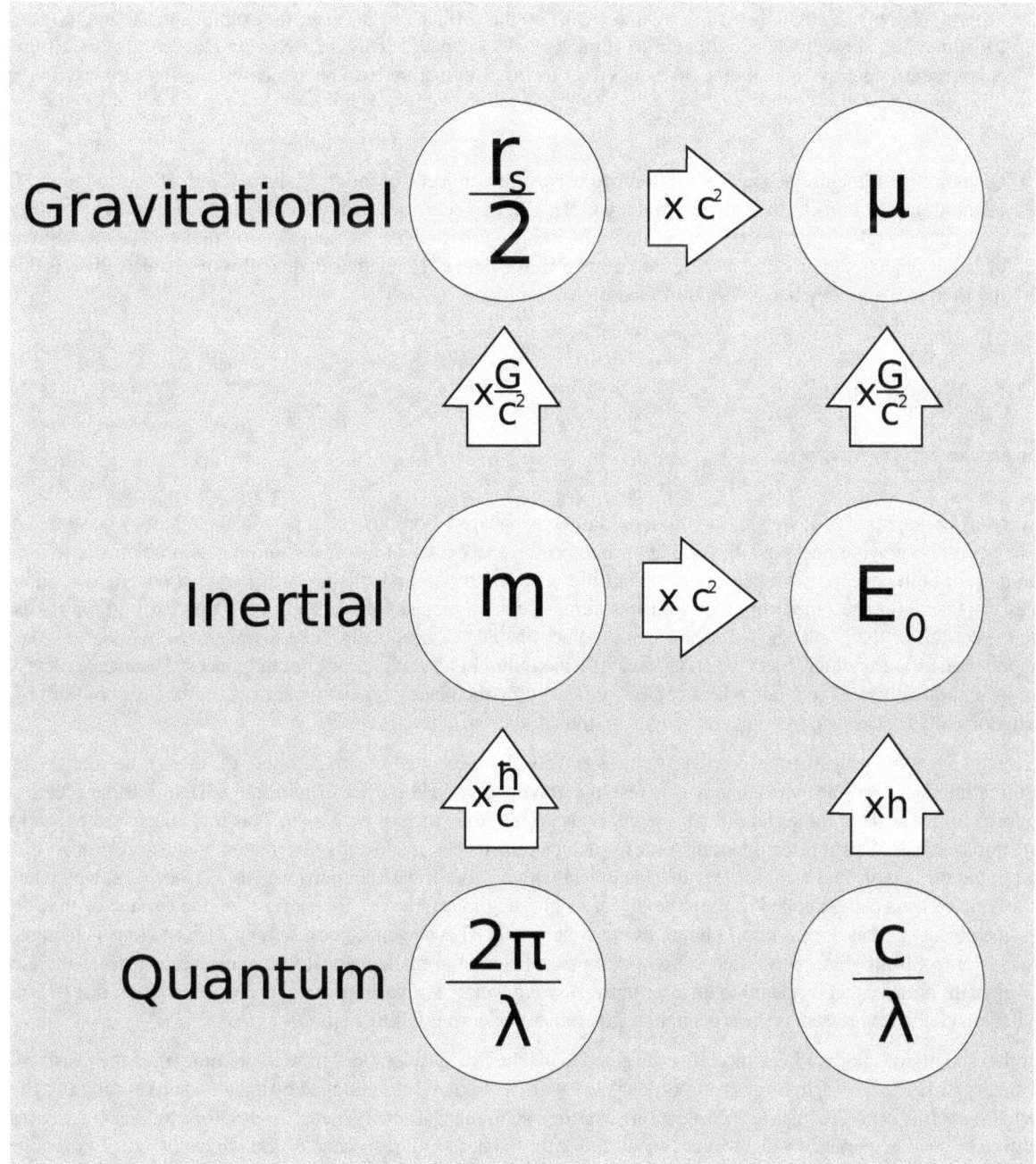

The relation between properties of mass and their associated physical constants. Every massive object is believed to exhibit all five properties. However, due to extremely large or extremely small constants, it is generally impossible to verify more than two or three properties for any object.

The Schwarzschild radius (r_s) represents the ability of mass to cause curvature in space and time.

The standard gravitational parameter (μ) represents the ability of a massive body to exert Newtonian gravitational forces on other bodies.

Inertial mass (m) represents the Newtonian response of mass to forces.

Rest energy (E_0) represents the ability of mass to be converted into other forms of energy.

The Compton wavelength (λ) represents the quantum response of mass to local geometry.

- Curvature of spacetime is a relativistic manifestation of the existence of mass. Curvature is extremely weak and difficult to measure. For this reason, curvature was not discovered until after it was predicted by Einstein's theory of general relativity. Extremely precise atomic clocks on the surface of the earth, for example, are found to measure less time (run slower) when compared to similar clocks in space. This difference in elapsed time is a form of curvature called gravitational time dilation. Other forms of curvature have been measured using the Gravity Probe B satellite.

- Quantum mass manifests itself as a difference between an object's quantum frequency and its wave number. The quantum mass of an electron, the Compton wavelength, can be determined through various forms of spectroscopy and is closely related to the Rydberg constant, the Bohr radius, and the classical electron radius. The quantum mass of larger objects can be directly measured using a watt balance. In relativistic quantum mechanics, mass is one of the irreducible representation labels of the Poincaré group.

12.2.1 Weight vs. mass

Main article: Mass versus weight

In everyday usage, mass and "weight" are often used interchangeably. For instance, a person's weight may be stated as 75 kg. In a constant gravitational field, the weight of an object is proportional to its mass, and it is unproblematic to use the same unit for both concepts. But because of slight differences in the strength of the Earth's gravitational field at different places, the distinction becomes important for measurements with a precision better than a few percent, and for places far from the surface of the Earth, such as in space or on other planets. Conceptually, "mass" (measured in kilograms) refers to an intrinsic property of an object, whereas "weight" (measured in newtons) measures an object's resistance to deviating from its natural course of free fall, which can be influenced by the nearby gravitational field. No matter how strong the gravitational field, objects in free fall are weightless, though they still have mass.[4]

The force known as "weight" is proportional to mass and acceleration in all situations where the mass is accelerated away from free fall. For example, when a body is at rest in a gravitational field (rather than in free fall), it must be accelerated by a force from a scale or the surface of a planetary body such as the Earth or the Moon. This force keeps the object from going into free fall. Weight is the opposing force in such circumstances, and is thus determined by the acceleration of free fall. On the surface of the Earth, for example, an object with a mass of 50 kilograms weighs 491 newtons, which means that 491 newtons is being applied to keep the object from going into free fall. By contrast, on the surface of the Moon, the same object still has a mass of 50 kilograms but weighs only 81.5 newtons, because only 81.5 newtons is required to keep this object from going into a free fall on the moon. Restated in mathematical terms, on the surface of the Earth, the weight W of an object is related to its mass m by $W = mg$, where $g = 9.80665$ m/s^2 is the acceleration due to Earth's gravitational field, (expressed as the acceleration experienced by a free-falling object).

For other situations, such as when objects are subjected to mechanical accelerations from forces other than the resistance of a planetary surface, the weight force is proportional to the mass of an object multiplied by the total acceleration away from free fall, which is called the proper acceleration. Through such mechanisms, objects in elevators, vehicles, centrifuges, and the like, may experience weight forces many times those caused by resistance to the effects of gravity on objects, resulting from planetary surfaces. In such cases, the generalized equation for weight W of an object is related to its mass m by the equation $W = -ma$, where a is the proper acceleration of the object caused by all influences other than gravity. (Again, if gravity is the only influence, such as occurs when an object falls freely, its weight will be zero).

Macroscopically, mass is associated with matter, although matter is not, ultimately, as clearly defined a concept as mass. On the subatomic scale, not only fermions, the particles often associated with matter, but also some bosons, the particles that act as force carriers, have rest mass. Another problem for easy definition is that much of the rest mass of ordinary matter derives from the invariant mass contributed to matter by particles and kinetic energies which have no rest mass themselves (only 1% of the rest mass of matter is accounted for by the rest mass of its fermionic quarks and electrons). From a fundamental physics perspective, mass is the number describing under which the representation of the little group of the Poincaré group a particle transforms. In the Standard Model of particle physics, this symmetry is described as arising as a consequence of a coupling of particles with rest mass to a postulated additional field, known as the Higgs field.

The total mass of the observable universe is estimated at between 10^{52} kg and 10^{53} kg, corresponding to the rest mass of between 10^{79} and 10^{80} protons.

12.2.2   Inertial vs. gravitational mass

Although inertial mass, passive gravitational mass and active gravitational mass are conceptually distinct, no experiment has ever unambiguously demonstrated any difference between them. In classical mechanics, Newton's third law implies that active and passive gravitational mass must always be identical (or at least proportional), but the classical theory offers no compelling reason why the gravitational mass has to equal the inertial mass. That it does is merely an empirical fact.

Albert Einstein developed his general theory of relativity starting from the assumption that this correspondence between inertial and (passive) gravitational mass is not accidental: that no experiment will ever detect a difference between them (the weak version of the equivalence principle). However, in the resulting theory, gravitation is not a force and thus not subject to Newton's third law, so "the equality of inertial and *active* gravitational mass [...] remains as puzzling as ever".[5]

The equivalence of inertial and gravitational masses is sometimes referred to as the "Galilean equivalence principle" or the "weak equivalence principle". The most important consequence of this equivalence principle applies to freely falling objects. Suppose we have an object with inertial and gravitational masses m and M, respectively. If the only force acting on the object comes from a gravitational field g, combining Newton's second law and the gravitational law yields the acceleration

$$a = \frac{M}{m}g.$$

This says that the ratio of gravitational to inertial mass of any object is equal to some constant K if and only if all objects fall at the same rate in a given gravitational field. This phenomenon is referred to as the "universality of free-fall". (In addition, the constant K can be taken to be 1 by defining our units appropriately.)

The first experiments demonstrating the universality of free-fall were conducted by Galileo. It is commonly stated that Galileo obtained his results by dropping objects from the Leaning Tower of Pisa, but this is most likely apocryphal; actually, he performed his experiments with balls rolling down nearly frictionless inclined planes to slow the motion and increase the timing accuracy. Increasingly precise experiments have been performed, such as those performed by Loránd Eötvös,[6] using the torsion balance pendulum, in 1889. As of 2008, no deviation from universality, and thus from Galilean equivalence, has ever been found, at least to the precision 10^{-12}. More precise experimental efforts are still being carried out.

The universality of free-fall only applies to systems in which gravity is the only acting force. All other forces, especially friction and air resistance, must be absent or at least negligible. For example, if a hammer and a feather are dropped from the same height through the air on Earth, the feather will take much longer to reach the ground; the feather is not really in *free*-fall because the force of air resistance upwards against the feather is comparable to the downward force of gravity. On the other hand, if the experiment is performed in a vacuum, in which there is no air resistance, the hammer and the feather should hit the ground at exactly the same time (assuming the acceleration of both objects towards each other, and of the ground towards both objects, for its own part, is negligible). This can easily be done in a high school laboratory by dropping the objects in transparent tubes that have the air removed with a vacuum pump. It is even more dramatic when done in an environment that naturally has a vacuum, as David Scott did on the surface of the Moon during Apollo 15.

A stronger version of the equivalence principle, known as the *Einstein equivalence principle* or the *strong equivalence principle*, lies at the heart of the general theory of relativity. Einstein's equivalence principle states that within sufficiently small regions of space-time, it is impossible to distinguish between a uniform acceleration and a uniform gravitational field. Thus, the theory postulates that the force acting on a massive object caused by a gravitational field is a result of the object's tendency to move in a straight line (in other words its inertia) and should therefore be a function of its inertial mass and the strength of the gravitational field.

12.2.3   Origin of mass

Main article: Mass generation mechanism

In theoretical physics, a mass generation mechanism is a theory which attempts to explain the origin of mass from the most fundamental laws of physics. To date, a number of different models have been proposed which advocate different views at the origin of mass. The problem is complicated by the fact that the notion of mass is strongly related to the gravitational interaction but a theory of the latter has not been yet reconciled with the currently popular model of particle physics, known as the Standard Model.

12.3 Pre-Newtonian concepts

12.3.1 Weight as an amount

Main article: weight

The concept of amount is very old and predates recorded history. Humans, at some early era, realized that the weight of

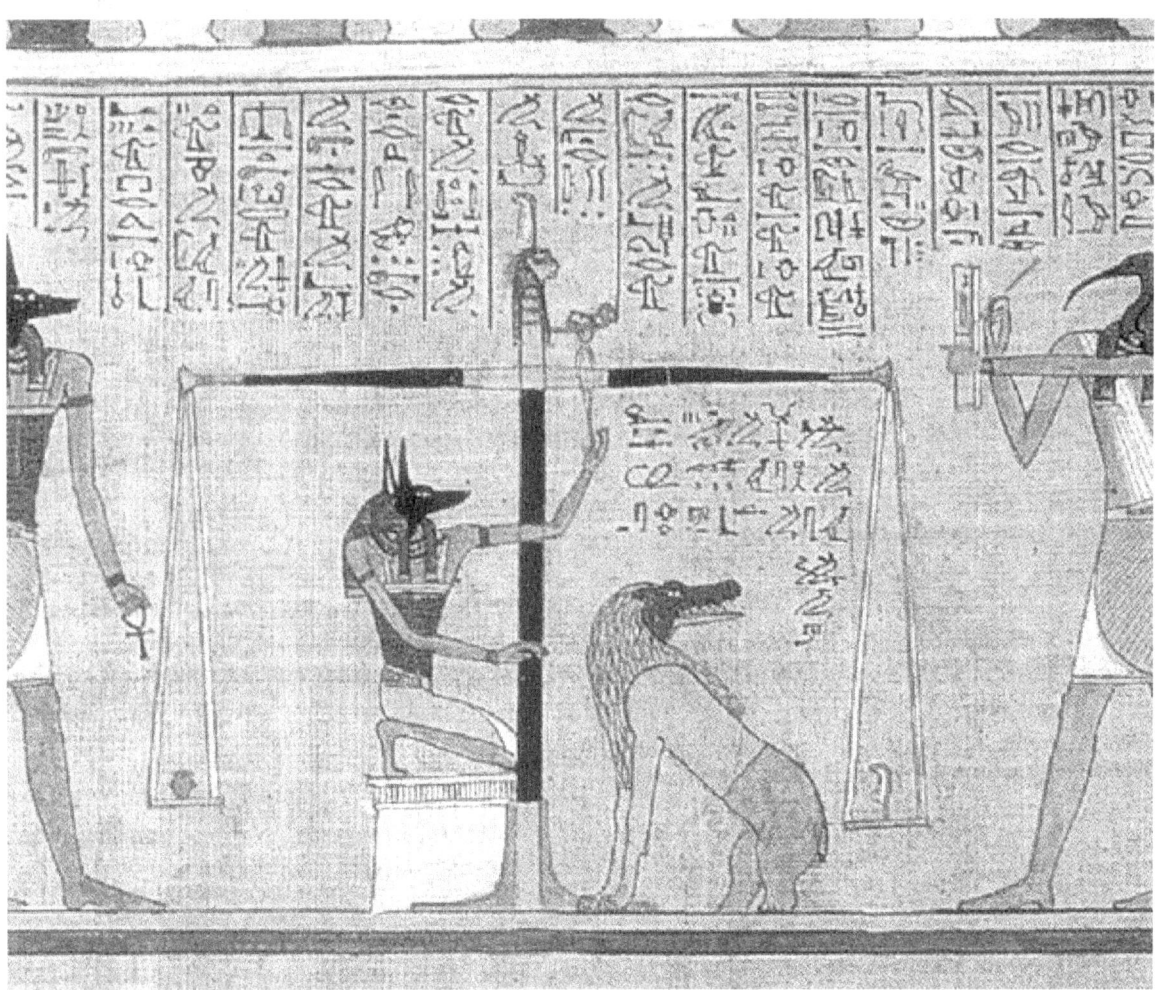

Depiction of early balance scales in the Papyrus of Hunefer (dated to the 19th dynasty, ca. 1285 BC). The scene shows Anubis weighing the heart of Hunefer.

a collection of similar objects was directly proportional to the number of objects in the collection:

$$W_n \propto n,$$

where W is the weight of the collection of similar objects and n is the number of objects in the collection. Proportionality, by definition, implies that two values have a constant ratio:

$$\frac{W_n}{n} = \frac{W_m}{m} \text{ , or equivalently } \frac{W_n}{W_m} = \frac{n}{m}.$$

An early use of this relationship is a balance scale, which balances the force of one object's weight against the force of another object's weight. The two sides of a balance scale are close enough that the objects experience similar gravitational fields. Hence, if they have similar masses then their weights will also be similar. This allows the scale, by comparing weights, to also compare masses.

Consequently, historical weight standards were often defined in terms of amounts. The Romans, for example, used the carob seed (carat or siliqua) as a measurement standard. If an object's weight was equivalent to 1728 carob seeds, then the object was said to weigh one Roman pound. If, on the other hand, the object's weight was equivalent to 144 carob seeds then the object was said to weigh one Roman ounce (uncia). The Roman pound and ounce were both defined in terms of different sized collections of the same common mass standard, the carob seed. The ratio of a Roman ounce (144 carob seeds) to a Roman pound (1728 carob seeds) was:

$$\frac{\text{ounce}}{\text{pound}} = \frac{W_{144}}{W_{1728}} = \frac{144}{1728} = \frac{1}{12}.$$

12.3.2 Planetary motion

See also: Kepler's laws of planetary motion

In 1600 AD, Johannes Kepler sought employment with Tycho Brahe, who had some of the most precise astronomical data available. Using Brahe's precise observations of the planet Mars, Kepler spent the next five years developing his own method for characterizing planetary motion. In 1609, Johannes Kepler published his three laws of planetary motion, explaining how the planets orbit the Sun. In Kepler's final planetary model, he described planetary orbits as following elliptical paths with the Sun at a focal point of the ellipse. Kepler discovered that the square of the orbital period of each planet is directly proportional to the cube of the semi-major axis of its orbit, or equivalently, that the ratio of these two values is constant for all planets in the Solar System.[note 4]

On 25 August 1609, Galileo Galilei demonstrated his first telescope to a group of Venetian merchants, and in early January of 1610, Galileo observed four dim objects near Jupiter, which he mistook for stars. However, after a few days of observation, Galileo realized that these "stars" were in fact orbiting Jupiter. These four objects (later named the Galilean moons in honor of their discoverer) were the first celestial bodies observed to orbit something other than the Earth or Sun. Galileo continued to observe these moons over the next eighteen months, and by the middle of 1611 he had obtained remarkably accurate estimates for their periods.

12.3.3 Galilean free fall

Sometime prior to 1638, Galileo turned his attention to the phenomenon of objects in free fall, attempting to characterize these motions. Galileo was not the first to investigate Earth's gravitational field, nor was he the first to accurately describe its fundamental characteristics. However, Galileo's reliance on scientific experimentation to establish physical principles would have a profound effect on future generations of scientists. It is unclear if these were just hypothetical experiments used to illustrate a concept, or if they were real experiments performed by Galileo,[7] but the results obtained from these experiments were both realistic and compelling. A biography by Galileo's pupil Vincenzo Viviani stated that Galileo had dropped balls of the same material, but different masses, from the Leaning Tower of Pisa to demonstrate that their time of descent was independent of their mass.[note 5] In support of this conclusion, Galileo had advanced the following theoretical argument: He asked if two bodies of different masses and different rates of fall are tied by a string, does the combined system fall faster because it is now more massive, or does the lighter body in its slower fall hold back the heavier body? The only convincing resolution to this question is that all bodies must fall at the same rate.[8]

A later experiment was described in Galileo's *Two New Sciences* published in 1638. One of Galileo's fictional characters, Salviati, describes an experiment using a bronze ball and a wooden ramp. The wooden ramp was "12 cubits long, half a cubit wide and three finger-breadths thick" with a straight, smooth, polished groove. The groove was lined with

Galileo Galilei (1636)

"parchment, also smooth and polished as possible". And into this groove was placed "a hard, smooth and very round bronze ball". The ramp was inclined at various angles to slow the acceleration enough so that the elapsed time could be measured. The ball was allowed to roll a known distance down the ramp, and the time taken for the ball to move the known distance was measured. The time was measured using a water clock described as follows:

"a large vessel of water placed in an elevated position; to the bottom of this vessel was soldered a pipe of small diameter giving a thin jet of water, which we collected in a small glass during the time of each descent, whether for the whole length of the channel or for a part of its length; the water thus collected was weighed, after each descent, on a very accurate balance; the differences and ratios of these weights gave us the differences and ratios of the times, and this with such accuracy that although the operation was repeated many, many times, there was no appreciable discrepancy in the results."[9]

Galileo found that for an object in free fall, the distance that the object has fallen is always proportional to the square of the elapsed time:

$$\text{Distance} \propto \text{Time}^2$$

Galileo had shown that objects in free fall under the influence of the Earth's gravitational field have a constant acceleration, and Galileo's contemporary, Johannes Kepler, had shown that the planets follow elliptical paths under the influence of the Sun's gravitational mass. However, Galileo's free fall motions and Kepler's planetary motions remained distinct during Galileo's lifetime.

12.4 Newtonian mass

Robert Hooke had published his concept of gravitational forces in 1674, stating that all celestial bodies have an attraction or gravitating power towards their own centers, and also attract all the other celestial bodies that are within the sphere of their activity. He further stated that gravitational attraction increases by how much nearer the body wrought upon is to their own center.[10] In correspondence with Isaac Newton from 1679 and 1680, Hooke conjectures that gravitational forces might decrease according to the double of the distance between the two bodies.[11] Hooke urged Newton, who was a pioneer in the development of calculus, to work through the mathematical details of Keplerian orbits to determine if Hooke's hypothesis was correct. Newton's own investigations verified that Hooke was correct, but due to personal differences between the two men, Newton chose not to reveal this to Hooke. Isaac Newton kept quiet about his discoveries until 1684, at which time he told a friend, Edmond Halley, that he had solved the problem of gravitational orbits, but had misplaced the solution in his office.[12] After being encouraged by Halley, Newton decided to develop his ideas about gravity and publish all of his findings. In November 1684, Isaac Newton sent a document to Edmund Halley, now lost but presumed to have been titled *De motu corporum in gyrum* (Latin for "On the motion of bodies in an orbit").[13] Halley presented Newton's findings to the Royal Society of London, with a promise that a fuller presentation would follow. Newton later recorded his ideas in a three book set, entitled Philosophiæ Naturalis Principia Mathematica (Latin: "Mathematical Principles of Natural Philosophy"). The first was received by the Royal Society on 28 April 1685–6; the second on 2 March 1686–7; and the third on 6 April 1686–7. The Royal Society published Newton's entire collection at their own expense in May 1686–7.[14]:31

Isaac Newton had bridged the gap between Kepler's gravitational mass and Galileo's gravitational acceleration, resulting in the discovery of the following relationship which governed both of these:

$$\mathbf{g} = -\mu \frac{\hat{\mathbf{R}}}{|\mathbf{R}|^2}$$

where $\mathbf{g}$ is the apparent acceleration of a body as it passes through a region of space where gravitational fields exist, μ is the gravitational mass (standard gravitational parameter) of the body causing gravitational fields, and $\mathbf{R}$ is the radial coordinate (the distance between the centers of the two bodies).

By finding the exact relationship between a body's gravitational mass and its gravitational field, Newton provided a second method for measuring gravitational mass. The mass of the Earth can be determined using Kepler's method (from the orbit of Earth's Moon), or it can be determined by measuring the gravitational acceleration on the Earth's surface, and multiplying that by the square of the Earth's radius. The mass of the Earth is approximately three millionths of the mass of the Sun. To date, no other accurate method for measuring gravitational mass has been discovered.[15]

12.4.1 Newton's cannonball

Main article: Newton's cannonball

Newton's cannonball was a thought experiment used to bridge the gap between Galileo's gravitational acceleration and Kepler's elliptical orbits. It appeared in Newton's 1728 book *A Treatise of the System of the World*. According to Galileo's concept of gravitation, a dropped stone falls with constant acceleration down towards the Earth. However, Newton explains that when a stone is thrown horizontally (meaning sideways or perpendicular to Earth's gravity) it follows a curved path. "For a stone projected is by the pressure of its own weight forced out of the rectilinear path, which by the projection alone it should have pursued, and made to describe a curve line in the air; and through that crooked way is at last brought down to the ground. And the greater the velocity is with which it is projected, the farther it goes before it falls to the Earth."[14]:513 Newton further reasons that if an object were "projected in an horizontal direction from the top of a high mountain" with sufficient velocity, "it would reach at last quite beyond the circumference of the Earth, and return to the mountain from which it was projected."

12.4.2 Universal gravitational mass

In contrast to earlier theories (e.g. celestial spheres) which stated that the heavens were made of entirely different material, Newton's theory of mass was groundbreaking partly because it introduced universal gravitational mass: every object has gravitational mass, and therefore, every object generates a gravitational field. Newton further assumed that the strength of each object's gravitational field would decrease according to the square of the distance to that object. If a large collection of small objects were formed into a giant spherical body such as the Earth or Sun, Newton calculated the collection would create a gravitational field proportional to the total mass of the body,[14]:397 and inversely proportional to the square of the distance to the body's center.[14]:221 [note 6]

For example, according to Newton's theory of universal gravitation, each carob seed produces a gravitational field. Therefore, if one were to gather an immense number of carob seeds and form them into an enormous sphere, then the gravitational field of the sphere would be proportional to the number of carob seeds in the sphere. Hence, it should be theoretically possible to determine the exact number of carob seeds that would be required to produce a gravitational field similar to that of the Earth or Sun. In fact, by unit conversion it is a simple matter of abstraction to realize that any traditional mass unit can theoretically be used to measure gravitational mass.

Measuring gravitational mass in terms of traditional mass units is simple in principle, but extremely difficult in practice. According to Newton's theory all objects produce gravitational fields and it is theoretically possible to collect an immense number of small objects and form them into an enormous gravitating sphere. However, from a practical standpoint, the gravitational fields of small objects are extremely weak and difficult to measure. Newton's books on universal gravitation were published in the 1680s, but the first successful measurement of the Earth's mass in terms of traditional mass units, the Cavendish experiment, did not occur until 1797, over a hundred years later. Cavendish found that the Earth's density was 5.448 ± 0.033 times that of water. As of 2009, the Earth's mass in kilograms is only known to around five digits of accuracy, whereas its gravitational mass is known to over nine significant figures.

Given two objects A and B, of masses M_A and M_B, separated by a displacement $\mathbf{R}_{AB}$, Newton's law of gravitation states that each object exerts a gravitational force on the other, of magnitude

$$\mathbf{F}_{AB} = -GM_A M_B \frac{\hat{\mathbf{R}}_{AB}}{|\mathbf{R}_{AB}|^2}$$

where G is the universal gravitational constant. The above statement may be reformulated in the following way: if g is the magnitude at a given location in a gravitational field, then the gravitational force on an object with gravitational mass M is

$$F = Mg$$

This is the basis by which masses are determined by weighing. In simple spring scales, for example, the force F is proportional to the displacement of the spring beneath the weighing pan, as per Hooke's law, and the scales are calibrated to take g into account, allowing the mass M to be read off. Assuming the gravitational field is equivalent on both sides of the balance, a balance measures relative weight, giving the relative gravitation mass of each object.

12.4.3 Inertial mass

Inertial mass is the mass of an object measured by its resistance to acceleration. The simple classical mechanics definition of mass is slightly different than the definition in the theory of special relativity, but the essential meaning is the same.

In classical mechanics, according to Newton's second law, we say that a body has a mass m if, at any instant of time, it obeys the equation of motion

$$\mathbf{F} = m\mathbf{a},$$

where $\mathbf{F}$ is the resultant force acting on the body and $\mathbf{a}$ is the acceleration of the body's centre of mass.[note 7] For the moment, we will put aside the question of what "force acting on the body" actually means.

This equation illustrates how mass relates to the inertia of a body. Consider two objects with different masses. If we apply an identical force to each, the object with a bigger mass will experience a smaller acceleration, and the object with a smaller mass will experience a bigger acceleration. We might say that the larger mass exerts a greater "resistance" to changing its state of motion in response to the force.

However, this notion of applying "identical" forces to different objects brings us back to the fact that we have not really defined what a force is. We can sidestep this difficulty with the help of Newton's third law, which states that if one object exerts a force on a second object, it will experience an equal and opposite force. To be precise, suppose we have two objects of constant inertial masses m_1 and m_2. We isolate the two objects from all other physical influences, so that the only forces present are the force exerted on m_1 by m_2, which we denote $\mathbf{F}_{12}$, and the force exerted on m_2 by m_1, which we denote $\mathbf{F}_{21}$. Newton's second law states that

$$\mathbf{F}_{12} = m_1\mathbf{a}_1,$$
$$\mathbf{F}_{21} = m_2\mathbf{a}_2,$$

where $\mathbf{a}_1$ and $\mathbf{a}_2$ are the accelerations of m_1 and m_2, respectively. Suppose that these accelerations are non-zero, so that the forces between the two objects are non-zero. This occurs, for example, if the two objects are in the process of colliding with one another. Newton's third law then states that

$$\mathbf{F}_{12} = -\mathbf{F}_{21};$$

and thus

$$m_1 = m_2\frac{|\mathbf{a}_2|}{|\mathbf{a}_1|}.$$

If $|\mathbf{a}_1|$ is non-zero, the fraction is well-defined, which allows us to measure the inertial mass of m_1. In this case, m_2 is our "reference" object, and we can define its mass m as (say) 1 kilogram. Then we can measure the mass of any other object in the universe by colliding it with the reference object and measuring the accelerations.

Additionally, mass relates a body's momentum $\mathbf{p}$ to its linear velocity $\mathbf{v}$:

$$\mathbf{p} = m\mathbf{v}$$

and the body's kinetic energy K to its velocity:

$$K = \frac{1}{2}m|\mathbf{v}|^2$$

12.5 Atomic mass

Main article: Atomic mass unit

Typically, the mass of objects is measured in relation to that of the kilogram, which is defined as the mass of the *international prototype kilogram* (IPK), a platinum alloy cylinder stored in an environmentally-monitored safe secured in a vault at the International Bureau of Weights and Measures in France. However, the IPK is not convenient for measuring the masses of atoms and particles of similar scale, as it contains trillions of trillions of atoms, and has most certainly lost or gained a little mass over time despite the best efforts to prevent this. It is much easier to precisely compare an atom's mass to that of another atom, thus scientists developed the atomic mass unit. By definition, 1 u is exactly one twelfth of the mass of a carbon-12 atom, and by extension a carbon-12 atom has a mass of exactly 12 u.

12.6 Mass in relativity

12.6.1 Special relativity

Main article: Mass in special relativity

In special relativity, there are two kinds of mass: rest mass (invariant mass),[note 8] and relativistic mass (which increases with velocity). Rest mass is the Newtonian mass as measured by an observer moving along with the object. *Relativistic mass* is the total quantity of energy in a body or system divided by c^2. The two are related by the following equation:

$$m_{\text{relative}} = \gamma(m_{\text{rest}})$$

where γ is the Lorentz factor:

$$\gamma = \frac{1}{\sqrt{1 - v^2/c^2}}$$

The invariant mass of systems is the same for observers in all inertial frames, while the relativistic mass depends on the observer's frame of reference. In order to formulate the equations of physics such that mass values do not change between observers, it is convenient to use rest mass. The rest mass of a body is also related to its energy E and the magnitude of its momentum $\mathbf{p}$ by the relativistic energy-momentum equation:

$$(m_{\text{rest}})c^2 = \sqrt{E_{\text{total}}^2 - (|\mathbf{p}|c)^2}.$$

So long as the system is closed with respect to mass and energy, both kinds of mass are conserved in any given frame of reference. The conservation of mass holds even as some types of particles are converted to others. Particles of matter may be converted to types of energy (e.g. light, kinetic energy, the potential energy in magnetic, electric and other fields) but this does not affect the amount of mass. Although things like heat may not be matter, all types of energy still continue to exhibit mass.[note 9][16] Thus, mass and energy do not change into one another in relativity; rather, both are names for the same thing, and neither mass nor energy *appear* without the other.

Both rest and relativistic mass can be expressed as an energy by applying the well-known relationship $E = mc^2$, yielding rest energy and "relativistic energy" (total system energy) respectively:

$$E_{\text{rest}} = (m_{\text{rest}})c^2$$

$$E_{\text{total}} = (m_{\text{relative}})c^2$$

The "relativistic" mass and energy concepts are related to their "rest" counterparts, but they do not have the same value as their rest counterparts in systems where there is a net momentum. Because the relativistic mass is proportional to the energy, it has gradually fallen into disuse among physicists.[17] There is disagreement over whether the concept remains useful pedagogically.[18][19][20]

In bound systems, the binding energy must often be subtracted from the mass of the unbound system, because binding energy commonly leaves the system at the time it is bound. Mass is not conserved in this process because the system is not closed during the binding process. For example, the binding energy of atomic nuclei is often lost in the form of gamma rays when the nuclei are formed, leaving nuclides which have less mass than the free particles (nucleons) of which they are composed.

12.6.2   General relativity

Main article: Mass in general relativity

In general relativity, the equivalence principle is any of several related concepts dealing with the equivalence of gravitational and inertial mass. At the core of this assertion is Albert Einstein's idea that the gravitational force as experienced locally while standing on a massive body (such as the Earth) is the same as the *pseudo-force* experienced by an observer in a non-inertial (i.e. accelerated) frame of reference.

However, it turns out that it is impossible to find an objective general definition for the concept of invariant mass in general relativity. At the core of the problem is the non-linearity of the Einstein field equations, making it impossible to write the gravitational field energy as part of the stress–energy tensor in a way that is invariant for all observers. For a given observer, this can be achieved by the stress–energy–momentum pseudotensor.[21]

12.7   Mass in quantum physics

In classical mechanics, the inert mass of a particle appears in the Euler–Lagrange equation as a parameter m:

$$\frac{\mathrm{d}}{\mathrm{d}t}\left(\frac{\partial L}{\partial \dot{x}_i}\right) = m\,\ddot{x}_i$$

After quantization, replacing the position vector x with a wave function, the parameter m appears in the kinetic energy operator:

$$i\hbar\frac{\partial}{\partial t}\Psi(\mathbf{r},\, t) = \left(-\frac{\hbar^2}{2m}\nabla^2 + V(\mathbf{r})\right)\Psi(\mathbf{r},\, t)$$

In the ostensibly covariant (relativistically invariant) Dirac equation, and in natural units, this becomes:

$$(-i\gamma^\mu\partial_\mu + m)\psi = 0$$

...where the "mass" parameter m is now simply a constant associated with the quantum described by the wave function ψ.

In the Standard Model of particle physics as developed in the 1960s, there is the proposal that this term arises from the coupling of the field ψ to an additional field Φ, the so-called Higgs field. In the case of fermions, the Higgs mechanism results in the replacement of the term $m\psi$ in the Lagrangian with $G_\psi \overline{\psi} \phi \psi$. This shifts the explanandum of the value for the mass of each elementary particle to the value of the unknown couplings $G\psi$. The tentatively confirmed discovery of a massive Higgs boson is regarded as a strong confirmation of this theory. But there is indirect evidence for the reality of the Electroweak symmetry breaking as described by the Higgs mechanism, and the non-existence of Higgs bosons would indicate a "Higgsless" description of this mechanism.

12.7.1 Tachyonic particles and imaginary (complex) mass

Main articles: Tachyonic field and Tachyon § Mass

A tachyonic field, or simply tachyon, is a quantum field with an imaginary mass.[22] Although tachyons (particles that move faster than light) are a purely hypothetical concept not generally believed to exist,[22] [23] fields with imaginary mass have come to play an important role in modern physics[24][25][26][24] and are discussed in popular books on physics.[22][27] Under no circumstances do any excitations ever propagate faster than light in such theories – the presence or absence of a tachyonic mass has no effect whatsoever on the maximum velocity of signals (there is no violation of causality).[28] While the *field* may have imaginary mass, any physical particles do not; the "imaginary mass" shows that the system becomes unstable, and sheds the instability by undergoing a type of phase transition called tachyon condensation (closely related to second order phase transitions) that results in symmetry breaking in current models of particle physics.

The term "tachyon" was coined by Gerald Feinberg in a 1967 paper,[29] but it was soon realized that Feinberg's model in fact did not allow for superluminal speeds.[28] Instead, the imaginary mass creates an instability in the configuration:- any configuration in which one or more field excitations are tachyonic will spontaneously decay, and the resulting configuration contains no physical tachyons. This process is known as tachyon condensation. Well known examples include the condensation of the Higgs boson in particle physics, and ferromagnetism in condensed matter physics.

Although the notion of a tachyonic imaginary mass might seem troubling because there is no classical interpretation of an imaginary mass, the mass is not quantized. Rather, the scalar field is; even for tachyonic quantum fields, the field operators at spacelike separated points still commute (or anticommute), thus preserving causality. Therefore, information still does not propagate faster than light,[29] and solutions grow exponentially, but not superluminally (there is no violation of causality). Tachyon condensation drives a physical system that has reached a local limit and might naively be expected to produce physical tachyons, to an alternate stable state where no physical tachyons exist. Once the tachyonic field reaches the minimum of the potential, its quanta are not tachyons any more but rather are ordinary particles with a positive mass-squared.[30]

This is a special case of the general rule, where unstable massive particles are formally described as having a complex mass, with the real part being their mass in usual sense, and the imaginary part being the decay rate in natural units.[30] However, in quantum field theory, a particle (a "one-particle state") is roughly defined as a state which is constant over time; i.e., an eigenvalue of the Hamiltonian. An unstable particle is a state which is only approximately constant over time; If it exists long enough to be measured, it can be formally described as having a complex mass, with the real part of the mass greater than its imaginary part. If both parts are of the same magnitude, this is interpreted as a resonance appearing in a scattering process rather than particle, as it is considered not to exist long enough to be measured independently of the scattering process. In the case of a tachyon the real part of the mass is zero, and hence no concept of a particle can be attributed to it.

In a Lorentz invariant theory, the same formulas that apply to ordinary slower-than-light particles (sometimes called "bradyons" in discussions of tachyons) must also apply to tachyons. In particular the energy–momentum relation:

$$E^2 = p^2 c^2 + m^2 c^4$$

(where **p** is the relativistic momentum of the bradyon and **m** is its rest mass) should still apply, along with the formula for the total energy of a particle:

$$E = \frac{mc^2}{\sqrt{1 - \frac{v^2}{c^2}}}.$$

This equation shows that the total energy of a particle (bradyon or tachyon) contains a contribution from its rest mass (the "rest mass–energy") and a contribution from its motion, the kinetic energy. When v is larger than c, the denominator in the equation for the energy is "imaginary", as the value under the radical is negative. Because the total energy must be real, the numerator must *also* be imaginary: i.e. the rest mass **m** must be imaginary, as a pure imaginary number divided by another pure imaginary number is a real number.

12.8   See also

- Mass versus weight

- Effective mass (spring–mass system)

- Effective mass (solid-state physics)

- Gell-Mann–Okubo mass formula

- International System of Quantities

12.9   Notes

[1] When a distinction is necessary, M is used to denote the active gravitational mass and m the passive gravitational mass.

[2] Since the Avogadro constant NA is defined as the number of atoms in 12 g of carbon-12, it follows that 1 u is exactly $1/(10^3 NA)$ kg.

[3] The distinction between "active" and "passive" gravitational mass does not exist in the Newtonian view of gravity as found inclassical mechanics, and can safely be ignored by laypersons. In most practical applications, Newtonian gravity is used becauseit is usually sufficiently accurate, and is simpler than General Relativity; for example, NASA uses primarily Newtonian gravity todesign space missions, although " accuracies are routinely enhanced by accounting for tiny relativistic effects".2 .phpThe distinction between"active"and"passive" is very abstract,and applies to post-graduate level applications of GeneralRelativity to certain problems in cosmology, and is otherwise not used. There is, nevertheless, an important conceptual dis-tinction in Newtonian physics between "*inertial* mass" and "gravitational mass", although these quantities are identical; theconceptual distinction between these two fundamental definitions of mass is maintained for teaching purposes because theyinvolve two distinct methods of measurement. It was long considered anomalous that the two distinct measurements of mass(inertial and gravitational) gave the identical result. The observed property, noted by Galileo, according to which objects ofdifferent mass fall with the same rate of acceleration (ignoring air resistance), is an expression of the fact that inertial andgravitational mass are the same.

[4] This constant ratio was later shown to be a direct measure of the Sun's active gravitational mass; it has units of distance cubed per time squared, and is known as the standard gravitational parameter:

$$\mu = 4\pi^2 \frac{\text{distance}^3}{\text{time}^2} \propto \text{mass gravitational}$$

[5] At the time when Viviani asserts that the experiment took place, Galileo had not yet formulated the final version of his law of free fall. He had, however, formulated an earlier version which predicted that bodies *of the same material* falling through the same medium would fall at the same speed. See Drake, S. (1978). *Galileo at Work*. University of Chicago Press. pp. 19–20. ISBN 0-226-16226-5.

[6] These two properties are very useful, as they allow spherical collections of objects to be treated exactly like large individual objects.

[7] In its original form, Newton's second law is valid only for bodies of constant mass.

[8] It is possible to make a slight distinction between "rest mass" and "invariant mass". For a system of two or more particles, none of the particles are required be at rest with respect to the observer for the system as a whole to be at rest with respect to the observer. To avoid this confusion, some sources will use "rest mass" only for individual particles, and "invariant mass" for systems.

[9] For example, a nuclear bomb in an idealized super-strong box, sitting on a scale, would in theory show no change in mass when detonated (although the inside of the box would become much hotter). In such a system, the mass of the box would change only if energy were allowed to escape from the box as light or heat. However, in that case, the removed energy would take its associated mass with it. Letting heat out of such a system is simply a way to remove mass. Thus, mass, like energy, cannot be destroyed, but only moved from one place to another.

12.10 References

[1] "New Quantum Theory Separates Gravitational and Inertial Mass". MIT Technology Review. 14 Jun 2010. Retrieved 3 Dec 2013.

[2] Jacob Aron (10 Jan 2013). "Most fundamental clock ever could redefine kilogram". NewScientist. Retrieved 17 Dec 2013.

[3] W. Rindler (2006). *Relativity: Special, General, And Cosmological*. Oxford University Press. pp. 16–18. ISBN 0-19-856731-6.

[4] Kane, Gordon (September 4, 2008). "The Mysteries of Mass". *Scientific American* (Nature America, Inc.). pp. 32–39. Retrieved 2013-07-05.

[5] Rindler, W. (2006). *Relativity: Special, General, And Cosmological*. Oxford University Press. p. 22. ISBN 0-19-856731-6.

[6] Eötvös, R. V.; Pekár, D.; Fekete, E. (1922). "*Beiträge zum Gesetz der Proportionalität von Trägheit und Gravität*". *Annalen der Physik* **68**: 11. Bibcode:1922AnP...373...11E. doi:10.1002/andp.19223730903.

[7] Drake,S. (1979). "Galileo's Discovery of the Law of Free Fall".*Scientific American***228**(5): 84–92. Bibcode:1973SciAm.228e. doi:10.1038/scientificamerican0573-84.

[8] Galileo, G. (1632). *Dialogue Concerning the Two Chief World Systems*.

[9] Galileo, G. (1638). *Discorsi e Dimostrazioni Matematiche, Intorno à Due Nuove Scienze* **213**. Louis Elsevier., translated in Crew, H.; de Salvio, A., eds. (1954). *Mathematical Discourses and Demonstrations, Relating to Two New Sciences*. Dover Publications. ISBN 1-275-10057-0. and also available in Hawking, S., ed. (2002). *On the Shoulders of Giants*. Running Press. pp. 534–535. ISBN 0-7624-1348-4.

[10] Hooke, R. (1674). "An attempt to prove the motion of the earth from observations". Royal Society.

[11] Turnbull, H. W., ed. (1960). *Correspondence of Isaac Newton, Volume 2 (1676–1687)*. Cambridge University Press. p. 297.

[12] Hawking, S., ed. (2005). *Principia*. Running Press. pp. 15*ff*. ISBN 978-0-7624-2022-3.

[13] Whiteside, D. T., ed. (2008). *The Mathematical Papers of Isaac Newton, Volume VI (1684–1691)*. Cambridge University Press. ISBN 978-0-521-04585-8. Retrieved 12 March 2011.

[14] Sir Isaac Newton; N. W. Chittenden (1848). *Newton's Principia: The mathematical principles of natural philosophy*. D. Adee. Retrieved 12 March 2011.

[15] Cuk, M. (January 2003). "Curious About Astronomy: How do you measure a planet's mass?". *Ask an Astronomer*. Retrieved 2011-03-12.

[16] Taylor, E. F.; Wheeler, J. A. (1992). *Spacetime Physics*. W. H. Freeman. pp. 248–149. ISBN 0-7167-2327-1.

[17] G. Oas (2005). "On the Abuse and Use of Relativistic Mass". arXiv:physics/0504110 [physics.ed-ph].

[18] Okun,L.B. (1989). "The Concept of Mass"(PDF).*Physics Today***42**(6): 31–36. Bibcode:1989PhT....42f..31O.doi:10..

[19] Rindler, W.; Vandyck, M. A.; Murugesan, P.; Ruschin, S.; Sauter, C.; Okun, L. B. (1990). "Putting to Rest Mass Misconceptions" (PDF). *Physics Today* **43** (5): 13–14, 115, 117. Bibcode:1990PhT....43e..13R. doi:10.1063/1.2810555.

[20] Sandin,T.R. (1991). "In Defense of Relativistic Mass".*American Journal of Physics***59**(11): 1032. Bibcode:1991AmJPh.. doi:10.1119/1.16642.

[21] Misner, C. W.; Thorne, K. S.; Wheeler, J. A. (1973). *Gravitation*. W. H. Freeman. p. 466. ISBN 978-0-7167-0344-0.

[22] Lisa Randall, *Warped Passages: Unraveling the Mysteries of the Universe's Hidden Dimensions*, p.286: "People initially thought of tachyons as particles travelling faster than the speed of light...But we now know that a tachyon indicates an instability in a theory that contains it. Regrettably for science fiction fans, tachyons are not real physical particles that appear in nature."

[23] Tipler, Paul A.; Llewellyn, Ralph A. (2008). *Modern Physics* (5th ed.). New York: W.H. Freeman & Co. p. 54. ISBN 978-0-7167-7550-8. ... so existence of particles v > c ... Called tachyons ... would present relativity with serious ... problems of infinite creation energies and causality paradoxes.

[24] Kutasov, David; Marino, Marcos & Moore, Gregory W. (2000). "Some exact results on tachyon condensation in string field theory". *JHEP* **0010**: 045. arXiv EFI-2000-32, RUNHETC-2000-34.

[25] Sen, A. (2002). Rolling tachyon. *JHEP* **0204**, 048. Cited 720 times as of 2/2012.

[26] Gibbons, G.W. (2000). Cosmological evolution of the rolling tachyon. *Phys. Lett. B* **537**:1.

[27] Brian Greene, *The Elegant Universe*, Vintage Books (2000)

[28] Aharonov, Y.; Komar, A.; Susskind, L. (1969). "Superluminal Behavior, Causality, and Instability". *Phys. Rev.* (American Physical Society) **182** (5): 1400–1403. Bibcode:1969PhRv..182.1400A. doi:10.1103/PhysRev.182.1400.

[29] Feinberg,Gerald(1967). "Possibility of Faster-Than-Light Particles".*Physical Review***159**(5): 1089–1105. Bibcode:1967PhRv. doi:10.1103/PhysRev.159.1089.

[30] Michael E. Peskin and Daniel V. Schroeder (1995). *An Introduction to Quantum Field Theory*, Perseus books publishing.

12.11   External links

- Francisco Flores (6 Feb 2012). "The Equivalence of Mass and Energy". Stanford Encyclopedia of Philosophy. Retrieved 3 Dec 2013.

- Gordon Kane (27 Jun 2005). "The Mysteries of Mass". Scientific American. Retrieved 3 Dec 2013.

- L. B. Okun (15 Nov 2001). "Photons, Clocks, Gravity and the Concept of Mass" (PDF). Nuclear Physics. Retrieved 3 Dec 2013.

- Frank Wilczek (13 May 2001). "The Origin of Mass and the Feebleness of Gravity" (video). MIT Video. Retrieved 3 Dec 2013.

- John Baez; et al. (2012). "Does mass change with velocity?". Retrieved 3 Dec 2013.

- John Baez; et al. (2008). "What is the mass of a photon?". Retrieved 3 Dec 2013.

- David R. Williams (12 February 2008). "The Apollo 15 Hammer-Feather Drop". NASA. Retrieved 3 Dec 2013.

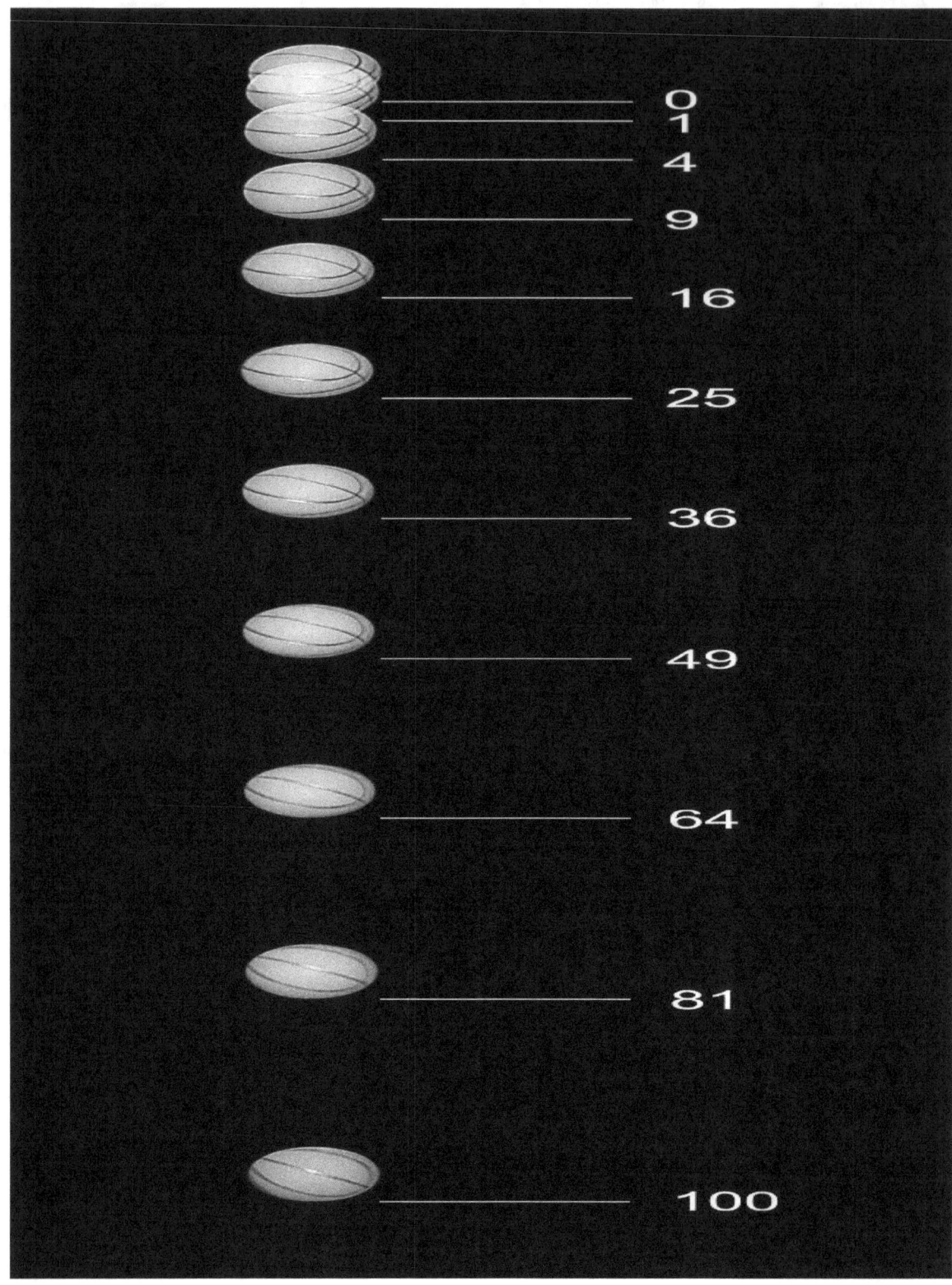

Distance traveled by a freely falling ball is proportional to the square of the elapsed time

Isaac Newton 1689

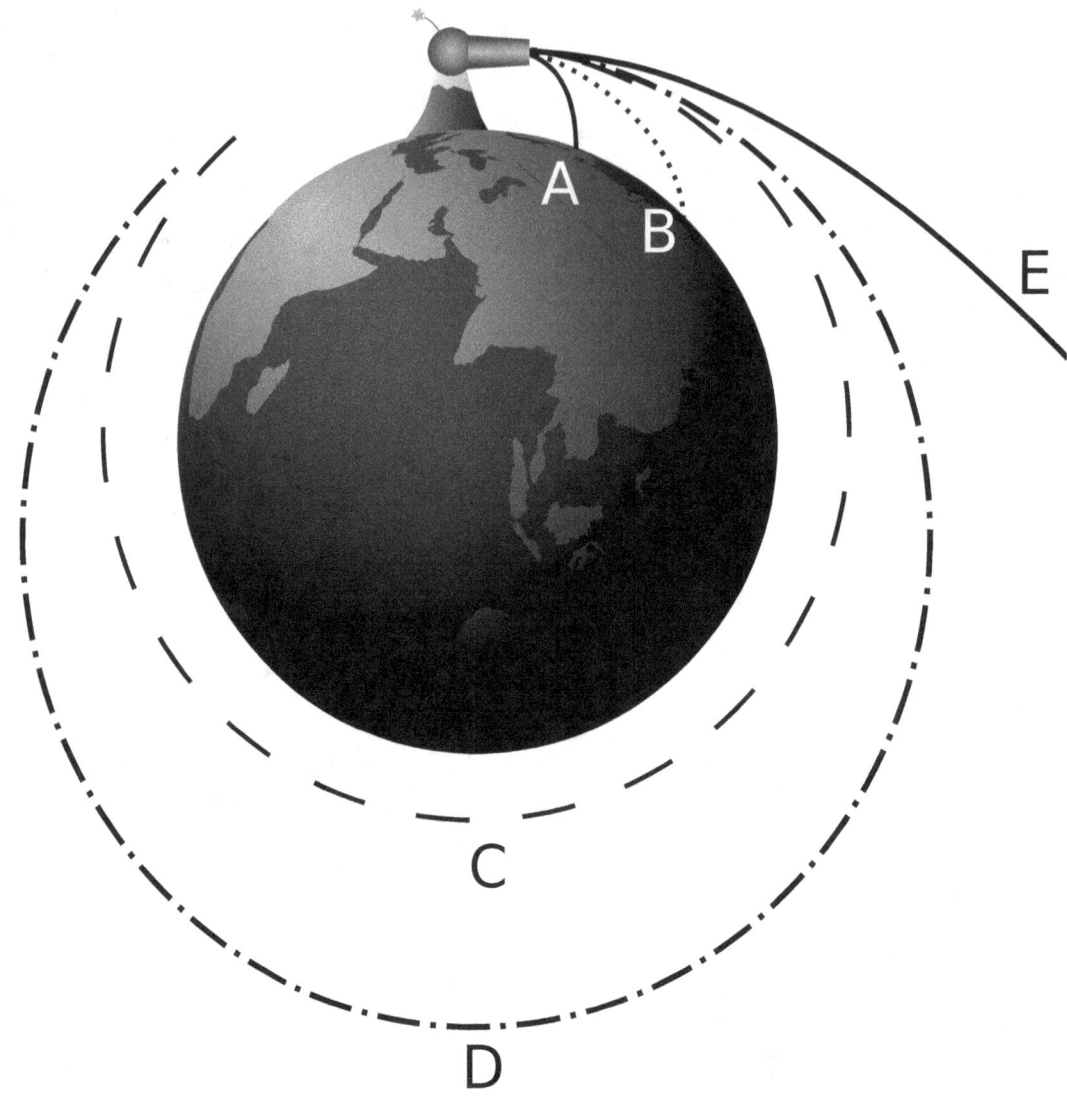

A cannon on top of a very high mountain shoots a cannonball horizontally. If the speed is low, the cannonball quickly falls back to Earth (A,B). At intermediate speeds, it will revolve around Earth along an elliptical orbit (C,D). At a sufficiently high speed, it will leave the Earth altogether (E).

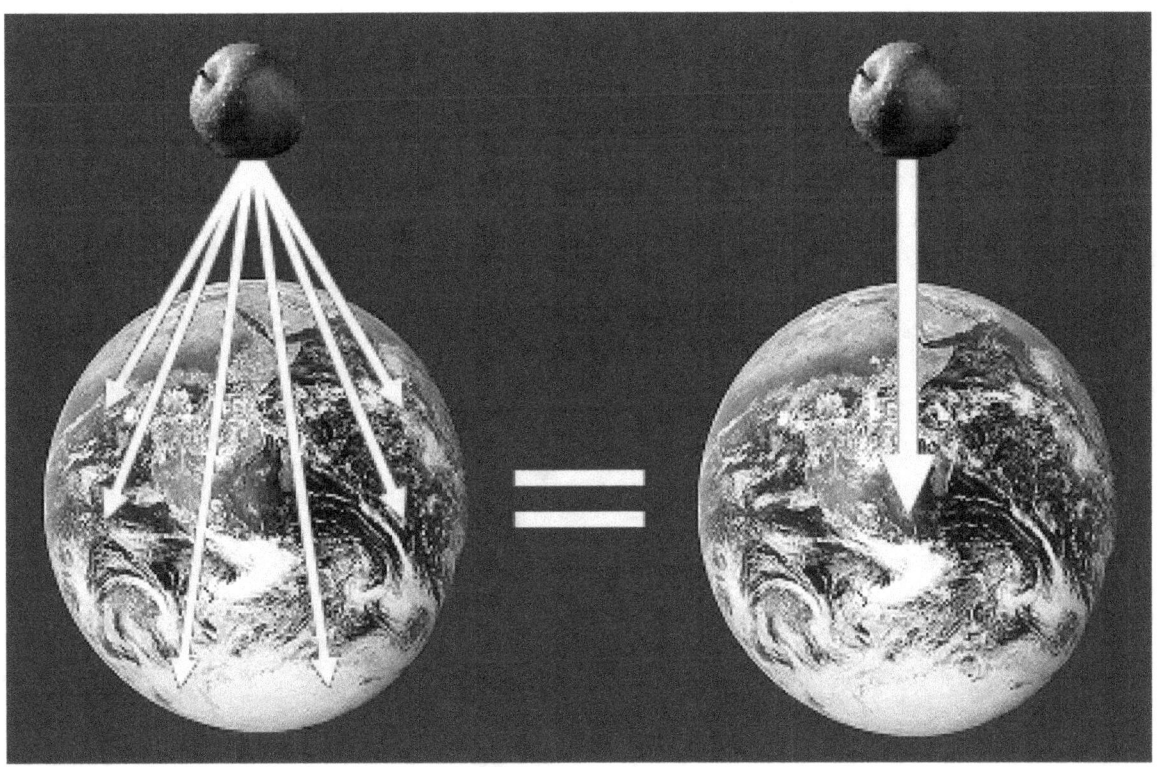

An apple experiences gravitational fields directed towards every part of the Earth; however, the sum total of these many fields produces a single gravitational field directed towards the Earth's center

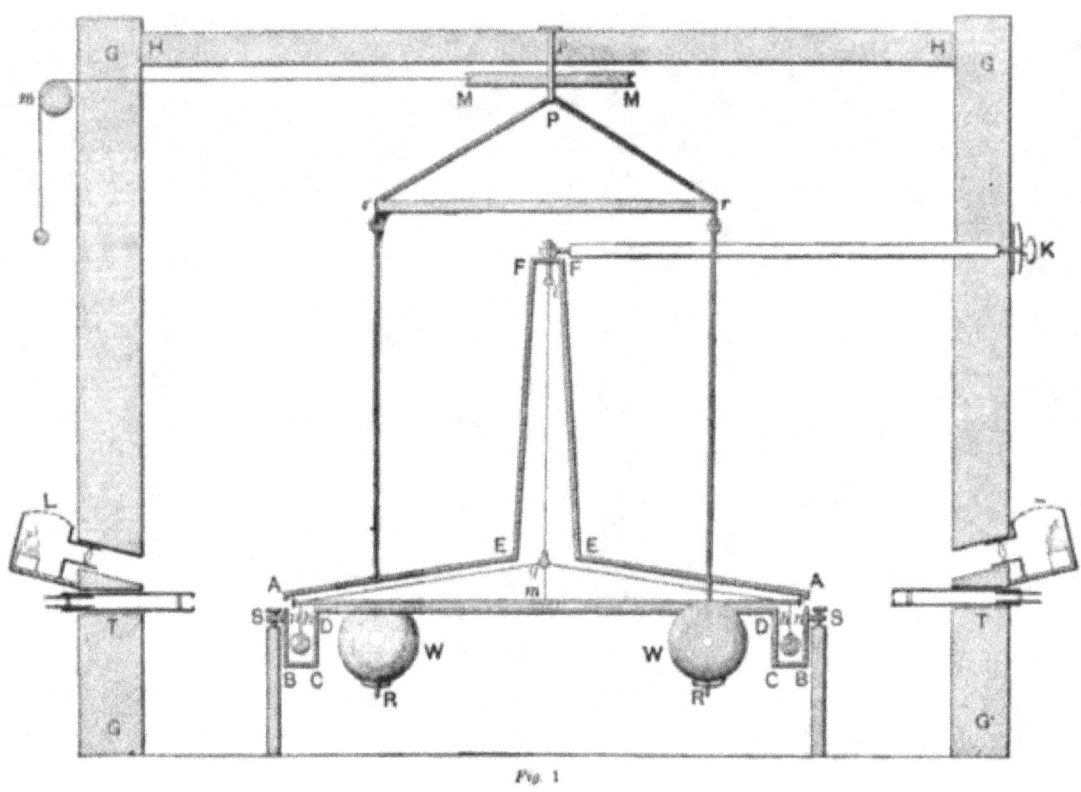

Fig. 1

Vertical section drawing of Cavendish's torsion balance instrument including the building in which it was housed. The large balls were hung from a frame so they could be rotated into position next to the small balls by a pulley from outside. Figure 1 of Cavendish's paper.

Massmeter, a device for measuring the inertial mass of an astronaut in weightlessness. The mass is calculated via the oscillation period for a spring with the astronaut attached (Tsiolkovsky State Museum of the History of Cosmonautics)

Chapter 13

Color charge

Color charge is a property of quarks and gluons that is related to the particles' strong interactions in the theory of quantum chromodynamics (QCD). The color charge of quarks and gluons is completely unrelated to visual perception of color,[1] because it is a property that has almost no manifestation at distances above the size of an atomic nucleus. The term *color* was chosen because the charge responsible for the strong force between particles can be analogized to the three primary colors of human vision: red, green, and blue.[2] Another color scheme is "red, yellow, and blue",[3] using paint, rather than light as the perceptible analogy.

Particles have corresponding antiparticles. A particle with red, green, or blue charge has a corresponding antiparticle in which the color charge must be the anticolor of red, green, and blue, respectively, for the color charge to be conserved in particle-antiparticle creation and annihilation. Particle physicists call these antired, antigreen, and antiblue. All three colors mixed together, or any one of these colors and its complement (or negative), is "colorless" or "white" and has a net color charge of zero. Free particles have a color charge of zero: baryons are composed of three quarks, but the individual quarks can have red, green, or blue charges, or negatives; mesons are made from a quark and antiquark, the quark can be any color, and the antiquark will have the negative of that color. This color charge differs from electromagnetic charges since electromagnetic charges have only one kind of value. Positive and negative electrical charges are the same kind of charge as they only differ by the sign.

Shortly after the existence of quarks was first proposed in 1964, Oscar W. Greenberg introduced the notion of color charge to explain how quarks could coexist inside some hadrons in otherwise identical quantum states without violating the Pauli exclusion principle. The theory of quantum chromodynamics has been under development since the 1970s and constitutes an important component of the Standard Model of particle physics.

13.1   Red, green, and blue

In QCD, a quark's color can take one of three values or charges, red, green, and blue. An antiquark can take one of three anticolors, called antired, antigreen, and antiblue (represented as cyan, magenta and yellow, respectively). Gluons are mixtures of two colors, such as red and antigreen, which constitutes their color charge. QCD considers eight gluons of the possible nine color–anticolor combinations to be unique; see *eight gluon colors* for an explanation.

The following illustrates the coupling constants for color-charged particles:

- The quark colors (red, green, blue) combine to be colorless

- The quark anticolors (antired, antigreen, antiblue) also combine to be colorless

- A hadron with 3 quarks (red, green, blue) before a color change

- Blue quark emits a blue-antigreen gluon

- Green quark has absorbed the blue-antigreen gluon and is now blue; color remains conserved

- An animation of the interaction inside a neutron. The gluons are represented as circles with the color charge in the center and the anti-color charge on the outside.

13.1.1 Field lines from color charges

Main article: Field (physics)

Analogous to an electric field and electric charges, the strong force acting between color charges can be depicted using field lines. However, the color field lines do not arc outwards from one charge to another as much, because they are pulled together tightly by gluons (within 1 fm).[4] This effect confines quarks within hadrons.

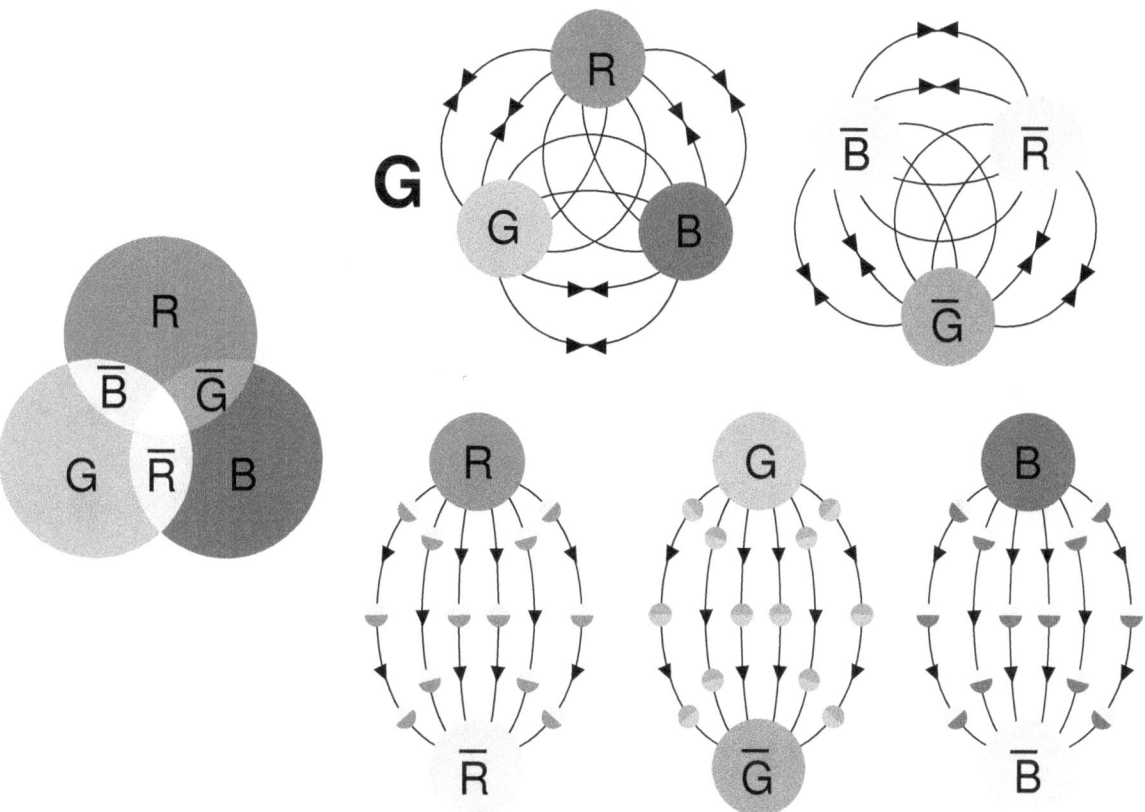

*Fields due to color charges, as in quarks (**G** is the gluon field strength tensor). These are "colorless" combinations. **Top:** Color charge has "ternary neutral states" as well as binary neutrality (analogous to electric charge). **Bottom:** Quark/antiquark combinations.[5][6]*

13.2 Coupling constant and charge

In a quantum field theory, a coupling constant and a charge are different but related notions. The coupling constant sets the magnitude of the force of interaction; for example, in quantum electrodynamics, the fine-structure constant is a coupling constant. The charge in a gauge theory has to do with the way a particle transforms under the gauge symmetry; i.e., its representation under the gauge group. For example, the electron has charge −1 and the positron has charge +1, implying that the gauge transformation has opposite effects on them in some sense. Specifically, if a local gauge transformation $\phi(x)$ is applied in electrodynamics, then one finds (using tensor index notation):

$$A_\mu \to A_\mu + \partial_\mu\phi(x)\ ,\ \psi \to \exp[iQ\phi(x)]\psi \text{ and } \overline{\psi} \to \exp[-iQ\phi(x)]\overline{\psi}$$

where A_μ is the photon field, and ψ is the electron field with $Q=-1$ (a bar over ψ denotes its antiparticle — the positron). Since QCD is a non-abelian theory, the representations, and hence the color charges, are more complicated. They are dealt with in the next section.

13.3 Quark and gluon fields and color charges

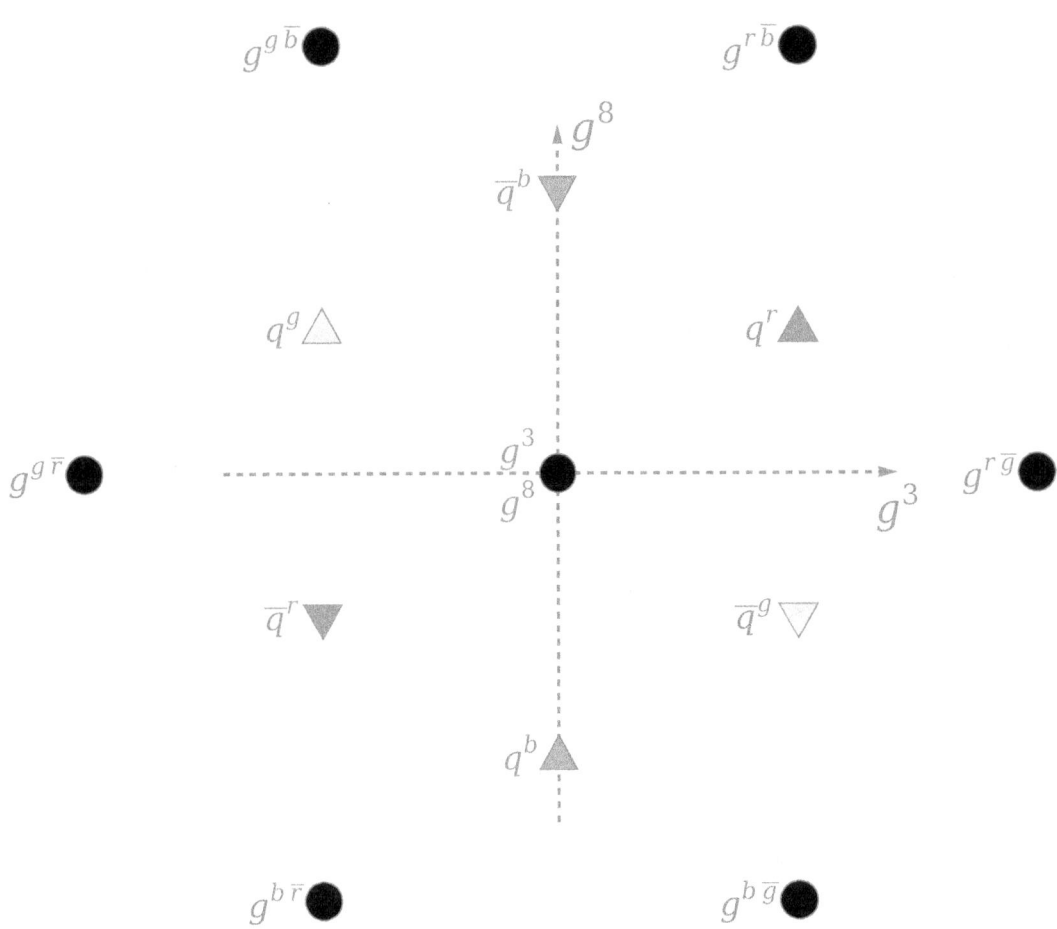

The pattern of strong charges for the three colors of quark, three antiquarks, and eight gluons (with two of zero charge overlapping).

In QCD the gauge group is the non-abelian group SU(3). The *running coupling* is usually denoted by α_s. Each flavor of quark belongs to the fundamental representation (**3**) and contains a triplet of fields together denoted by ψ. The antiquark field belongs to the complex conjugate representation (**3***) and also contains a triplet of fields. We can write

$$\psi = \begin{pmatrix} \psi_1 \\ \psi_2 \\ \psi_3 \end{pmatrix} \text{ and } \overline{\psi} = \begin{pmatrix} \overline{\psi}_1^* \\ \overline{\psi}_2^* \\ \overline{\psi}_3^* \end{pmatrix}.$$

The gluon contains an octet of fields (see gluon field), and belongs to the adjoint representation (**8**), and can be written using the Gell-Mann matrices as

$$\mathbf{A}_\mu = A_\mu^a \lambda_a.$$

(there is an implied summation over $a = 1, 2, \dots 8$). All other particles belong to the trivial representation (**1**) of color SU(3). The **color charge** of each of these fields is fully specified by the representations. Quarks have a color charge of red, green or blue and antiquarks have a color charge of antired, antigreen or antiblue. Gluons have a combination of two color charges (one of red, green or blue and one of antired, antigreen and antiblue) in a superposition of states which are given by the Gell-Mann matrices. All other particles have zero color charge. Mathematically speaking, the color charge of a particle is the value of a certain quadratic Casimir operator in the representation of the particle.

In the simple language introduced previously, the three indices "1", "2" and "3" in the quark triplet above are usually identified with the three colors. The colorful language misses the following point. A gauge transformation in color SU(3) can be written as $\psi \to U\,\psi$, where U is a 3×3 matrix which belongs to the group SU(3). Thus, after gauge transformation, the new colors are linear combinations of the old colors. In short, the simplified language introduced before is not gauge invariant.

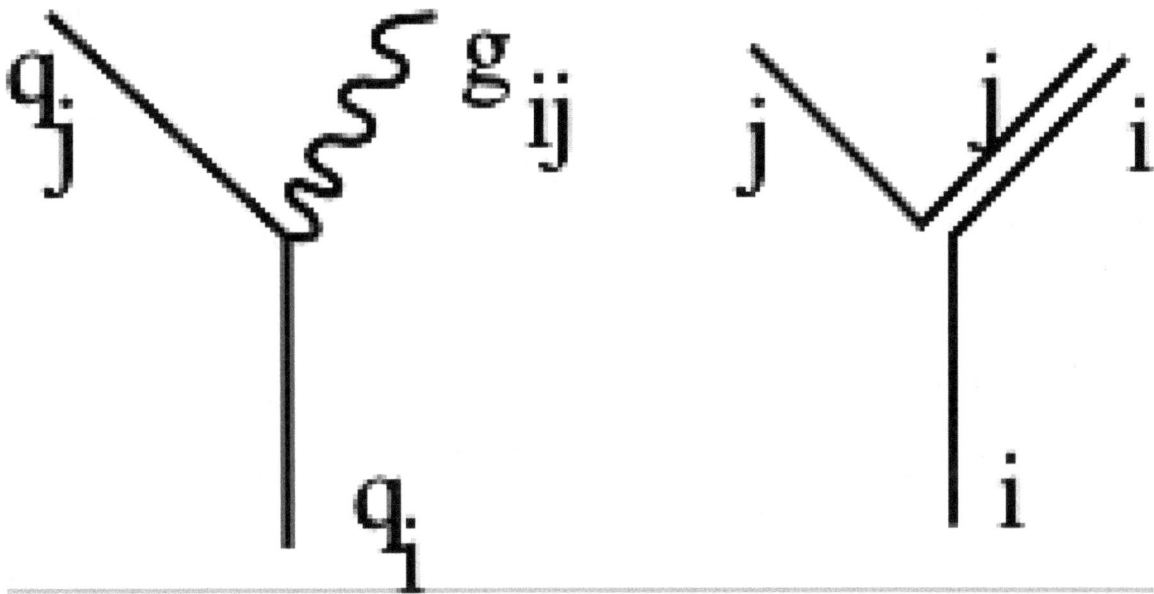

Color-line representation of QCD vertex

Color charge is conserved, but the book-keeping involved in this is more complicated than just adding up the charges, as is done in quantum electrodynamics. One simple way of doing this is to look at the interaction vertex in QCD and replace it by a color-line representation. The meaning is the following. Let ψ_i represent the i-th component of a quark field (loosely called the i-th color). The *color* of a gluon is similarly given by **A** which corresponds to the particular Gell-Mann matrix it is associated with. This matrix has indices i and j. These are the *color labels* on the gluon. At the interaction vertex one has $qi \to gi\,j + qj$. The **color-line** representation tracks these indices. Color charge conservation means that the ends of these color-lines must be either in the initial or final state, equivalently, that no lines break in the middle of a diagram.

Since gluons carry color charge, two gluons can also interact. A typical interaction vertex (called the three gluon vertex) for gluons involves $g + g \to g$. This is shown here, along with its color-line representation. The color-line diagrams can be

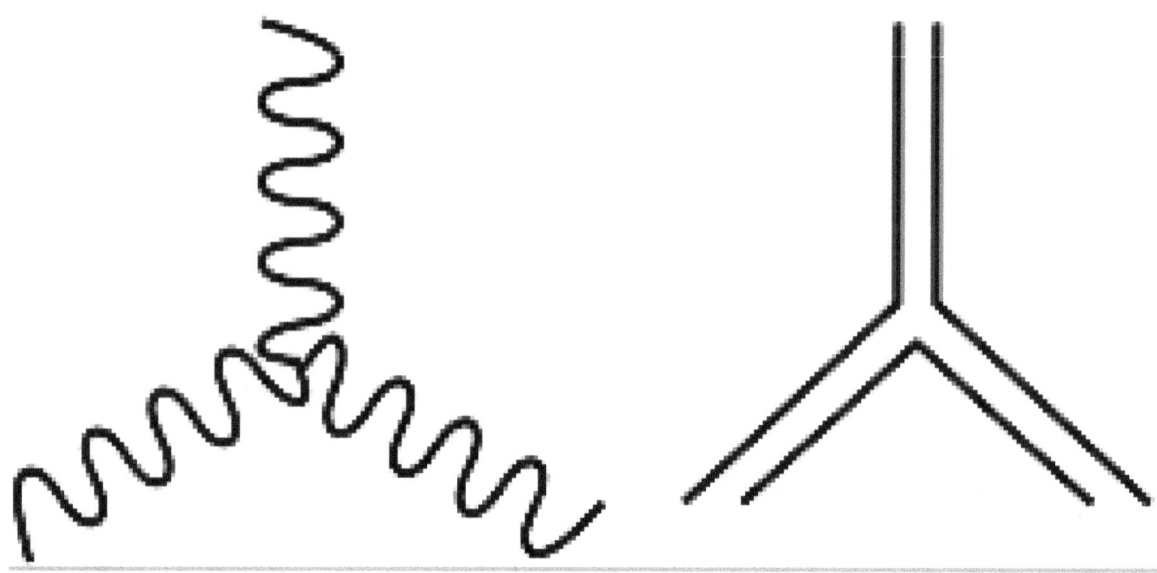

Color-line representation of 3-gluon vertex

restated in terms of conservation laws of color; however, as noted before, this is not a gauge invariant language. Note that in a typical non-abelian gauge theory the gauge boson carries the charge of the theory, and hence has interactions of this kind; for example, the W boson in the electroweak theory. In the electroweak theory, the W also carries electric charge, and hence interacts with a photon.

13.4 See also

- Color confinement
- Gluon field strength tensor

13.5 References

[1] Feynman, Richard (1985), *QED: The Strange Theory of Light and Matter*, Princeton University Press, p. 136, ISBN 0-691-08388-6, The idiot physicists, unable to come up with any wonderful Greek words anymore, call this type of polarization by the unfortunate name of 'color,' which has nothing to do with color in the normal sense.

[2] Close (2007)

[3] R. Penrose (2005). *The Road to Reality*. Vintage books. p. 648. ISBN 978-00994-40680.

[4] R. Resnick, R. Eisberg (1985), *Quantum Physics of Atoms, Molecules, Solids, Nuclei and Particles* (2nd ed.), John Wiley & Sons, p. 684, ISBN 978-0-471-87373-0

[5] Parker, C.B. (1994), *McGraw Hill Encyclopaedia of Physics* (2nd ed.), Mc Graw Hill, ISBN 0-07-051400-3

[6] M. Mansfield, C. O'Sullivan (2011), *Understanding Physics* (4th ed.), John Wiley & Sons, ISBN 978-0-47-0746370

13.6 Further reading

- Georgi, Howard (1999), *Lie algebras in particle physics*, Perseus Books Group, ISBN 0-7382-0233-9.

- Griffiths, David J. (1987), *Introduction to Elementary Particles*, New York: John Wiley & Sons, ISBN 0-471-60386-4.

- Christman, J. Richard (2001), "Colour and Charm" (PDF), *Project PHYSNET document MISN-0-283* External link in |work= (help).

- Hawking, Stephen (1998), *A Brief History of Time*, Bantam Dell Publishing Group, ISBN 978-0-553-10953-5.

- Close, Frank (2007), *The New Cosmic Onion*, Taylor & Francis, ISBN 1-58488-798-2.

Chapter 14

Spin (physics)

This article is about spin in quantum mechanics. For rotation in classical mechanics, see angular momentum.

In quantum mechanics and particle physics, **spin** is an intrinsic form of angular momentum carried by elementary particles, composite particles (hadrons), and atomic nuclei.[1][2]

Spin is one of two types of angular momentum in quantum mechanics, the other being *orbital angular momentum*. The orbital angular momentum operator is the quantum-mechanical counterpart to the classical notion of angular momentum: it arises when a particle executes a rotating or twisting trajectory (such as when an electron orbits a nucleus).[3][4] The existence of spin angular momentum is inferred from experiments, such as the Stern–Gerlach experiment, in which particles are observed to possess angular momentum that cannot be accounted for by orbital angular momentum alone.[5]

In some ways, spin is like a vector quantity; it has a definite magnitude, and it has a "direction" (but quantization makes this "direction" different from the direction of an ordinary vector). All elementary particles of a given kind have the same magnitude of spin angular momentum, which is indicated by assigning the particle a *spin quantum number*.[2]

The SI unit of spin is the joule-second, just as with classical angular momentum. In practice, however, it is written as a multiple of the reduced Planck constant $\hbar$, usually in natural units, where the $\hbar$ is omitted, resulting in a unitless number. Spin quantum numbers are unitless numbers by definition.

When combined with the spin-statistics theorem, the spin of electrons results in the Pauli exclusion principle, which in turn underlies the periodic table of chemical elements.

Wolfgang Pauli was the first to propose the concept of spin, but he did not name it. In 1925, Ralph Kronig, George Uhlenbeck and Samuel Goudsmit at Leiden University suggested a physical interpretation of particles spinning around their own axis. The mathematical theory was worked out in depth by Pauli in 1927. When Paul Dirac derived his relativistic quantum mechanics in 1928, electron spin was an essential part of it.

14.1 Quantum number

Main article: Spin quantum number

As the name suggests, spin was originally conceived as the rotation of a particle around some axis. This picture is correct so far as spin obeys the same mathematical laws as quantized angular momenta do. On the other hand, spin has some peculiar properties that distinguish it from orbital angular momenta:

- Spin quantum numbers may take half-integer values.

- Although the direction of its spin can be changed, an elementary particle cannot be made to spin faster or slower.

- The spin of a charged particle is associated with a magnetic dipole moment with a g-factor differing from 1. This could only occur classically if the internal charge of the particle were distributed differently from its mass.

The conventional definition of the **spin quantum number**, s, is $s = n/2$, where n can be any non-negative integer. Hence the allowed values of s are 0, 1/2, 1, 3/2, 2, etc. The value of s for an elementary particle depends only on the type of particle, and cannot be altered in any known way (in contrast to the *spin direction* described below). The spin angular momentum, S, of any physical system is quantized. The allowed values of S are:

$$S = \frac{h}{2\pi} \sqrt{s(s+1)} = \frac{h}{4\pi} \sqrt{n(n+2)},$$

where h is the Planck constant. In contrast, orbital angular momentum can only take on integer values of s; i.e., even-numbered values of n.

14.1.1 Fermions and bosons

Those particles with half-integer spins, such as 1/2, 3/2, 5/2, are known as fermions, while those particles with integer spins, such as 0, 1, 2, are known as bosons. The two families of particles obey different rules and *broadly* have different roles in the world around us. A key distinction between the two families is that fermions obey the Pauli exclusion principle; that is, there cannot be two identical fermions simultaneously having the same quantum numbers (meaning, roughly, having the same position, velocity and spin direction). In contrast, bosons obey the rules of Bose–Einstein statistics and have no such restriction, so they may "bunch together" even if in identical states. Also, composite particles can have spins different from the particles which comprise them. For example, a helium atom can have spin 0 and therefore can behave like a boson even though the quarks and electrons which make it up are all fermions.

This has profound practical applications:

- Quarks and leptons (including electrons and neutrinos), which make up what is classically known as matter, are all fermions with spin 1/2. The common idea that "matter takes up space" actually comes from the Pauli exclusion principle acting on these particles to prevent the fermions that make up matter from being in the same quantum state. Further compaction would require electrons to occupy the same energy states, and therefore a kind of pressure (sometimes known as degeneracy pressure of electrons) acts to resist the fermions being overly close. It is also this pressure which prevents stars collapsing inwardly, and which, when it finally gives way under immense gravitational pressure in a dying massive star, triggers inward collapse and the dramatic explosion into a supernova.

  Elementary fermions with other spins (3/2, 5/2 etc.) are not known to exist, as of 2014.

- Elementary particles which are thought of as carrying forces are all bosons with spin 1. They include the photon which carries the electromagnetic force, the gluon (strong force), and the W and Z bosons (weak force). The ability of bosons to occupy the same quantum state is used in the laser, which aligns many photons having the same quantum number (the same direction and frequency), superfluid liquid helium resulting from helium-4 atoms being bosons, and superconductivity where pairs of electrons (which individually are fermions) act as single composite bosons.

  Elementary bosons with other spins (0, 2, 3 etc.) were not historically known to exist, although they have received considerable theoretical treatment and are well established within their respective mainstream theories. In particular theoreticians have proposed the graviton (predicted to exist by some quantum gravity theories) with spin 2, and the Higgs boson (explaining electroweak symmetry breaking) with spin 0. Since 2013 the Higgs boson with spin 0 has been considered proven to exist. It is the first scalar particle (spin 0) known to exist in nature.

Theoretical and experimental studies have shown that the spin possessed by elementary particles cannot be explained by postulating that they are made up of even smaller particles rotating about a common center of mass analogous to a classical electron radius; as far as can be presently determined, these elementary particles have no inner structure. The spin of an elementary particle is therefore seen as a truly intrinsic physical property, akin to the particle's electric charge and rest mass.

14.1.2 Spin-statistics theorem

The proof that particles with half-integer spin (fermions) obey Fermi–Dirac statistics and the Pauli Exclusion Principle, and particles with integer spin (bosons) obey Bose–Einstein statistics, occupy "symmetric states", and thus can share quantum states, is known as the spin-statistics theorem. The theorem relies on both quantum mechanics and the theory of special relativity, and this connection between spin and statistics has been called "one of the most important applications of the special relativity theory".[6]

14.2 Magnetic moments

Main article: Spin magnetic moment

Particles with spin can possess a magnetic dipole moment, just like a rotating electrically charged body in classical electrodynamics. These magnetic moments can be experimentally observed in several ways, e.g. by the deflection of particles by inhomogeneous magnetic fields in a Stern–Gerlach experiment, or by measuring the magnetic fields generated by the particles themselves.

The intrinsic magnetic moment $\boldsymbol{\mu}$ of a spin-1/2 particle with charge q, mass m, and spin angular momentum $\mathbf{S}$, is[7]

$$\boldsymbol{\mu} = \frac{g_s q}{2m}\mathbf{S}$$

where the dimensionless quantity g_s is called the spin g-factor. For exclusively orbital rotations it would be 1 (assuming that the mass and the charge occupy spheres of equal radius).

The electron, being a charged elementary particle, possesses a nonzero magnetic moment. One of the triumphs of the theory of quantum electrodynamics is its accurate prediction of the electron g-factor, which has been experimentally determined to have the value −2.0023193043622(15), with the digits in parentheses denoting measurement uncertainty in the last two digits at one standard deviation.[8] The value of 2 arises from the Dirac equation, a fundamental equation connecting the electron's spin with its electromagnetic properties, and the correction of 0.002319304... arises from the electron's interaction with the surrounding electromagnetic field, including its own field.[9] Composite particles also possess magnetic moments associated with their spin. In particular, the neutron possesses a non-zero magnetic moment despite being electrically neutral. This fact was an early indication that the neutron is not an elementary particle. In fact, it is made up of quarks, which are electrically charged particles. The magnetic moment of the neutron comes from the spins of the individual quarks and their orbital motions.

Neutrinos are both elementary and electrically neutral. The minimally extended Standard Model that takes into account non-zero neutrino masses predicts neutrino magnetic moments of:[10][11][12]

$$\mu_\nu \approx 3 \times 10^{-19} \mu_B \frac{m_\nu}{\text{eV}}$$

where the μ_ν are the neutrino magnetic moments, m_ν are the neutrino masses, and μ_B is the Bohr magneton. New physics above the electroweak scale could, however, lead to significantly higher neutrino magnetic moments. It can be shown in a model independent way that neutrino magnetic moments larger than about 10^{-14} μ_B are unnatural, because they would also lead to large radiative contributions to the neutrino mass. Since the neutrino masses cannot exceed about 1 eV, these radiative corrections must then be assumed to be fine tuned to cancel out to a large degree.[13]

The measurement of neutrino magnetic moments is an active area of research. As of 2001, the latest experimental results have put the neutrino magnetic moment at less than 1.2×10^{-10} times the electron's magnetic moment.

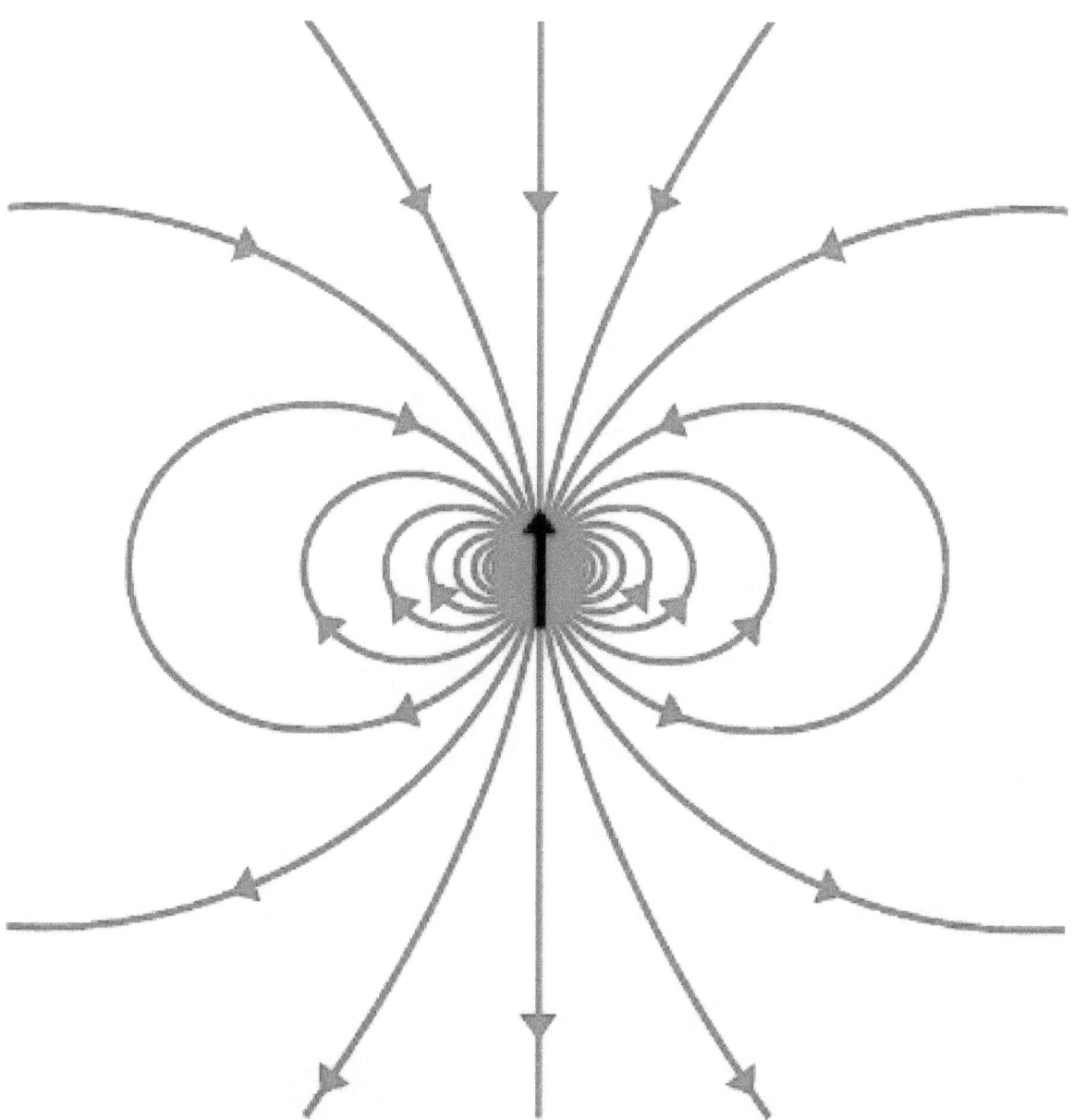

Schematic diagram depicting the spin of the neutron as the black arrow and magnetic field lines associated with the neutron magnetic moment. The neutron has a negative magnetic moment. While the spin of the neutron is upward in this diagram, the magnetic field lines at the center of the dipole are downward.

In ordinary materials, the magnetic dipole moments of individual atoms produce magnetic fields that cancel one another, because each dipole points in a random direction. Ferromagnetic materials below their Curie temperature, however, exhibit magnetic domains in which the atomic dipole moments are locally aligned, producing a macroscopic, non-zero magnetic field from the domain. These are the ordinary "magnets" with which we are all familiar.

In paramagnetic materials, the magnetic dipole moments of individual atoms spontaneously align with an externally applied magnetic field. In diamagnetic materials, on the other hand, the magnetic dipole moments of individual atoms spontaneously align oppositely to any externally applied magnetic field, even if it requires energy to do so.

The study of the behavior of such "spin models" is a thriving area of research in condensed matter physics. For instance, the Ising model describes spins (dipoles) that have only two possible states, up and down, whereas in the Heisenberg model the spin vector is allowed to point in any direction. These models have many interesting properties, which have led to interesting results in the theory of phase transitions.

14.3 Direction

Further information: Angular momentum operator

14.3.1 Spin projection quantum number and multiplicity

In classical mechanics, the angular momentum of a particle possesses not only a magnitude (how fast the body is rotating), but also a direction (either up or down on the axis of rotation of the particle). Quantum mechanical spin also contains information about direction, but in a more subtle form. Quantum mechanics states that the component of angular momentum measured along any direction can only take on the values [14]

$$S_i = \hbar s_i, \quad s_i \in \{-s, -(s-1), \ldots, s-1, s\}$$

where Si is the spin component along the i-axis (either x, y, or z), si is the spin projection quantum number along the i-axis, and s is the principal spin quantum number (discussed in the previous section). Conventionally the direction chosen is the z-axis:

$$S_z = \hbar s_z, \quad s_z \in \{-s, -(s-1), \ldots, s-1, s\}$$

where Sz is the spin component along the z-axis, sz is the spin projection quantum number along the z-axis.

One can see that there are $2s+1$ possible values of s_z. The number "$2s + 1$" is the multiplicity of the spin system. For example, there are only two possible values for a spin-1/2 particle: $s_z = +1/2$ and $s_z = -1/2$. These correspond to quantum states in which the spin is pointing in the +z or −z directions respectively, and are often referred to as "spin up" and "spin down". For a spin-3/2 particle, like a delta baryon, the possible values are +3/2, +1/2, −1/2, −3/2.

14.3.2 Vector

For a given quantum state, one could think of a spin vector $\langle S \rangle$ whose components are the expectation values of the spin components along each axis, i.e., $\langle S \rangle = [\langle S_x \rangle, \langle S_y \rangle, \langle S_z \rangle]$. This vector then would describe the "direction" in which the spin is pointing, corresponding to the classical concept of the axis of rotation. It turns out that the spin vector is not very useful in actual quantum mechanical calculations, because it cannot be measured directly: sx, sy and sz cannot possess simultaneous definite values, because of a quantum uncertainty relation between them. However, for statistically large collections of particles that have been placed in the same pure quantum state, such as through the use of a Stern–Gerlach apparatus, the spin vector does have a well-defined experimental meaning: It specifies the direction in ordinary space in which a subsequent detector must be oriented in order to achieve the maximum possible probability (100%) of detecting every particle in the collection. For spin-1/2 particles, this maximum probability drops off smoothly as the angle between the spin vector and the detector increases, until at an angle of 180 degrees—that is, for detectors oriented in the opposite direction to the spin vector—the expectation of detecting particles from the collection reaches a minimum of 0%.

As a qualitative concept, the spin vector is often handy because it is easy to picture classically. For instance, quantum mechanical spin can exhibit phenomena analogous to classical gyroscopic effects. For example, one can exert a kind of "torque" on an electron by putting it in a magnetic field (the field acts upon the electron's intrinsic magnetic dipole moment—see the following section). The result is that the spin vector undergoes precession, just like a classical gyroscope. This phenomenon is known as electron spin resonance (ESR). The equivalent behaviour of protons in atomic nuclei is used in nuclear magnetic resonance (NMR) spectroscopy and imaging.

Mathematically, quantum mechanical spin states are described by vector-like objects known as spinors. There are subtle differences between the behavior of spinors and vectors under coordinate rotations. For example, rotating a spin-1/2

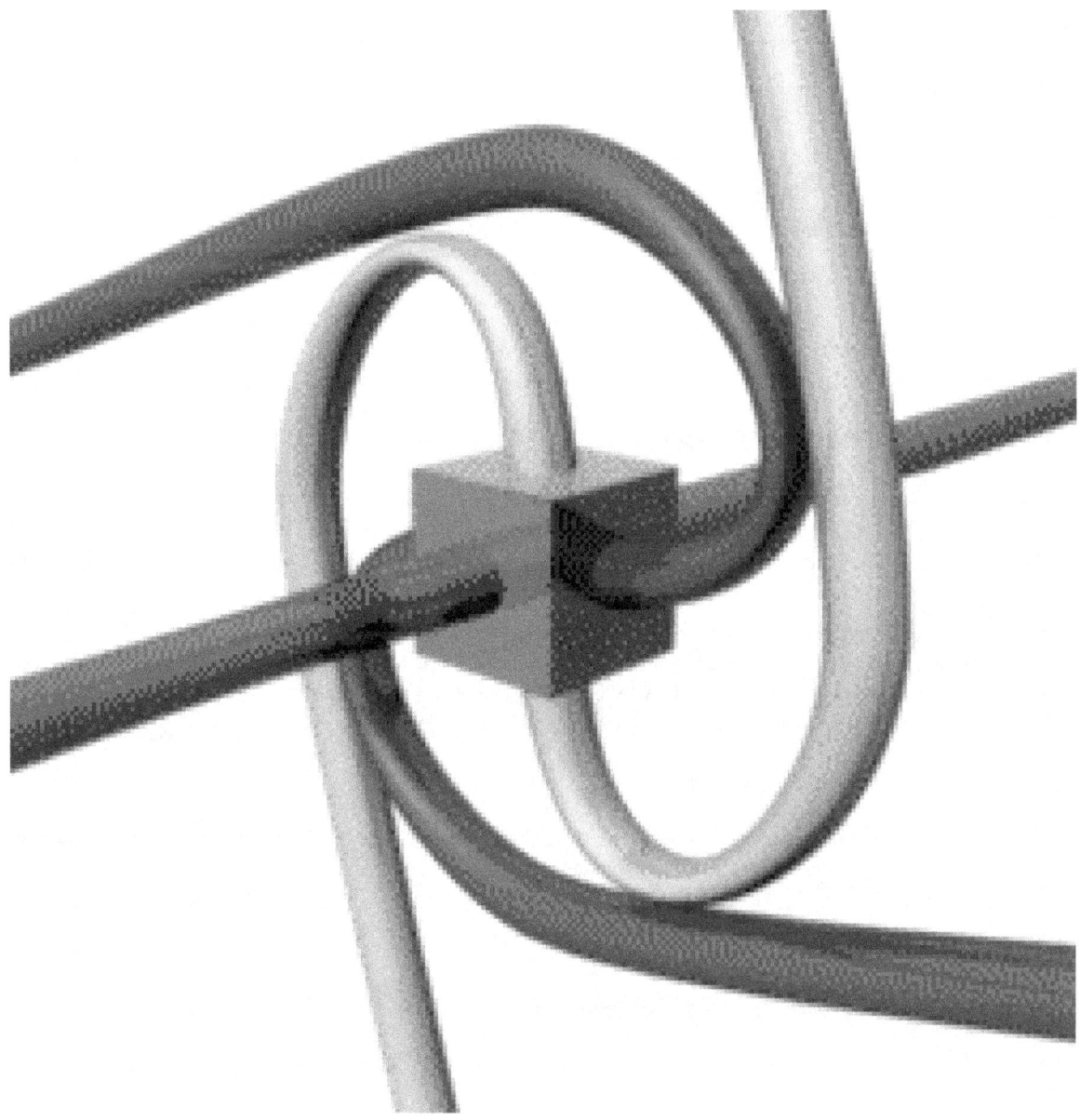

A single point in space can spin continuously without becoming tangled. Notice that after a 360 degree rotation, the spiral flips between clockwise and counterclockwise orientations. It returns to its original configuration after spinning a full 720 degrees.

particle by 360 degrees does not bring it back to the same quantum state, but to the state with the opposite quantum phase; this is detectable, in principle, with interference experiments. To return the particle to its exact original state, one needs a 720 degree rotation. A spin-zero particle can only have a single quantum state, even after torque is applied. Rotating a spin-2 particle 180 degrees can bring it back to the same quantum state and a spin-4 particle should be rotated 90 degrees to bring it back to the same quantum state. The spin 2 particle can be analogous to a straight stick that looks the same even after it is rotated 180 degrees and a spin 0 particle can be imagined as sphere which looks the same after whatever angle it is turned through.

14.4   Mathematical formulation

14.4.1 Operator

Spin obeys commutation relations analogous to those of the orbital angular momentum:

$$[S_i, S_j] = i\hbar\epsilon_{ijk}S_k$$

where ϵ_{ijk} is the Levi-Civita symbol. It follows (as with angular momentum) that the eigenvectors of S^2 and S_z (expressed as kets in the total S basis) are:

$$S^2|s, m\rangle = \hbar^2 s(s+1)|s, m\rangle$$
$$S_z|s, m\rangle = \hbar m|s, m\rangle.$$

The spin raising and lowering operators acting on these eigenvectors give:

$$S_\pm|s, m\rangle = \hbar\sqrt{s(s+1) - m(m \pm 1)}|s, m \pm 1\rangle \text{, where } S_\pm = S_x \pm iS_y.$$

But unlike orbital angular momentum the eigenvectors are not spherical harmonics. They are not functions of θ and φ. There is also no reason to exclude half-integer values of s and m.

In addition to their other properties, all quantum mechanical particles possess an intrinsic spin (though it may have the intrinsic spin 0, too). The spin is quantized in units of the reduced Planck constant, such that the state function of the particle is, say, not $\psi = \psi(\mathbf{r})$, but $\psi = \psi(\mathbf{r}, \sigma)$ where σ is out of the following discrete set of values:

$$\sigma \in \{-s\hbar, -(s-1)\hbar, \cdots, +(s-1)\hbar, +s\hbar\}.$$

One distinguishes bosons (integer spin) and fermions (half-integer spin). The total angular momentum conserved in interaction processes is then the *sum* of the orbital angular momentum and the spin.

14.4.2 Pauli matrices

The quantum mechanical operators associated with spin-$\frac{1}{2}$ observables are:

$$\hat{\mathbf{S}} = \frac{\hbar}{2}\sigma$$

where in Cartesian components:

$$S_x = \frac{\hbar}{2}\sigma_x, \quad S_y = \frac{\hbar}{2}\sigma_y, \quad S_z = \frac{\hbar}{2}\sigma_z.$$

For the special case of spin-1/2 particles, σx, σy and σz are the three Pauli matrices, given by:

$$\sigma_x = \begin{pmatrix} 0 & 1 \\ 1 & 0 \end{pmatrix} \quad \sigma_y = \begin{pmatrix} 0 & -i \\ i & 0 \end{pmatrix} \quad \sigma_z = \begin{pmatrix} 1 & 0 \\ 0 & -1 \end{pmatrix}.$$

14.4.3 Pauli exclusion principle

For systems of N identical particles this is related to the Pauli exclusion principle, which states that by interchanges of any two of the N particles one must have

$$\psi(\cdots \mathbf{r}_i, \sigma_i \cdots \mathbf{r}_j, \sigma_j \cdots) = (-1)^{2s} \psi(\cdots \mathbf{r}_j, \sigma_j \cdots \mathbf{r}_i, \sigma_i \cdots).$$

Thus, for bosons the prefactor $(-1)^{2s}$ will reduce to $+1$, for fermions to -1. In quantum mechanics all particles are either bosons or fermions. In some speculative relativistic quantum field theories "supersymmetric" particles also exist, where linear combinations of bosonic and fermionic components appear. In two dimensions, the prefactor $(-1)^{2s}$ can be replaced by any complex number of magnitude 1 such as in the Anyon.

The above permutation postulate for N-particle state functions has most-important consequences in daily life, e.g. the periodic table of the chemists or biologists.

14.4.4   Rotations

See also: symmetries in quantum mechanics

As described above, quantum mechanics states that components of angular momentum measured along any direction can only take a number of discrete values. The most convenient quantum mechanical description of particle's spin is therefore with a set of complex numbers corresponding to amplitudes of finding a given value of projection of its intrinsic angular momentum on a given axis. For instance, for a spin 1/2 particle, we would need two numbers $a_{\pm 1/2}$, giving amplitudes of finding it with projection of angular momentum equal to $\hbar/2$ and $-\hbar/2$, satisfying the requirement

$$\left| a_{\frac{1}{2}} \right|^2 + \left| a_{-\frac{1}{2}} \right|^2 = 1.$$

For a generic particle with spin s, we would need $2s + 1$ such parameters. Since these numbers depend on the choice of the axis, they transform into each other non-trivially when this axis is rotated. It's clear that the transformation law must be linear, so we can represent it by associating a matrix with each rotation, and the product of two transformation matrices corresponding to rotations A and B must be equal (up to phase) to the matrix representing rotation AB. Further, rotations preserve the quantum mechanical inner product, and so should our transformation matrices:

$$\sum_{m=-j}^{j} a_m^* b_m = \sum_{m=-j}^{j} \left(\sum_{n=-j}^{j} U_{nm} a_n \right)^* \left(\sum_{k=-j}^{j} U_{km} b_k \right)$$

$$\sum_{n=-j}^{j} \sum_{k=-j}^{j} U_{np}^* U_{kq} = \delta_{pq}.$$

Mathematically speaking, these matrices furnish a unitary projective representation of the rotation group SO(3). Each such representation corresponds to a representation of the covering group of SO(3), which is SU(2).[15] There is one n-dimensional irreducible representation of SU(2) for each dimension, though this representation is n-dimensional real for odd n and n-dimensional complex for even n (hence of real dimension $2n$). For a rotation by angle θ in the plane with normal vector $\hat{\boldsymbol{\theta}}$, U can be written

$$U = e^{-\frac{i}{\hbar} \boldsymbol{\theta} \cdot \mathbf{S}},$$

where $\boldsymbol{\theta} = \theta \hat{\boldsymbol{\theta}}$ is a and $\mathbf{S}$ is the vector of spin operators.

(Click "show" at right to see a proof or "hide" to hide it.)

Working in the coordinate system where $\hat{\boldsymbol{\theta}} = \hat{z}$, we would like to show that S_x and S_y are rotated into each other by the angle θ. Starting with S_x. Using units where $\hbar = 1$:

$$S_x \to U^\dagger S_x U = e^{i\theta S_z} S_x e^{-i\theta S_z}$$

$$= S_x + (i\theta)[S_z, S_x] + \left(\frac{1}{2!}\right)(i\theta)^2[S_z, [S_z, S_x]] + \left(\frac{1}{3!}\right)(i\theta)^3[S_z, [S_z, [S_z, S_x]]] + \cdots$$

Using the spin operator commutation relations, we see that the commutators evaluate to iS_y for the odd terms in the series, and to S_x for all of the even terms. Thus:

$$U^\dagger S_x U = S_x \left[1 - \frac{\theta^2}{2!} + \ldots\right] - S_y \left[\theta - \frac{\theta^3}{3!} \cdots\right]$$

$$= S_x \cos\theta - S_y \sin\theta$$

as expected. Note that since we only relied on the spin operator commutation relations, this proof holds for any dimension (i.e., for any principal spin quantum number s).[16]

A generic rotation in 3-dimensional space can be built by compounding operators of this type using Euler angles:

$$\mathcal{R}(\alpha, \beta, \gamma) = e^{-i\alpha S_x} e^{-i\beta S_y} e^{-i\gamma S_z}$$

An irreducible representation of this group of operators is furnished by the Wigner D-matrix:

$$D^s_{m'm}(\alpha, \beta, \gamma) \equiv \langle sm'|\mathcal{R}(\alpha, \beta, \gamma)|sm\rangle = e^{-im'\alpha} d^s_{m'm}(\beta) e^{-im\gamma},$$

where

$$d^s_{m'm}(\beta) = \langle sm'|e^{-i\beta s_y}|sm\rangle$$

is Wigner's small d-matrix. Note that for $\gamma = 2\pi$ and $\alpha = \beta = 0$; i.e., a full rotation about the z-axis, the Wigner D-matrix elements become

$$D^s_{m'm}(0, 0, 2\pi) = d^s_{m'm}(0) e^{-im2\pi} = \delta_{m'm}(-1)^{2m}.$$

Recalling that a generic spin state can be written as a superposition of states with definite m, we see that if s is an integer, the values of m are all integers, and this matrix corresponds to the identity operator. However, if s is a half-integer, the values of m are also all half-integers, giving $(-1)^{2m} = -1$ for all m, and hence upon rotation by 2π the state picks up a minus sign. This fact is a crucial element of the proof of the spin-statistics theorem.

14.4.5 Lorentz transformations

We could try the same approach to determine the behavior of spin under general Lorentz transformations, but we would immediately discover a major obstacle. Unlike SO(3), the group of Lorentz transformations SO(3,1) is non-compact and therefore does not have any faithful, unitary, finite-dimensional representations.

In case of spin 1/2 particles, it is possible to find a construction that includes both a finite-dimensional representation and a scalar product that is preserved by this representation. We associate a 4-component Dirac spinor ψ with each particle. These spinors transform under Lorentz transformations according to the law

$$\psi' = \exp\left(\frac{1}{8}\omega_{\mu\nu}[\gamma_\mu, \gamma_\nu]\right)\psi$$

where γ_μ are gamma matrices and $\omega_{\mu\nu}$ is an antisymmetric 4×4 matrix parametrizing the transformation. It can be shown that the scalar product

$$\langle \psi | \phi \rangle = \bar{\psi}\phi = \psi^\dagger \gamma_0 \phi$$

is preserved. It is not, however, positive definite, so the representation is not unitary.

14.4.6 Metrology along the x, y, and z axes

Each of the (Hermitian) Pauli matrices has two eigenvalues, +1 and −1. The corresponding normalized eigenvectors are:

$$\psi_{x+} = \frac{1}{\sqrt{2}}\begin{pmatrix} 1 \\ 1 \end{pmatrix}, \quad \psi_{x-} = \frac{1}{\sqrt{2}}\begin{pmatrix} 1 \\ -1 \end{pmatrix},$$
$$\psi_{y+} = \frac{1}{\sqrt{2}}\begin{pmatrix} 1 \\ i \end{pmatrix}, \quad \psi_{y-} = \frac{1}{\sqrt{2}}\begin{pmatrix} 1 \\ -i \end{pmatrix},$$
$$\psi_{z+} = \begin{pmatrix} 1 \\ 0 \end{pmatrix}, \quad \psi_{z-} = \begin{pmatrix} 0 \\ 1 \end{pmatrix}.$$

By the postulates of quantum mechanics, an experiment designed to measure the electron spin on the x, y or z axis can only yield an eigenvalue of the corresponding spin operator (Sx, Sy or Sz) on that axis, i.e. $\hbar/2$ or $-\hbar/2$. The quantum state of a particle (with respect to spin), can be represented by a two component spinor:

$$\psi = \begin{pmatrix} a + bi \\ c + di \end{pmatrix}.$$

When the spin of this particle is measured with respect to a given axis (in this example, the x-axis), the probability that its spin will be measured as $\hbar/2$ is just $|\langle \psi_{x+} | \psi \rangle|^2$. Correspondingly, the probability that its spin will be measured as $-\hbar/2$ is just $|\langle \psi_{x-} | \psi \rangle|^2$. Following the measurement, the spin state of the particle will collapse into the corresponding eigenstate. As a result, if the particle's spin along a given axis has been measured to have a given eigenvalue, all measurements will yield the same eigenvalue (since $|\langle \psi_{x+} | \psi_{x+} \rangle|^2 = 1$, etc), provided that no measurements of the spin are made along other axes.

14.4.7 Metrology along an arbitrary axis

The operator to measure spin along an arbitrary axis direction is easily obtained from the Pauli spin matrices. Let $u = (ux, uy, uz)$ be an arbitrary unit vector. Then the operator for spin in this direction is simply

$$S_u = \frac{\hbar}{2}(u_x \sigma_x + u_y \sigma_y + u_z \sigma_z)$$

The operator Su has eigenvalues of $\pm\hbar/2$, just like the usual spin matrices. This method of finding the operator for spin in an arbitrary direction generalizes to higher spin states, one takes the dot product of the direction with a vector of the three operators for the three x, y, z axis directions.

A normalized spinor for spin-1/2 in the (ux, uy, uz) direction (which works for all spin states except spin down where it will give 0/0), is:

$$\frac{1}{\sqrt{2 + 2u_z}}\begin{pmatrix} 1 + u_z \\ u_x + iu_y \end{pmatrix}.$$

The above spinor is obtained in the usual way by diagonalizing the σ_u matrix and finding the eigenstates corresponding to the eigenvalues. In quantum mechanics, vectors are termed "normalized" when multiplied by a normalizing factor, which results in the vector having a length of unity.

14.4.8 Compatibility of metrology

Since the Pauli matrices do not commute, measurements of spin along the different axes are incompatible. This means that if, for example, we know the spin along the x-axis, and we then measure the spin along the y-axis, we have invalidated our previous knowledge of the x-axis spin. This can be seen from the property of the eigenvectors (i.e. eigenstates) of the Pauli matrices that:

$$| \langle \psi_{x\pm} | \psi_{y\pm} \rangle |^2 = | \langle \psi_{x\pm} | \psi_{z\pm} \rangle |^2 = | \langle \psi_{y\pm} | \psi_{z\pm} \rangle |^2 = \frac{1}{2}.$$

So when physicists measure the spin of a particle along the x-axis as, for example, $\hbar/2$, the particle's spin state collapses into the eigenstate $| \psi_{x+} \rangle$. When we then subsequently measure the particle's spin along the y-axis, the spin state will now collapse into either $| \psi_{y+} \rangle$ or $| \psi_{y-} \rangle$, each with probability 1/2. Let us say, in our example, that we measure $-\hbar/2$. When we now return to measure the particle's spin along the x-axis again, the probabilities that we will measure $\hbar/2$ or $-\hbar/2$ are each 1/2 (i.e. they are $| \langle \psi_{x+} | \psi_{y-} \rangle |^2$ and $| \langle \psi_{x-} | \psi_{y-} \rangle |^2$ respectively). This implies that the original measurement of the spin along the x-axis is no longer valid, since the spin along the x-axis will now be measured to have either eigenvalue with equal probability.

14.4.9 Higher spins

The spin-1/2 operator $\mathbf{S} = \hbar/2\boldsymbol{\sigma}$ form the fundamental representation of SU(2). By taking Kronecker products of this representation with itself repeatedly, one may construct all higher irreducible representations. That is, the resulting spin operators for higher spin systems in three spatial dimensions, for arbitrarily large s, can be calculated using this spin operator and ladder operators.

The resulting spin matrices for spin 1 are:

$$S_x = \frac{\hbar}{\sqrt{2}} \begin{pmatrix} 0 & 1 & 0 \\ 1 & 0 & 1 \\ 0 & 1 & 0 \end{pmatrix}$$

$$S_y = \frac{\hbar}{\sqrt{2}} \begin{pmatrix} 0 & -i & 0 \\ i & 0 & -i \\ 0 & i & 0 \end{pmatrix}$$

$$S_z = \hbar \begin{pmatrix} 1 & 0 & 0 \\ 0 & 0 & 0 \\ 0 & 0 & -1 \end{pmatrix}$$

for spin 3/2 they are

$$S_x = \frac{\hbar}{2} \begin{pmatrix} 0 & \sqrt{3} & 0 & 0 \\ \sqrt{3} & 0 & 2 & 0 \\ 0 & 2 & 0 & \sqrt{3} \\ 0 & 0 & \sqrt{3} & 0 \end{pmatrix}$$

$$S_y = \frac{\hbar}{2} \begin{pmatrix} 0 & -i\sqrt{3} & 0 & 0 \\ i\sqrt{3} & 0 & -2i & 0 \\ 0 & 2i & 0 & -i\sqrt{3} \\ 0 & 0 & i\sqrt{3} & 0 \end{pmatrix}$$

$$S_z = \frac{\hbar}{2} \begin{pmatrix} 3 & 0 & 0 & 0 \\ 0 & 1 & 0 & 0 \\ 0 & 0 & -1 & 0 \\ 0 & 0 & 0 & -3 \end{pmatrix}$$

and for spin 5/2 they are

$$S_x = \frac{\hbar}{2}\begin{pmatrix} 0 & \sqrt{5} & 0 & 0 & 0 & 0 \\ \sqrt{5} & 0 & 2\sqrt{2} & 0 & 0 & 0 \\ 0 & 2\sqrt{2} & 0 & 3 & 0 & 0 \\ 0 & 0 & 3 & 0 & 2\sqrt{2} & 0 \\ 0 & 0 & 0 & 2\sqrt{2} & 0 & \sqrt{5} \\ 0 & 0 & 0 & 0 & \sqrt{5} & 0 \end{pmatrix}$$

$$S_y = \frac{\hbar}{2}\begin{pmatrix} 0 & -i\sqrt{5} & 0 & 0 & 0 & 0 \\ i\sqrt{5} & 0 & -2i\sqrt{2} & 0 & 0 & 0 \\ 0 & 2i\sqrt{2} & 0 & -3i & 0 & 0 \\ 0 & 0 & 3i & 0 & -2i\sqrt{2} & 0 \\ 0 & 0 & 0 & 2i\sqrt{2} & 0 & -i\sqrt{5} \\ 0 & 0 & 0 & 0 & i\sqrt{5} & 0 \end{pmatrix}$$

$$S_z = \frac{\hbar}{2}\begin{pmatrix} 5 & 0 & 0 & 0 & 0 & 0 \\ 0 & 3 & 0 & 0 & 0 & 0 \\ 0 & 0 & 1 & 0 & 0 & 0 \\ 0 & 0 & 0 & -1 & 0 & 0 \\ 0 & 0 & 0 & 0 & -3 & 0 \\ 0 & 0 & 0 & 0 & 0 & -5 \end{pmatrix}.$$

The generalization of these matrices for arbitrary s is

$$(S_x)_{ab} = \frac{\hbar}{2}(\delta_{a,b+1} + \delta_{a+1,b})\sqrt{(s+1)(a+b-1)-ab}$$
$$(S_y)_{ab} = \frac{\hbar}{2i}(\delta_{a,b+1} - \delta_{a+1,b})\sqrt{(s+1)(a+b-1)-ab} \quad 1 \le a,b \le 2s+1$$
$$(S_z)_{ab} = \hbar(s+1-a)\delta_{a,b} = \hbar(s+1-b)\delta_{a,b}.$$

Also useful in the quantum mechanics of multiparticle systems, the general Pauli group *Gn* is defined to consist of all *n*-fold tensor products of Pauli matrices.

The analog formula of Euler's formula in terms of the Pauli matrices:

$$e^{i\theta(\hat{\mathbf{n}}\cdot\boldsymbol{\sigma})} = I\cos\theta + i(\hat{\mathbf{n}}\cdot\boldsymbol{\sigma})\sin\theta$$

for higher spins is tractable, but less simple.[17]

14.5 Parity

In tables of the spin quantum number *s* for nuclei or particles, the spin is often followed by a "+" or "−". This refers to the parity with "+" for even parity (wave function unchanged by spatial inversion) and "−" for odd parity (wave function negated by spatial inversion). For example, see the isotopes of bismuth.

14.6 Applications

Spin has important theoretical implications and practical applications. Well-established *direct* applications of spin include:

- Nuclear magnetic resonance (NMR) spectroscopy in chemistry;

- Electron spin resonance spectroscopy in chemistry and physics;

- Magnetic resonance imaging (MRI) in medicine, a type of applied NMR, which relies on proton spin density;

- Giant magnetoresistive (GMR) drive head technology in modern hard disks.

Electron spin plays an important role in magnetism, with applications for instance in computer memories. The manipulation of *nuclear spin* by radiofrequency waves (nuclear magnetic resonance) is important in chemical spectroscopy and medical imaging.

Spin-orbit coupling leads to the fine structure of atomic spectra, which is used in atomic clocks and in the modern definition of the second. Precise measurements of the g-factor of the electron have played an important role in the development and verification of quantum electrodynamics. *Photon spin* is associated with the polarization of light.

An emerging application of spin is as a binary information carrier in spin transistors. The original concept, proposed in 1990, is known as Datta-Das spin transistor.[18] Electronics based on spin transistors are referred to as spintronics. The manipulation of spin in dilute magnetic semiconductor materials, such as metal-doped ZnO or TiO_2 imparts a further degree of freedom and has the potential to facilitate the fabrication of more efficient electronics.[19]

There are many *indirect* applications and manifestations of spin and the associated Pauli exclusion principle, starting with the periodic table of chemistry.

14.7 History

Spin was first discovered in the context of the emission spectrum of alkali metals. In 1924 Wolfgang Pauli introduced what he called a "two-valued quantum degree of freedom" associated with the electron in the outermost shell. This allowed him to formulate the Pauli exclusion principle, stating that no two electrons can share the same quantum state at the same time.

The physical interpretation of Pauli's "degree of freedom" was initially unknown. Ralph Kronig, one of Landé's assistants, suggested in early 1925 that it was produced by the self-rotation of the electron. When Pauli heard about the idea, he criticized it severely, noting that the electron's hypothetical surface would have to be moving faster than the speed of light in order for it to rotate quickly enough to produce the necessary angular momentum. This would violate the theory of relativity. Largely due to Pauli's criticism, Kronig decided not to publish his idea.

In the autumn of 1925, the same thought came to two Dutch physicists, George Uhlenbeck and Samuel Goudsmit at Leiden University. Under the advice of Paul Ehrenfest, they published their results. It met a favorable response, especially after Llewellyn Thomas managed to resolve a factor-of-two discrepancy between experimental results and Uhlenbeck and Goudsmit's calculations (and Kronig's unpublished results). This discrepancy was due to the orientation of the electron's tangent frame, in addition to its position.

Mathematically speaking, a fiber bundle description is needed. The tangent bundle effect is additive and relativistic; that is, it vanishes if c goes to infinity. It is one half of the value obtained without regard for the tangent space orientation, but with opposite sign. Thus the combined effect differs from the latter by a factor two (Thomas precession).

Despite his initial objections, Pauli formalized the theory of spin in 1927, using the modern theory of quantum mechanics invented by Schrödinger and Heisenberg. He pioneered the use of Pauli matrices as a representation of the spin operators, and introduced a two-component spinor wave-function.

Pauli's theory of spin was non-relativistic. However, in 1928, Paul Dirac published the Dirac equation, which described the relativistic electron. In the Dirac equation, a four-component spinor (known as a "Dirac spinor") was used for the electron wave-function. In 1940, Pauli proved the *spin-statistics theorem*, which states that fermions have half-integer spin and bosons integer spin.

In retrospect, the first direct experimental evidence of the electron spin was the Stern–Gerlach experiment of 1922. However, the correct explanation of this experiment was only given in 1927.[20]

Wolfgang Pauli

14.8   See also

- Einstein–de Haas effect

- Spin-orbital

- Chirality (physics)

- Dynamic nuclear polarisation

- Helicity (particle physics)

- Pauli equation

- Pauli–Lubanski pseudovector

- Rarita–Schwinger equation

- Representation theory of SU(2)

- Spin-½

- Spin-flip

- Spin isomers of hydrogen

- Spin tensor

- Spin wave

- Spin engineering

- Yrast

- Zitterbewegung

14.9 References

[1] Merzbacher, Eugen (1998). *Quantum Mechanics* (3rd ed.). pp. 372–3.

[2] Griffiths, David (2005). *Introduction to Quantum Mechanics* (2nd ed.). pp. 183–4.

[3] "Angular Momentum Operator Algebra", class notes by Michael Fowler

[4] *A modern approach to quantum mechanics*, by Townsend, p. 31 and p. 80

[5] Eisberg, Robert; Resnick, Robert (1985). *Quantum Physics of Atoms, Molecules, Solids, Nuclei, and Particles* (2nd ed.). pp. 272–3.

[6] Pauli,Wolfgang(1940). "The Connection Between Spin and Statistics"(PDF).*Phys.Rev***58**(8): 716–722. Bibcode:1940. doi:10.1103/PhysRev.58.716.

[7] Physics of Atoms and Molecules, B.H. Bransden, C.J.Joachain, Longman, 1983, ISBN 0-582-44401-2

[8] "CODATA Value: electron g factor". *The NIST Reference on Constants, Units, and Uncertainty*. NIST. 2006. Retrieved 2013-11-15.

[9] R.P. Feynman (1985). "Electrons and Their Interactions".*QED:The Strange Theory of Light and Matter*. Princeton, Jersey: Princeton University Press. p. 115. ISBN 0-691-08388-6.

> "After some years, it was discovered that this value [−g/2] was not exactly 1, but slightly more—something like 1.00116. This correction was worked out for the first time in 1948 by Schwinger as j*j divided by 2 pi [*sic*] [where *j* is the square root of the fine-structure constant], and was due to an alternative way the electron can go from place to place: instead of going directly from one point to another, the electron goes along for a while and suddenly emits a photon; then (horrors!) it absorbs its own photon."

[10] W.J. Marciano, A.I. Sanda (1977). "Exotic decays of the muon and heavy leptons in gauge theories". *Physics Letters* **B67** (3): 303–305. Bibcode:1977PhLB...67..303M. doi:10.1016/0370-2693(77)90377-X.

[11] B.W. Lee, R.E. Shrock (1977). "Natural suppression of symmetry violation in gauge theories: Muon- and electron-lepton-number nonconservation".*Physical Review***D16**(5): 1444–1473. Bibcode:1977PhRvD..16.1444L.doi:10.1103/Phys.1444.

[12] K. Fujikawa, R. E. Shrock (1980). "Magnetic Moment of a Massive Neutrino and Neutrino-Spin Rotation". *Physical Review Letters* **45** (12): 963–966. Bibcode:1980PhRvL..45..963F. doi:10.1103/PhysRevLett.45.963.

[13] N.F. Bell; Cirigliano, V.; Ramsey-Musolf, M.; Vogel, P.; Wise, Mark; et al. (2005). "How Magnetic is the Dirac Neutrino?". *Physical Review Letters***95**(15): 151802. arXiv:hep-ph/0504134. Bibcode:2005PhRvL..95o1802B.doi:10.1.PMID16241715.

[14] Quanta: A handbook of concepts, P.W. Atkins, Oxford University Press, 1974, ISBN 0-19-855493-1

[15] B.C. Hall (2013). *Quantum Theory for Mathematicians*. Springer. pp. 354–358.

[16] *Modern Quantum Mechanics*, by J. J. Sakurai, p159

[17] Curtright, T L; Fairlie, D B; Zachos, C K (2014). "A compact formula for rotations as spin matrix polynomials". *SIGMA* **10**: 084. doi:10.3842/SIGMA.2014.084.

[18] Datta. S and B. Das (1990). "Electronic analog of the electrooptic modulator". *Applied Physics Letters* **56** (7): 665–667. Bibcode:1990ApPhL..56..665D. doi:10.1063/1.102730.

[19] Assadi, M.H.N; Hanaor, D.A.H (2013). "Theoretical study on copper's energetics and magnetism in TiO$_2$ polymorphs" (PDF). *Journal of Applied Physics* **113** (23): 233913. doi:10.1063/1.4811539.

[20] B. Friedrich, D. Herschbach (2003). "Stern and Gerlach: How a Bad Cigar Helped Reorient Atomic Physics". *Physics Today* **56** (12): 53. Bibcode:2003PhT....56l..53F. doi:10.1063/1.1650229.

14.10   Further reading

- Cohen-Tannoudji, Claude; Diu, Bernard; Laloë, Franck (2006). *Quantum Mechanics* (2 volume set ed.). John Wiley & Sons. ISBN 978-0-471-56952-7.

- Condon, E. U.; Shortley, G. H. (1935). "Especially Chapter 3". *The Theory of Atomic Spectra*. Cambridge University Press. ISBN 0-521-09209-4.

- Hipple, J. A.; Sommer, H.; Thomas, H.A. (1949). *A precise method of determining the faraday by magnetic resonance.* doi:10.1103/PhysRev.76.1877.2.https://www.academia.edu/6483539/John_A._Hipple_1911-1985as_knowledge

- Edmonds, A. R. (1957). *Angular Momentum in Quantum Mechanics*. Princeton University Press. ISBN 0-691-07912-9.

- Jackson, John David (1998). *Classical Electrodynamics* (3rd ed.). John Wiley & Sons. ISBN 978-0-471-30932-1.

- Serway, Raymond A.; Jewett, John W. (2004). *Physics for Scientists and Engineers* (6th ed.). Brooks/Cole. ISBN 0-534-40842-7.

- Thompson, William J. (1994). *Angular Momentum: An Illustrated Guide to Rotational Symmetries for Physical Systems*. Wiley. ISBN 0-471-55264-X.

- Tipler, Paul (2004). *Physics for Scientists and Engineers: Mechanics, Oscillations and Waves, Thermodynamics* (5th ed.). W. H. Freeman. ISBN 0-7167-0809-4.

- Sin-Itiro Tomonaga, The Story of Spin, 1997

14.11   External links

- "Spintronics. Feature Article" in *Scientific American*, June 2002.

- Goudsmit on the discovery of electron spin.

- *Nature*: "Milestones in 'spin' since 1896."

- ECE 495N Lecture 36: Spin Online lecture by S. Datta

Chapter 15

Up quark

The **up quark** or **u quark** (symbol: u) is the lightest of all quarks, a type of elementary particle, and a major constituent of matter. It, along with the down quark, forms the neutrons (one up quark, two down quarks) and protons (two up quarks, one down quark) of atomic nuclei. It is part of the first generation of matter, has an electric charge of $+\frac{2}{3}$ e and a bare mass of 1.8–3.0 MeV/c^2. Like all quarks, the up quark is an elementary fermion with spin-$\frac{1}{2}$, and experiences all four fundamental interactions: gravitation, electromagnetism, weak interactions, and strong interactions. The antiparticle of the up quark is the **up antiquark** (sometimes called *antiup quark* or simply *antiup*), which differs from it only in that some of its properties have equal magnitude but opposite sign.

Its existence (along with that of the down and strange quarks) was postulated in 1964 by Murray Gell-Mann and George Zweig to explain the *Eightfold Way* classification scheme of hadrons. The up quark was first observed by experiments at the Stanford Linear Accelerator Center in 1968.

15.1   History

In the beginnings of particle physics (first half of the 20th century), hadrons such as protons, neutrons and pions were thought to be elementary particles. However, as new hadrons were discovered, the 'particle zoo' grew from a few particles in the early 1930s and 1940s to several dozens of them in the 1950s. The relationships between each of them were unclear until 1961, when Murray Gell-Mann[2] and Yuval Ne'eman[3] (independently of each other) proposed a hadron classification scheme called the *Eightfold Way*, or in more technical terms, SU(3) flavor symmetry.

This classification scheme organized the hadrons into isospin multiplets, but the physical basis behind it was still unclear. In 1964, Gell-Mann[4] and George Zweig[5][6] (independently of each other) proposed the quark model, then consisting only of up, down, and strange quarks.[7] However, while the quark model explained the Eightfold Way, no direct evidence of the existence of quarks was found until 1968 at the Stanford Linear Accelerator Center.[8][9] Deep inelastic scattering experiments indicated that protons had substructure, and that protons made of three more-fundamental particles explained the data (thus confirming the quark model).[10]

At first people were reluctant to describe the three bodies as quarks, instead preferring Richard Feynman's parton description,[11][12][13] but over time the quark theory became accepted (see *November Revolution*).[14]

15.2   Mass

Despite being extremely common, the bare mass of the up quark is not well determined, but probably lies between 1.8 and 3.0 MeV/c^2.[1] Lattice QCD calculations give a more precise value: 2.01±0.14 MeV/c^2.[15]

When found in mesons (particles made of one quark and one antiquark) or baryons (particles made of three quarks), the 'effective mass' (or 'dressed' mass) of quarks becomes greater because of the binding energy caused by the gluon field between each quark (see mass–energy equivalence).The bare mass of up quarks is so light, it cannot be straightforwardly calculated because relativistic effects have to be taken into account.

15.3 See also

- Down quark

- Isospin

- Quark model

- Quantum Mechanics

15.4 References

[1] J. Beringer *et al.* (Particle Data Group) (2012). "PDGLive Particle Summary 'Quarks (u, d, s, c, b, t, b', t', Free)'" (PDF). Particle Data Group. Retrieved 2013-02-21.

[2] M. Gell-Mann (2000) [1964]. "The Eightfold Way: A theory of strong interaction symmetry". In M. Gell-Mann, Y. Ne'eman. *The Eightfold Way.* Westview Press. p. 11. ISBN 0-7382-0299-1.
Original: M. Gell-Mann (1961). "The Eightfold Way: A theory of strong interaction symmetry". *Synchrotron Laboratory Report CTSL-20* (California Institute of Technology)

[3] Y. Ne'eman (2000) [1964]. "Derivation of strong interactions from gauge invariance". In M. Gell-Mann, Y. Ne'eman. *The Eightfold Way.* Westview Press. ISBN 0-7382-0299-1.
Original Y. Ne'eman (1961). "Derivation of strong interactions from gauge invariance". *Nuclear Physics* **26** (2): 222 N.doi:10.1016/0029-5582(61)90134-1.

[4] M. Gell-Mann (1964). "A Schematic Model of Baryons and Mesons". *Physics Letters* **8** (3): 214–215. Bibcode:1964PhL......8..214G. doi:10.1016/S0031-9163(64)92001-3.

[5] G. Zweig (1964). "An SU(3) Model for Strong Interaction Symmetry and its Breaking". *CERN Report No.8181/Th 8419.*

[6] G. Zweig (1964). "An SU(3) Model for Strong Interaction Symmetry and its Breaking: II". *CERN Report No.8419/Th 8412.*

[7] B. Carithers, P. Grannis (1995). "Discovery of the Top Quark" (PDF). *Beam Line* (SLAC) **25** (3): 4–16. Retrieved 2008-09-23.

[8] E. D. Bloom; Coward, D.; Destaebler, H.; Drees, J.; Miller, G.; Mo, L.; Taylor, R.; Breidenbach, M.; et al. (1969). "High-Energy Inelastic *e–p* Scattering at 6° and 10°". *Physical Review Letters* **23** (16): 930–934. Bibcode:1969PhRvL..23..930B. doi:10.1103/PhysRevLett.23.930.

[9] M. Breidenbach; Friedman, J.; Kendall, H.; Bloom, E.; Coward, D.; Destaebler, H.; Drees, J.; Mo, L.; Taylor, R.; et al. (1969). "Observed Behavior of Highly Inelastic Electron–Proton Scattering". *Physical Review Letters* **23** (16): 935–939. Bibcode:1969PhRvL..23..935B. doi:10.1103/PhysRevLett.23.935.

[10] J. I. Friedman. "The Road to the Nobel Prize". Hue University. Retrieved 2008-09-29.

[11] R.P.Feynman(1969). "Very High-Energy Collisions of Hadrons".*Physical Review Letters***23**(24): 1415–1417. Bibcode:19. doi:10.1103/PhysRevLett.23.1415.

[12] S. Kretzer; Lai, H.; Olness, Fredrick; Tung, W.; et al. (2004). "CTEQ6 Parton Distributions with Heavy Quark Mass Effects". *Physical Review D***69**(11): 114005. arXiv:hep-ph/0307022. Bibcode:2004PhRvD..69k4005K.doi:10.1103/PhysRevD.69..

[13] D. J. Griffiths (1987). *Introduction to Elementary Particles.* John Wiley & Sons. p. 42. ISBN 0-471-60386-4.

[14] M. E. Peskin, D. V. Schroeder (1995). *An introduction to quantum field theory.* Addison–Wesley. p. 556. ISBN 0-201-50397-2.

[15] Cho, Adrian (April 2010). "Mass of the Common Quark Finally Nailed Down". Science Magazine.

15.5   Further reading

- A. Ali, G. Kramer; Kramer (2011). "JETS and QCD: A historical review of the discovery of the quark and gluon jets and its impact on QCD". *European Physical Journal H* **36** (2): 245. arXiv:1012.2288. Bibcode:2011EPJ A.doi:10.1140/epjh/e2011-10047-1.

- R. Nave. "Quarks". *HyperPhysics*. Georgia State University, Department of Physics and Astronomy. Retrieved 2008-06-29.

- A. Pickering (1984). *Constructing Quarks*. University of Chicago Press. pp. 114–125. ISBN 0-226-66799-5.

Chapter 16

Down quark

The **down quark** or **d quark** (symbol: d) is the second-lightest of all quarks, a type of elementary particle, and a major constituent of matter. Together with the up quark, it forms the neutrons (one up quark, two down quarks) and protons (two up quarks, one down quark) of atomic nuclei. It is part of the first generation of matter, has an electric charge of $-\frac{1}{3}$ e and a bare mass of 4.8+0.5
−0.3 MeV/c^2.[1] Like all quarks, the down quark is an elementary fermion with spin-½,and experiences all 4 fundamental interactions: gravitation, electromagnetism, weak interactions, and strong interactions. The antiparticle of the down quark is the **down antiquark** (sometimes called *antidown quark* or simply *antidown*), which differs from it only in that some of its properties have equal magnitude but opposite sign.

Its existence (along with that of the up and strange quarks) was postulated in 1964 by Murray Gell-Mann and George Zweig to explain the *Eightfold Way* classification scheme of hadrons. The down quark was first observed by experiments at the Stanford Linear Accelerator Center in 1968.

16.1 History

In the beginnings of particle physics (first half of the 20th century), hadrons such as protons, neutrons, and pions were thought to be elementary particles. However, as new hadrons were discovered, the 'particle zoo' grew from a few particles in the early 1930s and 1940s to several dozens of them in the 1950s. The relationships between each of them was unclear until 1961, when Murray Gell-Mann[2] and Yuval Ne'eman[3] (independently of each other) proposed a hadron classification scheme called the *Eightfold Way*, or in more technical terms, SU(3) flavor symmetry.

This classification scheme organized the hadrons into isospin multiplets, but the physical basis behind it was still unclear. In 1964, Gell-Mann[4] and George Zweig[5][6] (independently of each other) proposed the quark model, then consisting only of up, down, and strange quarks.[7] However, while the quark model explained the Eightfold Way, no direct evidence of the existence of quarks was found until 1968 at the Stanford Linear Accelerator Center.[8][9] Deep inelastic scattering experiments indicated that protons had substructure, and that protons made of three more-fundamental particles explained the data (thus confirming the quark model).[10]

At first people were reluctant to identify the three-bodies as quarks, instead preferring Richard Feynman's parton description,[11][12][13] but over time the quark theory became accepted (see *November Revolution*).[14]

16.2 Mass

Despite being extremely common, the bare mass of the down quark is not well determined, but probably lies between 4.5 and 5.3$10^0$ MeV/c^2.[1] Lattice QCD calculations give a more precise value: 4.79±0.16 MeV/c^2.[15]

When found in mesons (particles made of one quark and one antiquark) or baryons (particles made of three quarks), the 'effective mass' (or 'dressed' mass) of quarks becomes greater because of the binding energy caused by the gluon field between quarks (see mass–energy equivalence). For example, the effective mass of down quarks in a proton is around 330 MeV/c^2. Because the bare mass of down quarks is so small, it cannot be straightforwardly calculated because relativistic effects have to be taken into account.

16.3   See also

- Up quark

- Isospin

- Quark model

16.4   References

[1] J. Beringer (Particle Data Group); et al. (2013). "PDGLive Particle Summary 'Quarks (u, d, s, c, b, t, b′, t′, Free)'" (PDF). Particle Data Group. Retrieved 2013-07-23.

[2] M. Gell-Mann (2000) [1964]. "The Eightfold Way: A theory of strong interaction symmetry". In M. Gell-Mann, Y. Ne'eman. *The Eightfold Way*. Westview Press. p. 11. ISBN 0-7382-0299-1.
Original: M. Gell-Mann (1961). "The Eightfold Way: A theory of strong interaction symmetry". *Synchrotron Laboratory Report CTSL-20* (California Institute of Technology).

[3] Y. Ne'eman (2000) [1964]. "Derivation of strong interactions from gauge invariance". In M. Gell-Mann, Y. Ne'eman. *The Eightfold Way*. Westview Press. ISBN 0-7382-0299-1.
Original Y. Ne'eman (1961). "Derivation of strong interactions from gauge invariance". *Nuclear Physics* **26** (2): 222 N.doi:10.1016/0029-5582(61)90134-1.

[4] M.Gell-Mann(1964). "A Schematic Model of Baryons and Mesons".*Physics Letters***8**(3): 214–215. Bibcode:1964PhL......8..G. doi:10.1016/S0031-9163(64)92001-3.

[5] G. Zweig (1964). "An SU(3) Model for Strong Interaction Symmetry and its Breaking". *CERN Report No.8181/Th 8419*.

[6] G. Zweig (1964). "An SU(3) Model for Strong Interaction Symmetry and its Breaking: II". *CERN Report No.8419/Th 8412*.

[7] B. Carithers, P. Grannis (1995). "Discovery of the Top Quark" (PDF). *Beam Line* (SLAC) **25** (3): 4–16. Retrieved 2008-09-23.

[8] E. D. Bloom; Coward, D.; Destaebler, H.; Drees, J.; Miller, G.; Mo, L.; Taylor, R.; Breidenbach, M.; et al. (1969). "High-Energy Inelastic *e–p* Scattering at 6° and 10°". *Physical Review Letters* **23** (16): 930–934. Bibcode:1969PhRvL..23..930B. doi:10.1103/PhysRevLett.23.930.

[9] M. Breidenbach; Friedman, J.; Kendall, H.; Bloom, E.; Coward, D.; Destaebler, H.; Drees, J.; Mo, L.; Taylor, R.; et al. (1969). "Observed Behavior of Highly Inelastic Electron–Proton Scattering". *Physical Review Letters* **23** (16): 935–939. Bibcode:1969PhRvL..23..935B. doi:10.1103/PhysRevLett.23.935.

[10] J. I. Friedman. "The Road to the Nobel Prize". Hue University. Retrieved 2008-09-29.

[11] R.P.Feynman(1969). "Very High-Energy Collisions of Hadrons".*Physical Review Letters***23**(24): 1415–1417. Bibcode:1969P. doi:10.1103/PhysRevLett.23.1415.

[12] S. Kretzer; Lai, H.; Olness, Fredrick; Tung, W.; et al. (2004). "CTEQ6 Parton Distributions with Heavy Quark Mass Effects". *Physical Review D***69**(11): 114005. arXiv:hep-ph/0307022. Bibcode:2004PhRvD..69k4005K.doi:10.1103/PhysRevD.69..

[13] D. J. Griffiths (1987). *Introduction to Elementary Particles*. John Wiley & Sons. p. 42. ISBN 0-471-60386-4.

[14] M. E. Peskin, D. V. Schroeder (1995). *An introduction to quantum field theory*. Addison–Wesley. p. 556. ISBN 0-201-50397-2.

[15] Cho, Adrian (April 2010). "Mass of the Common Quark Finally Nailed Down". Science Magazine.

16.5   Further reading

- A. Ali, G. Kramer; Kramer (2011). "JETS and QCD: A historical review of the discovery of the quark and gluon jets and its impact on QCD". *European Physical Journal H* **36** (2): 245. arXiv:1012.2288. Bibcode:2011EPJH .doi:10.1140/epjh/e2011-10047-1.

- R. Nave. "Quarks". *HyperPhysics*. Georgia State University, Department of Physics and Astronomy. Retrieved 2008-06-29.

- A. Pickering (1984). *Constructing Quarks*. University of Chicago Press. pp. 114–125. ISBN 0-226-66799-5.

three more-fundamental particles explained the data (thus confirming the quark model).[13]

At first people were reluctant to identify the three-bodies as quarks, instead preferring Richard Feynman's parton description, [14][15][16]but over time the quark theory became accepted(see*November Revolution*).[17]

17.2 See also

- Quark model

- Strange matter

- Strangeness production

- Strangelet

- Strange star

17.3 References

[1] J. Beringer *et al.* (Particle Data Group) (2012). "PDGLive Particle Summary 'Quarks (u, d, s, c, b, t, b′, t′, Free)'" (PDF). Particle Data Group. Retrieved 2012-11-30.

[2] M. Gell-Mann (1953). "Isotopic Spin and New Unstable Particles". *Physical Review* **92** (3): 833. Bibcode:1953PhRv...92..833G. doi:10.1103/PhysRev.92.833.

[3] G. Johnson (2000). *Strange Beauty: Murray Gell-Mann and the Revolution in Twentieth-Century Physics*. Random House. p. 119. ISBN 0-679-43764-9. By the end of the summer... [Gell-Mann] completed his first paper, "Isotopic Spin and Curious Particles" and send it of to *Physical Review*. The editors hated the title, so he amended it to "Strange Particles". They wouldn't go for that either—never mind that almost everybody used the term—suggesting instead "Isotopic Spin and New Unstable Particles".

[4] K. Nishijima, Kazuhiko (1955). "Charge Independence Theory of V Particles". *Progress of Theoretical Physics* **13** (3): 285. Bibcode:1955PThPh..13..285N. doi:10.1143/PTP.13.285.

[5] M. Gell-Mann (2000) [1964]. "The Eightfold Way: A theory of strong interaction symmetry". In M. Gell-Mann, Y. Ne'eman. *The Eightfold Way*. Westview Press. p. 11. ISBN 0-7382-0299-1.
Original: M. Gell-Mann (1961). "The Eightfold Way: A theory of strong interaction symmetry". *Synchrotron Laboratory Report CTSL-20* (California Institute of Technology)

[6] Y. Ne'eman (2000) [1964]. "Derivation of strong interactions from gauge invariance". In M. Gell-Mann, Y. Ne'eman. *The Eightfold Way*. Westview Press. ISBN 0-7382-0299-1.
Original Y. Ne'eman (1961). "Derivation of strong interactions from gauge invariance". *Nuclear Physics* **26** (2): 222 N.doi:10.1016/0029-5582(61)90134-1.

[7] M.Gell-Mann(1964). "A Schematic Model of Baryons and Mesons".*Physics Letters***8**(3): 214–215. Bibcode:1964PhL. doi:10.1016/S0031-9163(64)92001-3.

[8] G. Zweig (1964). "An SU(3) Model for Strong Interaction Symmetry and its Breaking". *CERN Report No.8181/Th 8419*.

[9] G. Zweig (1964). "An SU(3) Model for Strong Interaction Symmetry and its Breaking: II". *CERN Report No.8419/Th 8412*.

[10] B. Carithers, P. Grannis (1995). "Discovery of the Top Quark" (PDF). *Beam Line* (SLAC) **25** (3): 4–16. Retrieved 2008-09-23.

[11] E. D. Bloom; Coward, D.; Destaebler, H.; Drees, J.; Miller, G.; Mo, L.; Taylor, R.; Breidenbach, M.; et al. (1969). "High-Energy Inelastic *e–p* Scattering at 6° and 10°". *Physical Review Letters* **23** (16): 930–934. Bibcode:1969PhRvL..23..930B. doi:10.1103/PhysRevLett.23.930.

[12] M. Breidenbach; Friedman, J.; Kendall, H.; Bloom, E.; Coward, D.; Destaebler, H.; Drees, J.; Mo, L.; Taylor, R.; et al. (1969). "Observed Behavior of Highly Inelastic Electron–Proton Scattering". *Physical Review Letters* **23** (16): 935–939. Bibcode:1969PhRvL..23..935B. doi:10.1103/PhysRevLett.23.935.

[13] J. I. Friedman. "The Road to the Nobel Prize". Hue University. Retrieved 2008-09-29.

[14] R.P.Feynman(1969). "Very High-Energy Collisions of Hadrons".*Physical Review Letters***23**(24): 1415–1417. Bibcode:1969. doi:10.1103/PhysRevLett.23.1415.

[15] S. Kretzer; Lai, H.; Olness, Fredrick; Tung, W.; et al. (2004). "CTEQ6 Parton Distributions with Heavy Quark Mass Effects". *Physical Review D***69**(11): 114005. arXiv:hep-th/0307022. Bibcode:2004PhRvD..69k4005K.doi:10.1103/PhysRevD.69..

[16] D. J. Griffiths (1987). *Introduction to Elementary Particles*. John Wiley & Sons. p. 42. ISBN 0-471-60386-4.

[17] M. E. Peskin, D. V. Schroeder (1995). *An introduction to quantum field theory*. Addison–Wesley. p. 556. ISBN 0-201-50397-2.

17.4   Further reading

- R. Nave. "Quarks". *HyperPhysics*. Georgia State University, Department of Physics and Astronomy. Retrieved 2008-06-29.

- A. Pickering (1984). *Constructing Quarks*. University of Chicago Press. pp. 114–125. ISBN 0-226-66799-5.

Chapter 18

Charm quark

The **charm quark** or **c quark** (from its symbol, c) is the third most massive of all quarks, a type of elementary particle. Charm quarks are found in hadrons, which are subatomic particles made of quarks. Example of hadrons containing charm quarks include the J/ψ meson (J/ψ), D mesons (D), charmed Sigma baryons (Σ c), and other charmed particles.

It, along with the strange quark is part of the second generation of matter, and has an electric charge of $+\frac{2}{3}$ e and a bare mass of $1.29 {+0.05 \atop -0.11}$ GeV/c^2.[1] Like all quarks, the charm quark is an elementary fermion with spin-$\frac{1}{2}$, and experiences all four fundamental interactions: gravitation, electromagnetism, weak interactions, and strong interactions. The antiparticle of the charm quark is the **charm antiquark** (sometimes called *anticharm quark* or simply *anticharm*), which differs from it only in that some of its properties have equal magnitude but opposite sign.

The existence of a fourth quark had been speculated by a number of authors around 1964 (for instance by James Bjorken and Sheldon Glashow[4]), but its prediction is usually credited to Sheldon Glashow, John Iliopoulos and Luciano Maiani in 1970 (see GIM mechanism).[5] The first charmed particle (a particle containing a charm quark) to be discovered was the J/ψ meson. It was discovered by a team at the Stanford Linear Accelerator Center (SLAC), led by Burton Richter,[6] and one at the Brookhaven National Laboratory (BNL), led by Samuel Ting.[7]

The 1974 discovery of the J/ψ (and thus the charm quark) ushered in a series of breakthroughs which are collectively known as the *November Revolution*.

18.1   Hadrons containing charm quarks

Main articles: List of baryons and list of mesons

Some of the hadrons containing charm quarks include:

- D mesons contain a charm quark (or its antiparticle) and an up or down quark.

- D
  s mesons contain a charm quark and a strange quark.

- There are many charmonium states, for example the J/ψ particle. These consist of a charm quark and its antiparticle.

- Charmed baryons have been observed, and are named in analogy with strange baryons (e.g. Λ+ c).

145

18.2   See also

- Quark model

18.3   Notes

[1] K. Nakamura *et al.* (Particle Data Group) (2011). "PDGLive Particle Summary 'Quarks (u, d, s, c, b, t, b′, t′, Free)'" (PDF). Particle Data Group. Retrieved 2011-08-08.

[2] Carl Rod Nave. "Transformation of Quark Flavors by the Weak Interaction". Retrieved 2010-12-06. The c quark has about 5% probability of decaying into a d quark instead of an s quark.

[3] K. Nakamura; et al. (2010). "Review of Particles Physics: The CKM Quark-Mixing Matrix" (PDF). *J. Phys. G* **37** (75021): 150.

[4] B.J.Bjorken,S.L.Glashow;Glashow(1964). "Elementary particles and SU(4)".*Physics Letters***11**(3): 255–257. Bibcode:. doi:10.1016/0031-9163(64)90433-0.

[5] S.L. Glashow, J. Iliopoulos, L. Maiani; Iliopoulos; Maiani (1970). "Weak Interactions with Lepton–Hadron Symmetry". *Physical Review D* **2** (7): 1285–1292. Bibcode:1970PhRvD...2.1285G. doi:10.1103/PhysRevD.2.1285.

[6] J.-E. Augustin; Boyarski, A.; Breidenbach, M.; Bulos, F.; Dakin, J.; Feldman, G.; Fischer, G.; Fryberger, D.; Hanson, G.; Jean-Marie, B.; Larsen, R.; Lüth, V.; Lynch, H.; Lyon, D.; Morehouse, C.; Paterson, J.; Perl, M.; Richter, B.; Rapidis, P.; Schwitters, R.; Tanenbaum, W.; Vannucci, F.; Abrams, G.; Briggs, D.; Chinowsky, W.; Friedberg, C.; Goldhaber, G.; Hollebeek, R.; Kadyk, J.; Lulu, B. (1974). "Discovery of a Narrow Resonance in e^+e^- Annihilation". *Physical Review Letters* **33** (23): 1406. Bibcode:1974PhRvL..33.1406A. doi:10.1103/PhysRevLett.33.1406.

[7] J.J.Aubert;et al. (1974). "Experimental Observation of a Heavy Particle*J*".*Physical Review Letters***33**(23): 1404. Bibcode:1974A. doi:10.1103/PhysRevLett.33.1404.

18.4   Further reading

- R. Nave. "Quarks". *HyperPhysics*. Georgia State University, Department of Physics and Astronomy. Retrieved 2008-06-29.

- A. Pickering (1984). *Constructing Quarks*. University of Chicago Press. pp. 114–125. ISBN 0-226-66799-5.

Chapter 19

Top quark

The **top quark**, also known as the **t quark** (symbol: t) or **truth quark**, is an elementary particle and a fundamental constituent of matter. Like all quarks, the top quark is an elementary fermion with spin-$\frac{1}{2}$, and experiences all four fundamental interactions: gravitation, electromagnetism, weak interactions, and strong interactions. It has an electric charge of $+\frac{2}{3}\,e$,[2] and is the most massive of all observed elementary particles. It has a mass of 173.34 ± 0.27 (stat) $\pm$ 0.71 (syst)10^0 GeV/c^2,[1] which is about the same mass as an atom of tungsten. The antiparticle of the top quark is the **top antiquark** (symbol: t, sometimes called *antitop quark* or simply *antitop*), which differs from it only in that some of its properties have equal magnitude but opposite sign.

The top quark interacts primarily by the strong interaction, but can only decay through the weak force. It decays to a W boson and either a bottom quark (most frequently), a strange quark, or, on the rarest of occasions, a down quark. The Standard Model predicts its mean lifetime to be roughly 5×10^{-25} s.[3] This is about a twentieth of the timescale for strong interactions, and therefore it does not form hadrons, giving physicists a unique opportunity to study a "bare" quark (all other quarks hadronize, meaning that they combine with other quarks to form hadrons, and can only be observed as such). Because it is so massive, the properties of the top quark allow predictions to be made of the mass of the Higgs boson under certain extensions of the Standard Model (see Mass and coupling to the Higgs boson below). As such, it is extensively studied as a means to discriminate between competing theories.

Its existence (and that of the bottom quark) was postulated in 1973 by Makoto Kobayashi and Toshihide Maskawa to explain the observed CP violations in kaon decay,[4] and was discovered in 1995 by the CDF[5] and DØ[6] experiments at Fermilab. Kobayashi and Maskawa won the 2008 Nobel Prize in Physics for the prediction of the top and bottom quark, which together form the third generation of quarks.[7]

19.1   History

In 1973, Makoto Kobayashi and Toshihide Maskawa predicted the existence of a third generation of quarks to explain observed CP violations in kaon decay.[4] The names top and bottom were introduced by Haim Harari in 1975,[8][9] to match the names of the first generation of quarks (up and down) reflecting the fact that the two were the 'up' and 'down' component of a weak isospin doublet.[10] The top quark was sometimes called *truth quark* in the past, but over time *top quark* became the predominant use.[11]

The proposal of Kobayashi and Maskawa heavily relied on the GIM mechanism put forward by Sheldon Lee Glashow, John Iliopoulos and Luciano Maiani,[12] which predicted the existence of the then still unobserved charm quark. When in November 1974 teams at Brookhaven National Laboratory (BNL) and the Stanford Linear Accelerator Center (SLAC) simultaneously announced the discovery of the J/ψ meson, it was soon after identified as a bound state of the missing charm quark with its antiquark. This discovery allowed the GIM mechanism to become part of the Standard Model.[13] With the acceptance of the GIM mechanism, Kobayashi and Maskawa's prediction also gained in credibility. Their case was further strengthened by the discovery of the tau by Martin Lewis Perl's team at SLAC between 1974 and 1978.[14] This announced a third generation of leptons, breaking the new symmetry between leptons and quarks introduced by the

GIM mechanism. Restoration of the symmetry implied the existence of a fifth and sixth quark.

It was in fact not long until a fifth quark, the bottom, was discovered by the E288 experiment team, led by Leon Lederman at Fermilab in 1977.[15][16][17] This strongly suggested that there must also be a sixth quark, the top, to complete the pair. It was known that this quark would be heavier than the bottom, requiring more energy to create in particle collisions, but the general expectation was that the sixth quark would soon be found. However, it took another 18 years before the existence of the top was confirmed.[18]

Early searches for the top quark at SLAC and DESY (in Hamburg) came up empty-handed. When, in the early eighties, the Super Proton Synchrotron (SPS) at CERN discovered the W boson and the Z boson, it was again felt that the discovery of the top was imminent. As the SPS gained competition from the Tevatron at Fermilab there was still no sign of the missing particle, and it was announced by the group at CERN that the top mass must be at least 41 GeV/c^2. After a race between CERN and Fermilab to discover the top, the accelerator at CERN reached its limits without creating a single top, pushing the lower bound on its mass up to 77 GeV/c^2.[18]

The Tevatron was (until the start of LHC operation at CERN in 2009) the only hadron collider powerful enough to produce top quarks. In order to be able to confirm a future discovery, a second detector, the DØ detector, was added to the complex (in addition to the Collider Detector at Fermilab (CDF) already present). In October 1992, the two groups found their first hint of the top, with a single creation event that appeared to contain the top. In the following years, more evidence was collected and on April 22, 1994, the CDF group submitted their paper presenting tentative evidence for the existence of a top quark with a mass of about 175 GeV/c^2. In the meantime, DØ had found no more evidence than the suggestive event in 1992. A year later, on March 2, 1995, after having gathered more evidence and a reanalysis of the DØ data (who had been searching for a much lighter top), the two groups jointly reported the discovery of the top with a certainty of 99.9998% at a mass of 176±18 GeV/c^2.[5][6][18]

In the years leading up to the top quark discovery, it was realized that certain precision measurements of the electroweak vector boson masses and couplings are very sensitive to the value of the top quark mass. These effects become much larger for higher values of the top mass and therefore could indirectly see the top quark even if it could not be directly produced in any experiment at the time. The largest effect from the top quark mass was on the T parameter and by 1994 the precision of these indirect measurements had led to a prediction of the top quark mass to be between 145 GeV/c^2 and 185 GeV/c^2. It is the development of techniques that ultimately allowed such precision calculations that led to Gerardus 't Hooft and Martinus Veltman winning the Nobel Prize in physics in 1999.[19][20]

19.2   Properties

- At the final Tevatron energy of 1.96 TeV, top–antitop pairs were produced with a cross section of about 7 picobarns (pb).[21] The Standard Model prediction (at next-to-leading order with m_t = 175 GeV/c^2) is 6.7–7.5 pb.

- The W bosons from top quark decays carry polarization from the parent particle, hence pose themselves as a unique probe to top polarization.

- In the Standard Model, the top quark is predicted to have a spin quantum number of $^1/_2$ and electric charge $+^2/_3$. A first measurement of the top quark charge has been published, resulting in approximately 90% confidence limit that the top quark charge is indeed $+^2/_3$.[22]

19.3   Production

Because top quarks are very massive, large amounts of energy are needed to create one. The only way to achieve such high energies is through high energy collisions. These occur naturally in the Earth's upper atmosphere as cosmic rays collide with particles in the air, or can be created in a particle accelerator. In 2011, after the Tevatron ceased operations, the Large Hadron Collider at CERN became the only accelerator that generates a beam of sufficient energy to produce top quarks, with a center-of-mass energy of 7 TeV.

There are multiple processes that can lead to the production of a top quark. The most common is production of a top–antitop pair via strong interactions. In a collision, a highly energetic gluon is created, which subsequently decays into a top and antitop. This process was responsible for the majority of the top events at Tevatron and was the process observed when the top was first discovered in 1995.[23] It is also possible to produce pairs of top–antitop through the decay of an intermediate photon or Z-boson. However, these processes are predicted to be much rarer and have a virtually identical experimental signature in a hadron collider like Tevatron.

A distinctly different process is the production of single tops via weak interaction. This can happen in two ways (called channels): either an intermediate W-boson decays into a top and antibottom quark ("s-channel") or a bottom quark (probably created in a pair through the decay of a gluon) transforms to top quark by exchanging a W-boson with an up or down quark ("t-channel"). The first evidence for these processes was published by the DØ collaboration in December 2006,[24] and in March 2009 the CDF[25] and DØ[23] collaborations released twin papers with the definitive observation of these processes. The main significance of measuring these production processes is that their frequency is directly proportional to the | V_{tb} |² component of the CKM matrix.

19.4 Decay

The only known way that a top quark can decay is through the weak interaction producing a W-boson and a down-type quark (down, strange, or bottom). Because of its enormous mass, the top quark is extremely short-lived with a predicted lifetime of only 5×10^{-25} s.[3] As a result top quarks do not have time to form hadrons before they decay, as other quarks do. This provides physicists with the unique opportunity to study the behavior of a "bare" quark.

In particular, it is possible to directly determine the branching ratio $\Gamma(W^+b)$ / $\Gamma(W^+q$ (q = b,s,d)). The best current determination of this ratio is 0.91±0.04.[26] Since this ratio is equal to | V_{tb} |² according to the Standard Model, this gives another way of determining the CKM element | V_{tb} |, or in combination with the determination of | V_{tb} | from single top production provides tests for the assumption that the CKM matrix is unitary.[27]

The Standard Model also allows more exotic decays, but only at one loop level, meaning that they are extremely suppressed. In particular, it is possible for a top quark to decay into another up-type quark (an up or a charm) by emitting a photon or a Z-boson.[28] Searches for these exotic decay modes have provided no evidence for their existence in accordance with expectations from the Standard Model. The branching ratios for these decays have been determined to be less than 5.9 in 1,000 for photonic decay and less than 2.1 in 1,000 for Z-boson decay at 95% confidence.[26]

19.5 Mass and coupling to the Higgs boson

The Standard Model describes fermion masses through the Higgs mechanism. The Higgs boson has a Yukawa coupling to the left- and right-handed top quarks. After electroweak symmetry breaking (when the Higgs acquires a vacuum expectation value), the left- and right-handed components mix, becoming a mass term.

$$\mathcal{L} = y_t h q u^c \to \frac{y_t v}{\sqrt{2}}(1 + h^0/v) u u^c$$

The top quark Yukawa coupling has a value of

$$y_t = \sqrt{2} m_t / v \simeq 1$$

where v = 246 GeV is the value of the Higgs vacuum expectation value.

19.5.1 Yukawa couplings

See also: Beta function (physics)

In the Standard Model, all of the quark and lepton Yukawa couplings are small compared to the top quark Yukawa coupling. Understanding this hierarchy in the fermion masses is an open problem in theoretical physics. Yukawa couplings are not constants and their values change depending on the energy scale (distance scale) at which they are measured. The dynamics of Yukawa couplings are determined by the renormalization group equation.

One of the prevailing views in particle physics is that the size of the top quark Yukawa coupling is determined by the renormalization group, leading to the "quasi-infrared fixed point."

The Yukawa couplings of the up, down, charm, strange and bottom quarks, are hypothesized to have small values at the extremely high energy scale of grand unification, 10^{15} GeV. They increase in value at lower energy scales, at which the quark masses are generated by the Higgs. The slight growth is due to corrections from the QCD coupling. The corrections from the Yukawa couplings are negligible for the lower mass quarks.

If, however, a quark Yukawa coupling has a large value at very high energies, its Yukawa corrections will evolve and cancel against the QCD corrections. This is known as a (quasi-) infrared fixed point. No matter what the initial starting value of the coupling is, if it is sufficiently large it will reach this fixed point value. The corresponding quark mass is then predicted.

The top quark Yukawa coupling lies very near the infrared fixed point of the Standard Model. The renormalization group equation is:

$$\mu \frac{\partial}{\partial \mu} y_t \approx \frac{y_t}{16\pi^2} \left(\frac{9}{2} y_t^2 - 8g_3^2 - \frac{9}{4} g_2^2 - \frac{17}{20} g_1^2 \right),$$

where g_3 is the color gauge coupling, g_2 is the weak isospin gauge coupling, and g_1 is the weak hypercharge gauge coupling. This equation describes how the Yukawa coupling changes with energy scale μ. Solutions to this equation for large initial values y_t cause the right-hand side of the equation to quickly approach zero, locking y_t to the QCD coupling g_3. The value of the fixed point is fairly precisely determined in the Standard Model, leading to a top quark mass of 230 GeV. However, if there is more than one Higgs doublet, the mass value will be reduced by Higgs mixing angle effects in an unpredicted way.

In the minimal supersymmetric extension of the Standard Model (MSSM), there are two Higgs doublets and the renormalization group equation for the top quark Yukawa coupling is slightly modified:

$$\mu \frac{\partial}{\partial \mu} y_t \approx \frac{y_t}{16\pi^2} \left(6y_t^2 + y_b^2 - \frac{16}{3} g_3^2 - 3g_2^2 - \frac{13}{15} g_1^2 \right),$$

where y_b is the bottom quark Yukawa coupling. This leads to a fixed point where the top mass is smaller, 170–200 GeV. The uncertainty in this prediction arises because the bottom quark Yukawa coupling can be amplified in the MSSM. Some theorists believe this is supporting evidence for the MSSM.

The quasi-infrared fixed point has subsequently formed the basis of top quark condensation theories of electroweak symmetry breaking in which the Higgs boson is composite at *extremely* short distance scales, composed of a pair of top and antitop quarks.

19.6   See also

- CDF experiment

- Topness

- Top quark condensate

- Topcolor

- Quark model

19.7 References

[1] The ATLAS, CDF, CMS, D0 Collaborations (2014). "First combination of Tevatron and LHC measurements of the top-quark mass". Retrieved 2014-03-19.

[2] S. Willenbrock (2003). "The Standard Model and the Top Quark". In H.B Prosper and B. Danilov (eds.). *Techniques and Concepts of High-Energy Physics XII*. NATO Science Series **123**. Kluwer Academic. pp. 1–41. arXiv:hep-ph/0211067v3. ISBN 1-4020-1590-9.

[3] A.Quadt(2006). "Top quark physics at hadron colliders".*European Physical Journal C***48**(3): 835–1000. Bibcode:2006EPJC.. doi:10.1140/epjc/s2006-02631-6.

[4] M. Kobayashi, T. Maskawa (1973). "*CP*-Violation in the Renormalizable Theory of Weak Interaction". *Progress of Theoretical Physics* **49** (2): 652. Bibcode:1973PThPh..49..652K. doi:10.1143/PTP.49.652.

[5] F. Abe *et al.* (CDF Collaboration) (1995). "Observation of Top Quark Production in pp Collisions with the Collider Detector at Fermilab". *Physical Review Letters* **74** (14): 2626–2631. Bibcode:1995PhRvL..74.2626A. doi:10.1103/PhysRevLett.74.2626. PMID 10057978.

[6] S. Abachi *et al.* (DØ Collaboration) (1995). "Search for High Mass Top Quark Production in pp Collisions at $\sqrt{s}$ = 1.8 TeV". *Physical Review Letters* **74** (13): 2422–2426. Bibcode:1995PhRvL..74.2422A. doi:10.1103/PhysRevLett.74.2422.

[7] "2008 Nobel Prize in Physics". The Nobel Foundation. 2008. Retrieved 2009-09-11.

[8] H.Harari(1975). "A new quark model for hadrons".*Physics Letters B***57**(3): 265. Bibcode:1975PhLB...57..265H.doi:10.1016-2693(75)90072-6.

[9] K.W. Staley (2004). *The Evidence for the Top Quark*. Cambridge University Press. pp. 31–33. ISBN 978-0-521-82710-2.

[10] D.H. Perkins (2000). *Introduction to high energy physics*. Cambridge University Press. p. 8. ISBN 0-521-62196-8.

[11] F. Close (2006). *The New Cosmic Onion*. CRC Press. p. 133. ISBN 1-58488-798-2.

[12] S.L. Glashow, J. Iliopoulous, L. Maiani (1970). "Weak Interactions with Lepton–Hadron Symmetry". *Physical Review D* **2** (7): 1285–1292. Bibcode:1970PhRvD...2.1285G. doi:10.1103/PhysRevD.2.1285.

[13] A. Pickering (1999). *Constructing Quarks: A Sociological History of Particle Physics*. University of Chicago Press. pp. 253–254. ISBN 978-0-226-66799-7.

[14] M.L. Perl; et al. (1975). "Evidence for Anomalous Lepton Production in e+e− Annihilation". *Physical Review Letters* **35** (22): 1489. Bibcode:1975PhRvL..35.1489P. doi:10.1103/PhysRevLett.35.1489.

[15] "Discoveries at Fermilab – Discovery of the Bottom Quark" (Press release). Fermilab. 7 August 1977. Retrieved 2009-07-24.

[16] L.M. Lederman (2005). "Logbook: Bottom Quark". *Symmetry Magazine* **2** (8).

[17] S.W. Herb; et al. (1977). "Observation of a Dimuon Resonance at 9.5 GeV in 400-GeV Proton-Nucleus Collisions". *Physical Review Letters* **39** (5): 252. Bibcode:1977PhRvL..39..252H. doi:10.1103/PhysRevLett.39.252.

[18] T.M. Liss, P.L. Tipton (1997). "The Discovery of the Top Quark" (PDF). *Scientific American*: 54–59.

[19] "The Nobel Prize in Physics 1999". The Nobel Foundation. Retrieved 2009-09-10.

[20] "The Nobel Prize in Physics 1999, Press Release" (Press release). The Nobel Foundation. 12 October 1999. Retrieved 2009-09-10.

[21] D. Chakraborty (DØ and CDF collaborations) (2002). *Top quark and W/Z results from the Tevatron* (PDF). Rencontres de Moriond. p. 26.

[22] V.M. Abazov *et al.* (DØ Collaboration) (2007). "Experimental discrimination between charge 2*e*/3 top quark and charge 4*e*/3 exotic quark production scenarios". *Physical Review Letters* **98** (4): 041801. arXiv:hep-ex/0608044. Bibcode:2007PhRvL..98 .doi:10.1103/PhysRevLett.98.041801. PMID17358756.

[23] V.M. Abazov *et al.* (DØ Collaboration) (2009). "Observation of Single Top Quark Production". *Physical Review Letters* **103** (9). arXiv:0903.0850. Bibcode:2009PhRvL.103i2001A. doi:10.1103/PhysRevLett.103.092001.

[24] V.M. Abazov *et al.* (DØ Collaboration) (2007). "Evidence for production of single top quarks and first direct measurement of |V$_{tb}$|".*Physical Review Letters***98**(18): 181802. arXiv:hep-ex/0612052. Bibcode:2007PhRvL..98r1802A.doi:10.1103 .PMID17501561.

[25] T. Aaltonen *et al.* (CDF Collaboration) (2009). "First Observation of Electroweak Single Top Quark Production". *Physical Review Letters* **103** (9). arXiv:0903.0885. Bibcode:2009PhRvL.103i2002A. doi:10.1103/PhysRevLett.103.092002.

[26] J. Beringer *et al.* (Particle Data Group) (2012). "PDGLive Particle Summary 'Quarks (u, d, s, c, b, t, b', t', Free)'" (PDF). Particle Data Group. Retrieved 2013-07-23.

[27] V.M. Abazov *et al.* (DØ Collaboration) (2008). "Simultaneous measurement of the ratio B(t→Wb)/B(t→Wq) and the top-quark pair production cross section with the DØ detector at √s = 1.96 TeV". *Physical Review Letters* **100** (19): 192003. arXiv:0801.1326. Bibcode:2008PhRvL.100s2003A. doi:10.1103/PhysRevLett.100.192003.

[28] S. Chekanov *et al.* (ZEUS Collaboration) (2003). "Search for single-top production in ep collisions at HERA". *Physics Letters B* **559** (3–4): 153. arXiv:hep-ex/0302010. Bibcode:2003PhLB..559..153Z. doi:10.1016/S0370-2693(03)00333-2.

19.8   Further reading

- Frank Fiedler; for the D0; CDF Collaborations (June 2005). "Top Quark Production and Properties at the Tevatron". arXiv:hep-ex/0506005 [hep-ex].

- R. Nave. "Quarks". *HyperPhysics*. Georgia State University, Department of Physics and Astronomy. Retrieved 2008-06-29.

- A. Pickering (1984). *Constructing Quarks*. University of Chicago Press. pp. 114–125. ISBN 0-226-66799-5.

19.9   External links

- Top quark on arxiv.org

- Tevatron Electroweak Working Group

- Top quark information on Fermilab website

- Logbook pages from CDF and DZero collaborations' top quark discovery

- Scientific American article on the discovery of the top quark

- Public Homepage of Top Quark Analysis Results from DØ Collaboration at Fermilab

- Public Homepage of Top Quark Analysis Results from CDF Collaboration at Fermilab

- Harvard Magazine article about the 1994 top quark discovery

- 1999 Nobel Prize in Physics

Chapter 20

Bottom quark

The **bottom quark** or **b quark**, also known as the **beauty quark**, is a third-generation quark with a charge of $-\frac{1}{3}\,e$. Although all quarks are described in a similar way by quantum chromodynamics, the bottom quark's large bare mass (around 4.2 GeV/c^2,[3] a bit more than four times the mass of a proton), combined with low values of the CKM matrix elements V_{ub} and V_{cb}, gives it a distinctive signature that makes it relatively easy to identify experimentally (using a technique called B-tagging). Because three generations of quark are required for CP violation (see CKM matrix), mesons containing the bottom quark are the easiest particles to use to investigate the phenomenon; such experiments are being performed at the BaBar, Belle and LHCb experiments. The bottom quark is also notable because it is a product in almost all top quark decays, and is a frequent decay product for the Higgs boson.

The bottom quark was theorized in 1973 by physicists Makoto Kobayashi and Toshihide Maskawa to explain CP violation.[1] The name "bottom" was introduced in 1975 by Haim Harari.[4][5] The bottom quark was discovered in 1977 by the Fermilab E288 experiment team led by Leon M. Lederman, when collisions produced bottomonium.[2][6][7] Kobayashi and Maskawa won the 2008 Nobel Prize in Physics for their explanation of CP-violation.[8][9] On its discovery, there were efforts to name the bottom quark "beauty", but "bottom" became the predominant usage.

The bottom quark can decay into either an up quark or charm quark via the weak interaction. Both these decays are suppressed by the CKM matrix, making lifetimes of most bottom particles (~10^{-12} s) somewhat higher than those of charmed particles (~10^{-13} s), but lower than those of strange particles (from ~10^{-10} to ~10^{-8} s).

20.1 Hadrons containing bottom quarks

Main articles: list of baryons and list of mesons

Some of the hadrons containing bottom quarks include:

- B mesons contain a bottom quark (or its antiparticle) and an up or down quark.

- B
  c and B
  s mesons contain a bottom quark along with a charm quark or strange quark respectively.

- There are many bottomonium states, for example the Υ meson and χ_b(3P), the first particle discovered in LHC. These consist of a bottom quark and its antiparticle.

- Bottom baryons have been observed, and are named in analogy with strange baryons (e.g. Λ0
  b).

20.2   See also

- Quark model

20.3   References

[1] M. Kobayashi; T. Maskawa (1973). "CP-Violation in the Renormalizable Theory of Weak Interaction". *Progress of Theoretical Physics* **49** (2): 652–657. Bibcode:1973PThPh..49..652K. doi:10.1143/PTP.49.652.

[2] "Discoveries at Fermilab – Discovery of the Bottom Quark" (Press release). Fermilab. 7 August 1977. Retrieved 2009-07-24.

[3] J. Beringer (Particle Data Group); et al. (2012). "PDGLive Particle Summary 'Quarks (u, d, s, c, b, t, b′, t′, Free)'" (PDF). Particle Data Group. Retrieved 2012-12-18.

[4] H.Harari(1975). "A new quark model for hadrons".*Physics Letters B***57**(3): 265. Bibcode:1975PhLB...57..265H.doi:10.1016/ 2693(75)90072-6.

[5] K.W. Staley (2004). *The Evidence for the Top Quark*. Cambridge University Press. pp. 31–33. ISBN 978-0-521-82710-2.

[6] L.M. Lederman (2005). "Logbook: Bottom Quark". *Symmetry Magazine* **2** (8).

[7] S.W. Herb; Hom, D.; Lederman, L.; Sens, J.; Snyder, H.; Yoh, J.; Appel, J.; Brown, B.; Brown, C.; Innes, W.; Ueno, K.; Yamanouchi, T.; Ito, A.; Jöstlein, H.; Kaplan, D.; Kephart, R.; et al. (1977). "Observation of a Dimuon Resonance at 9.5 GeV in 400-GeV Proton-Nucleus Collisions". *Physical Review Letters* **39** (5): 252. Bibcode:1977PhRvL..39..252H. doi:10.1103/PhysRevLett.39.252.

[8] 2008 Physics Nobel Prize lecture by Makoto Kobayashi

[9] 2008 Physics Nobel Prize lecture by Toshihide Maskawa

20.4   Further reading

- L. Lederman (1978). "The Upsilon Particle". *Scientific American* **239** (4): 72. doi:10.1038/scientificamerican1078-72.

- R. Nave. "Quarks". *HyperPhysics*. Georgia State University, Department of Physics and Astronomy. Retrieved 2008-06-29.

- A. Pickering (1984). *Constructing Quarks*. University of Chicago Press. pp. 114–125. ISBN 0-226-66799-5.

- J. Yoh (1997). "The Discovery of the b Quark at Fermilab in 1977: The Experiment Coordinator's Story" (PDF). *Proceedings of Twenty Beautiful Years of Bottom Physics*. Fermilab. Retrieved 2009-07-24.

20.5   External links

- History of the discovery of the bottom quark / Upsilon meson

Chapter 21

Baryon

Not to be confused with Baryonyx.

A **baryon** is a composite subatomic particle made up of three quarks (as distinct from mesons, which are composed of one quark and one antiquark). Baryons and mesons belong to the hadron family of particles, which are the quark-based particles. The name "baryon" comes from the Greek word for "heavy" (βαρύς, *barys*), because, at the time of their naming, most known elementary particles had lower masses than the baryons.

As quark-based particles, baryons participate in the strong interaction, whereas leptons, which are not quark-based, do not. The most familiar baryons are the protons and neutrons that make up most of the mass of the visible matter in the universe. Electrons (the other major component of the atom) are leptons.

Each baryon has a corresponding antiparticle (antibaryon) where quarks are replaced by their corresponding antiquarks. For example, a proton is made of two up quarks and one down quark; and its corresponding antiparticle, the antiproton, is made of two up antiquarks and one down antiquark.

21.1   Background

Baryons are strongly interacting fermions that is, they experience the strong nuclear force and are described by Fermi–Dirac statistics, which apply to all particles obeying the Pauli exclusion principle. This is in contrast to the bosons, which do not obey the exclusion principle.

Baryons, along with mesons, are hadrons, meaning they are particles composed of quarks. Quarks have baryon numbers of $B = \frac{1}{3}$ and antiquarks have baryon number of $B = -\frac{1}{3}$. The term "baryon" usually refers to *triquarks*—baryons made of three quarks ($B = \frac{1}{3} + \frac{1}{3} + \frac{1}{3} = 1$).

Other exotic baryons have been proposed, such as pentaquarks—baryons made of four quarks and one antiquark ($B = \frac{1}{3} + \frac{1}{3} + \frac{1}{3} + \frac{1}{3} - \frac{1}{3} = 1$), but their existence is not generally accepted. In theory, heptaquarks (5 quarks, 2 antiquarks), nonaquarks (6 quarks, 3 antiquarks), etc. could also exist. Until recently, it was believed that some experiments showed the existence of pentaquarks—baryons made of four quarks and one antiquark.[1][2] The particle physics community as a whole did not view their existence as likely in 2006,[3] and in 2008, considered evidence to be overwhelmingly against the existence of the reported pentaquarks.[4] However, in July 2015, the LHCb experiment observed two resonances consistent with pentaquark states in the Λ0
b → J/ψK−
p decay, with a combined statistical significance of 15σ.[5][6]

21.2   Baryonic matter

Nearly all matter that may be encountered or experienced in everyday life is baryonic matter, which includes atoms of any sort, and provides those with the quality of mass. Non-baryonic matter, as implied by the name, is any sort of matter that is not composed primarily of baryons. Those might include neutrinos or free electrons, dark matter, such as supersymmetric particles, axions, or black holes.

The very existence of baryons is also a significant issue in cosmology because it is assumed that the Big Bang produced a state with equal amounts of baryons and antibaryons. The process by which baryons came to outnumber their antiparticles is called baryogenesis.

21.3   Baryogenesis

Main article: Baryogenesis

Experiments are consistent with the number of quarks in the universe being a constant and, to be more specific, the number of baryons being a constant ; in technical language, the total baryon number appears to be *conserved*. Within the prevailing Standard Model of particle physics, the number of baryons may change in multiples of three due to the action of sphalerons, although this is rare and has not been observed under experiment. Some grand unified theories of particle physics also predict that a single proton can decay, changing the baryon number by one; however, this has not yet been observed under experiment. The excess of baryons over antibaryons in the present universe is thought to be due to non-conservation of baryon number in the very early universe, though this is not well understood.

21.4   Properties

21.4.1   Isospin and charge

Main article: Isospin

 The concept of isospin was first proposed by Werner Heisenberg in 1932 to explain the similarities between protons and neutrons under the strong interaction.[7] Although they had different electric charges, their masses were so similar that physicists believed they were actually the same particle. The different electric charges were explained as being the result of some unknown excitation similar to spin. This unknown excitation was later dubbed *isospin* by Eugene Wigner in 1937.[8]

This belief lasted until Murray Gell-Mann proposed the quark model in 1964 (containing originally only the u, d, and s quarks).[9] The success of the isospin model is now understood to be the result of the similar masses of the u and d quarks. Since the u and d quarks have similar masses, particles made of the same number then also have similar masses. The exact specific u and d quark composition determines the charge, as u quarks carry charge $+2/3$ while d quarks carry charge $-1/3$. For example the four Deltas all have different charges (Δ++ (uuu), Δ+ (uud), Δ0 (udd), Δ− (ddd)), but have similar masses (~1,232 MeV/c^2) as they are each made of a combination of three u and d quarks. Under the isospin model, they were considered to be a single particle in different charged states.

The mathematics of isospin was modeled after that of spin. Isospin projections varied in increments of 1 just like those of spin, and to each projection was associated a "charged state". Since the "Delta particle" had four "charged states", it was said to be of isospin $I = 3/2$. Its "charged states" Δ++, Δ+, Δ0, and Δ−, corresponded to the isospin projections $I_3 = +3/2$, $I_3 = +1/2$, $I_3 = -1/2$, and $I_3 = -3/2$, respectively. Another example is the "nucleon particle". As there were two nucleon "charged states", it was said to be of isospin $1/2$. The positive nucleon N+ (proton) was identified with $I_3 = +1/2$ and the neutral nucleon N0 (neutron) with $I_3 = -1/2$.[10] It was later noted that the isospin projections were related to the up and down quark content of particles by the relation:

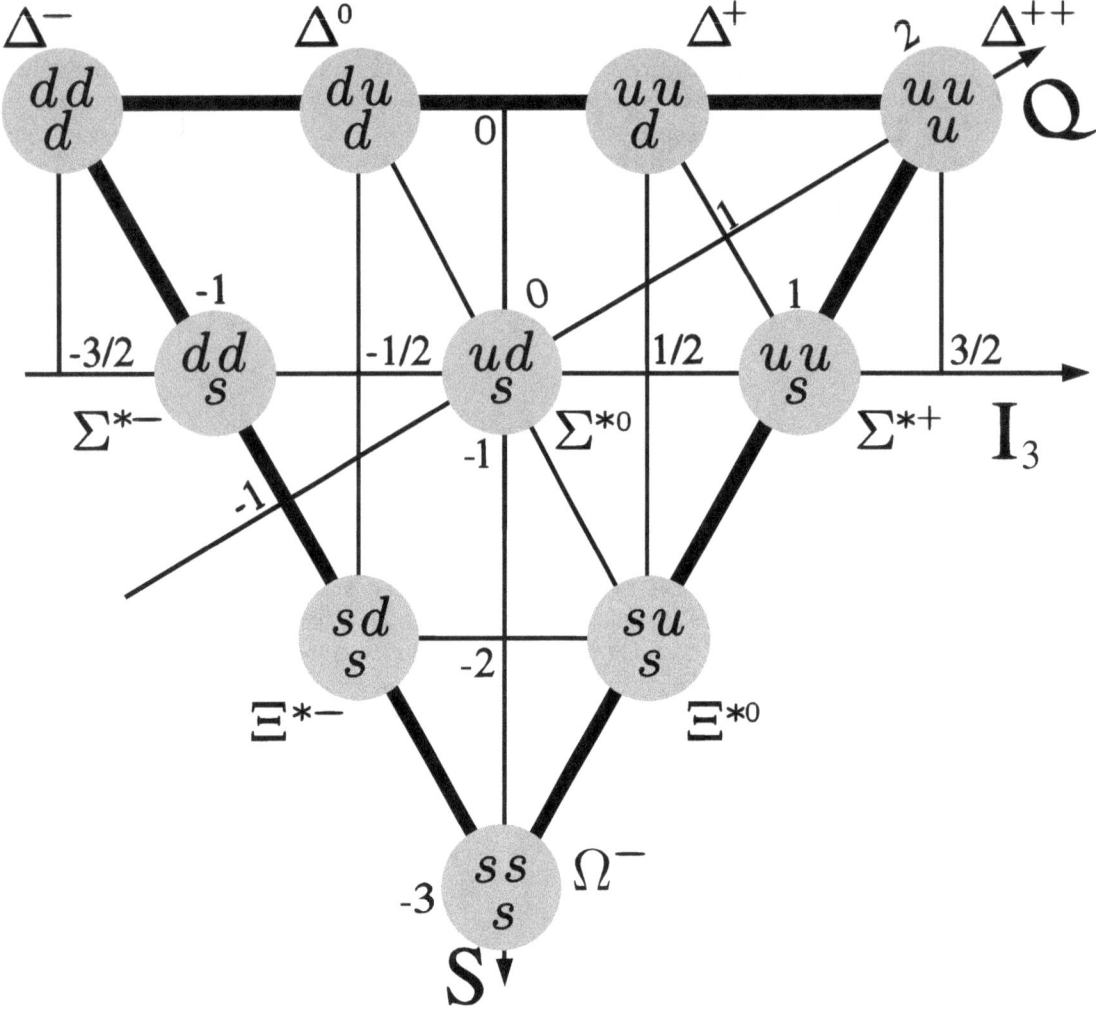

*Combinations of three **u**, **d** or **s** quarks forming baryons with a spin-$\frac{3}{2}$ form the* uds baryon decuplet

$$I_3 = \frac{1}{2}[(n_u - n_{\bar{u}}) - (n_d - n_{\bar{d}})],$$

where the n's are the number of up and down quarks and antiquarks.

In the "isospin picture", the four Deltas and the two nucleons were thought to be the different states of two particles. However in the quark model, Deltas are different states of nucleons (the N^{++} or N^- are forbidden by Pauli's exclusion principle). Isospin, although conveying an inaccurate picture of things, is still used to classify baryons, leading to unnatural and often confusing nomenclature.

21.4.2 Flavour quantum numbers

Main article: Flavour (particle physics) § Flavour quantum numbers

The strangeness flavour quantum number S (not to be confused with spin) was noticed to go up and down along with particle mass. The higher the mass, the lower the strangeness (the more s quarks). Particles could be described with

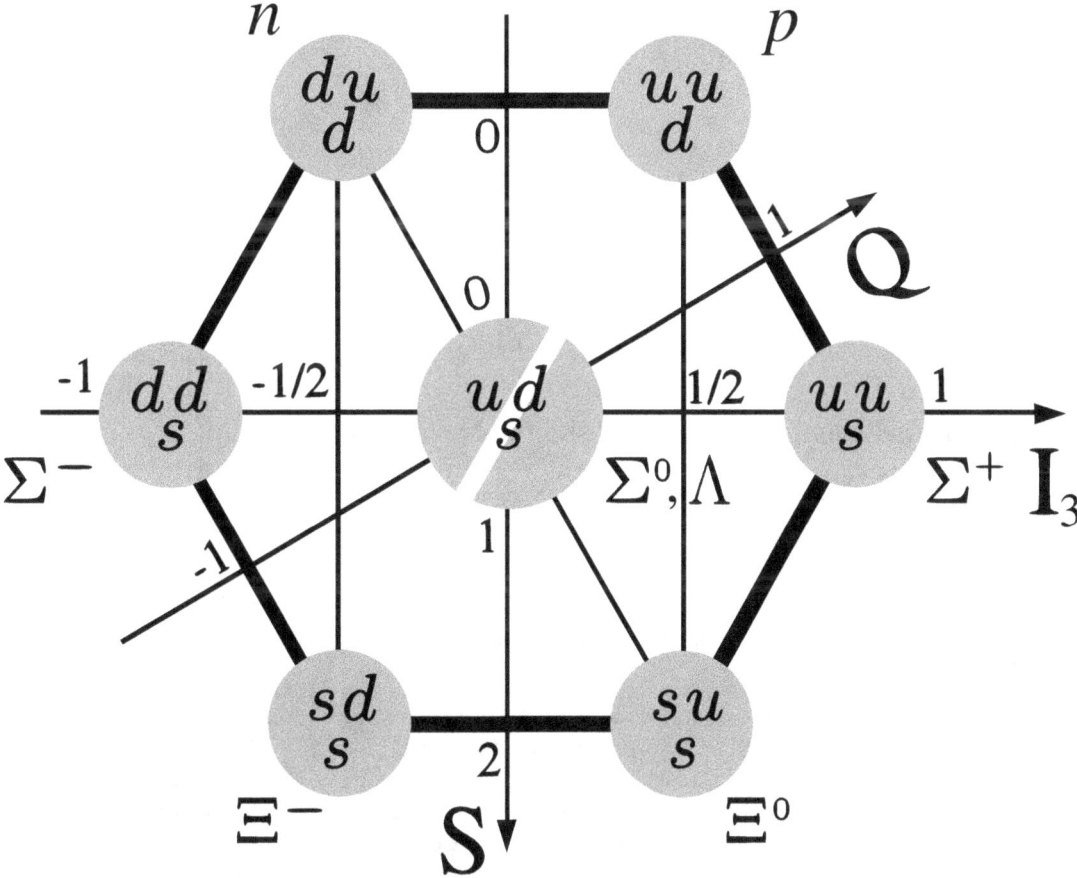

Combinations of three **u, d** *or s quarks forming baryons with a spin-*$\frac{1}{2}$ *form the* uds baryon octet

isospin projections (related to charge) and strangeness (mass) (see the uds octet and decuplet figures on the right). As other quarks were discovered, new quantum numbers were made to have similar description of udc and udb octets and decuplets. Since only the u and d mass are similar, this description of particle mass and charge in terms of isospin and flavour quantum numbers works well only for octet and decuplet made of one u, one d, and one other quark, and breaks down for the other octets and decuplets (for example, ucb octet and decuplet). If the quarks all had the same mass, their behaviour would be called *symmetric*, as they would all behave in exactly the same way with respect to the strong interaction. Since quarks do not have the same mass, they do not interact in the same way (exactly like an electron placed in an electric field will accelerate more than a proton placed in the same field because of its lighter mass), and the symmetry is said to be broken.

It was noted that charge (Q) was related to the isospin projection (I_3), the baryon number (B) and flavour quantum numbers (S, C, B', T) by the Gell-Mann–Nishijima formula:[10]

$$Q = I_3 + \frac{1}{2}(B + S + C + B' + T),$$

where S, C, B', and T represent the strangeness, charm, bottomness and topness flavour quantum numbers, respectively. They are related to the number of strange, charm, bottom, and top quarks and antiquark according to the relations:

$$S = -(n_s - n_{\bar{s}}),$$

$$C = +(n_c - n_{\bar{c}}),$$

$$B' = -(n_b - n_{\bar{b}}),$$

$$T = +(n_t - n_{\bar{t}}),$$

meaning that the Gell-Mann–Nishijima formula is equivalent to the expression of charge in terms of quark content:

$$Q = \frac{2}{3}[(n_u - n_{\bar{u}}) + (n_c - n_{\bar{c}}) + (n_t - n_{\bar{t}})] - \frac{1}{3}[(n_d - n_{\bar{d}}) + (n_s - n_{\bar{s}}) + (n_b - n_{\bar{b}})].$$

21.4.3   Spin, orbital angular momentum, and total angular momentum

Main articles: Spin (physics), Angular momentum operator, Quantum numbers and Clebsch–Gordan coefficients

Spin (quantum number S) is a vector quantity that represents the "intrinsic" angular momentum of a particle. It comes in increments of $\frac{1}{2}$ ℏ (pronounced "h-bar"). The ℏ is often dropped because it is the "fundamental" unit of spin, and it is implied that "spin 1" means "spin 1 ℏ". In some systems of natural units, ℏ is chosen to be 1, and therefore does not appear anywhere.

Quarks are fermionic particles of spin $\frac{1}{2}$ ($S = \frac{1}{2}$). Because spin projections vary in increments of 1 (that is 1 ℏ), a single quark has a spin vector of length $\frac{1}{2}$, and has two spin projections ($S_z = +\frac{1}{2}$ and $S_z = -\frac{1}{2}$). Two quarks can have their spins aligned, in which case the two spin vectors add to make a vector of length $S = 1$ and three spin projections ($S_z = +1$, $S_z = 0$, and $S_z = -1$). If two quarks have unaligned spins, the spin vectors add up to make a vector of length $S = 0$ and has only one spin projection ($S_z = 0$), etc. Since baryons are made of three quarks, their spin vectors can add to make a vector of length $S = \frac{3}{2}$, which has four spin projections ($S_z = +\frac{3}{2}$, $S_z = +\frac{1}{2}$, $S_z = -\frac{1}{2}$, and $S_z = -\frac{3}{2}$), or a vector of length $S = \frac{1}{2}$ with two spin projections ($S_z = +\frac{1}{2}$, and $S_z = -\frac{1}{2}$).[11]

There is another quantity of angular momentum, called the orbital angular momentum, (azimuthal quantum number L), that comes in increments of 1 ℏ, which represent the angular moment due to quarks orbiting around each other. The total angular momentum (total angular momentum quantum number J) of a particle is therefore the combination of intrinsic angular momentum (spin) and orbital angular momentum. It can take any value from $J = |L - S|$ to $J = |L + S|$, in increments of 1.

Particle physicists are most interested in baryons with no orbital angular momentum ($L = 0$), as they correspond to ground states—states of minimal energy. Therefore the two groups of baryons most studied are the $S = \frac{1}{2}$; $L = 0$ and $S = \frac{3}{2}$; $L = 0$, which corresponds to $J = \frac{1}{2}^{+}$ and $J = \frac{3}{2}^{+}$, respectively, although they are not the only ones. It is also possible to obtain $J = \frac{3}{2}^{+}$ particles from $S = \frac{1}{2}$ and $L = 2$, as well as $S = \frac{3}{2}$ and $L = 2$. This phenomenon of having multiple particles in the same total angular momentum configuration is called *degeneracy*. How to distinguish between these degenerate baryons is an active area of research in baryon spectroscopy.[12][13]

21.4.4   Parity

Main article: Parity (physics)

If the universe were reflected in a mirror, most of the laws of physics would be identical—things would behave the same way regardless of what we call "left" and what we call "right". This concept of mirror reflection is called *intrinsic parity* or *parity* (*P*). Gravity, the electromagnetic force, and the strong interaction all behave in the same way regardless of whether or not the universe is reflected in a mirror, and thus are said to conserve parity (P-symmetry). However, the weak interaction *does* distinguish "left" from "right", a phenomenon called parity violation (P-violation).

Based on this, one might think that, if the wavefunction for each particle (in more precise terms, the quantum field for each particle type) were simultaneously mirror-reversed, then the new set of wavefunctions would perfectly satisfy the laws of physics (apart from the weak interaction). It turns out that this is not quite true: In order for the equations to be satisfied, the wavefunctions of certain types of particles have to be multiplied by −1, in addition to being mirror-reversed. Such particle types are said to have *negative* or *odd* parity ($P = -1$, or alternatively $P = -$), while the other particles are said to have *positive* or *even* parity ($P = +1$, or alternatively $P = +$).

For baryons, the parity is related to the orbital angular momentum by the relation:[14]

$$P = (-1)^{L}.$$

As a consequence, baryons with no orbital angular momentum ($L = 0$) all have even parity ($P = +$).

21.5   Nomenclature

Baryons are classified into groups according to their isospin (I) values and quark (q) content. There are six groups of baryons—nucleon (N), Delta (Δ), Lambda (Λ), Sigma (Σ), Xi (Ξ), and Omega (Ω). The rules for classification are defined by the Particle Data Group. These rules consider the up (u), down (d) and strange (s) quarks to be *light* and the charm (c), bottom (b), and top (t) quarks to be *heavy*. The rules cover all the particles that can be made from three of each of the six quarks, even though baryons made of t quarks are not expected to exist because of the t quark's short lifetime. The rules do not cover pentaquarks.[15]

- Baryons with three u and/or d quarks are N's ($I = \frac{1}{2}$) or Δ's ($I = \frac{3}{2}$).

- Baryons with two u and/or d quarks are Λ's ($I = 0$) or Σ's ($I = 1$). If the third quark is heavy, its identity is given by a subscript.

- Baryons with one u or d quark are Ξ's ($I = \frac{1}{2}$). One or two subscripts are used if one or both of the remaining quarks are heavy.

- Baryons with no u or d quarks are Ω's ($I = 0$), and subscripts indicate any heavy quark content.

- Baryons that decay strongly have their masses as part of their names. For example, Σ^0 does not decay strongly, but $\Delta^{++}(1232)$ does.

It is also a widespread (but not universal) practice to follow some additional rules when distinguishing between some states that would otherwise have the same symbol.[10]

- Baryons in total angular momentum $J = \frac{3}{2}$ configuration that have the same symbols as their $J = \frac{1}{2}$ counterparts are denoted by an asterisk (*).

- Two baryons can be made of three different quarks in $J = \frac{1}{2}$ configuration. In this case, a prime (′) is used to distinguish between them.

  - *Exception*: When two of the three quarks are one up and one down quark, one baryon is dubbed Λ while the other is dubbed Σ.

Quarks carry charge, so knowing the charge of a particle indirectly gives the quark content. For example, the rules above say that a Λ+
c contains a c quark and some combination of two u and/or d quarks. The c quark has a charge of ($Q = +\frac{2}{3}$), therefore the other two must be a u quark ($Q = +\frac{2}{3}$), and a d quark ($Q = -\frac{1}{3}$) to have the correct total charge ($Q = +1$).

21.6 See also

- Eightfold way

- List of baryons

- List of particles

- Meson

- Timeline of particle discoveries

21.7 Notes

[1] H. Muir (2003)

[2] K. Carter (2003)

[3] W.-M. Yao *et al.* (2006): Particle listings – Θ^+

[4] C. Amsler *et al.* (2008): Pentaquarks

[5] LHCb (14 July 2015). "Observation of particles composed of five quarks, pentaquark-charmonium states, seen in $\Lambda_b^0 \to J/\psi pK^-$ decays.". *CERN website*. Retrieved 2015-07-14.

[6] R. Aaij et al. (LHCb collaboration) (2015). "Observation of J/ψp resonances consistent with pentaquark states in Λ0 b→J/ψK− p decays". *Physical Review Letters* **115** (7). Bibcode:2015PhRvL.115g2001A. doi:10.1103/PhysRevLett.115.072001.

[7] W. Heisenberg (1932)

[8] E. Wigner (1937)

[9] M. Gell-Mann (1964)

[10] S.S.M. Wong (1998a)

[11] R. Shankar (1994)

[12] H. Garcilazo *et al.* (2007)

[13] D.M. Manley (2005)

[14] S.S.M. Wong (1998b)

[15] C. Amsler *et al.* (2008): Naming scheme for hadrons

21.8 References

- C. Amsler *et al.* (Particle Data Group) (2008). "Review of Particle Physics". *Physics Letters B* **667** (1): 1–1340. Bibcode:2008PhLB..667....1P. doi:10.1016/j.physletb.2008.07.018.

- H. Garcilazo, J. Vijande, and A. Valcarce (2007). "Faddeev study of heavy-baryon spectroscopy". *Journal of Physics G* **34** (5): 961–976. doi:10.1088/0954-3899/34/5/014.

- K. Carter (2006). "The rise and fall of the pentaquark". Fermilab and SLAC. Retrieved 2008-05-27.

- W.-M. Yao *et al.*(Particle Data Group) (2006). "Review of Particle Physics". *Journal of Physics G* **33**: 1–1232. arXiv:astro-ph/0601168. Bibcode:2006JPhG...33....1Y. doi:10.1088/0954-3899/33/1/001.

• D.M. Manley (2005). "Status of baryon spectroscopy". *Journal of Physics: Conference Series* **5**: 230–237. Bibcode:2005JPhCS...9..230M. doi:10.1088/1742-6596/9/1/043.

• H. Muir (2003). "Pentaquark discovery confounds sceptics". New Scientist. Retrieved 2008-05-27.

• S.S.M. Wong (1998a). "Chapter 2—Nucleon Structure". *Introductory Nuclear Physics* (2nd ed.). New York (NY): John Wiley & Sons. pp. 21–56. ISBN 0-471-23973-9.

• S.S.M. Wong (1998b). "Chapter 3—The Deuteron". *Introductory Nuclear Physics* (2nd ed.). New York (NY): John Wiley & Sons. pp. 57–104. ISBN 0-471-23973-9.

• R. Shankar (1994). *Principles of Quantum Mechanics* (2nd ed.). New York (NY): Plenum Press. ISBN 0-306-44790-8.

• E. Wigner (1937). "On the Consequences of the Symmetry of the Nuclear Hamiltonian on the Spectroscopy of Nuclei". *Physical Review* **51** (2): 106–119. Bibcode:1937PhRv...51..106W. doi:10.1103/PhysRev.51.106.

• M.Gell-Mann(1964). "A Schematic of Baryons and Mesons".*Physics Letters***8**(3): 214–215. Bibcode:1964PhL.. doi:10.1016/S0031-9163(64)92001-3.

• W.Heisenberg(1932). "Über den Bau der Atomkerne I".*Zeitschrift für Physik*(in German)**77**: 1–11. Bibcode1H. doi:10.1007/BF01342433.

• W. Heisenberg (1932). "Über den Bau der Atomkerne II". *Zeitschrift für Physik* (in German) **78** (3–4): 156–164. Bibcode:1932ZPhy...78..156H. doi:10.1007/BF01337585.

• W. Heisenberg (1932). "Über den Bau der Atomkerne III". *Zeitschrift für Physik* (in German) **80** (9–10): 587–596. Bibcode:1933ZPhy...80..587H. doi:10.1007/BF01335696.

21.9 External links

• Particle Data Group—Review of Particle Physics (2008).

• Georgia State University—HyperPhysics

• Baryons made thinkable, an interactive visualisation allowing physical properties to be compared

Chapter 22

Meson

In particle physics, **mesons** (/ˈmiːzɒnz/ or /ˈmɛzɒnz/) are hadronic subatomic particles composed of one quark and one antiquark, bound together by the strong interaction. Because mesons are composed of sub-particles, they have a physical size, with a diameter of roughly one fermi, which is about $^2/_3$ the size of a proton or neutron. All mesons are unstable, with the longest-lived lasting for only a few hundredths of a microsecond. Charged mesons decay (sometimes through intermediate particles) to form electrons and neutrinos. Uncharged mesons may decay to photons.

Mesons are not produced by radioactive decay, but appear in nature only as short-lived products of very high-energy interactions in matter, between particles made of quarks. In cosmic ray interactions, for example, such particles are ordinary protons and neutrons. Mesons are also frequently produced artificially in high-energy particle accelerators that collide protons, anti-protons, or other particles.

In nature, the importance of lighter mesons is that they are the associated quantum-field particles that transmit the nuclear force, in the same way that photons are the particles that transmit the electromagnetic force. The higher energy (more massive) mesons were created momentarily in the Big Bang, but are not thought to play a role in nature today. However, such particles are regularly created in experiments, in order to understand the nature of the heavier types of quark that compose the heavier mesons.

Mesons are part of the hadron particle family, defined simply as particles composed of two quarks. The other members of the hadron family are the baryons: subatomic particles composed of three quarks rather than two. Some experiments show evidence of exotic mesons, which don't have the conventional valence quark content of one quark and one antiquark.

Because quarks have a spin of $^1/_2$, the difference in quark-number between mesons and baryons results in conventional two-quark mesons being bosons, whereas baryons are fermions.

Each type of meson has a corresponding antiparticle (antimeson) in which quarks are replaced by their corresponding antiquarks and vice versa. For example, a positive pion ($\pi+$) is made of one up quark and one down antiquark; and its corresponding antiparticle, the negative pion ($\pi-$), is made of one up antiquark and one down quark.

Because mesons are composed of quarks, they participate in both the weak and strong interactions. Mesons with net electric charge also participate in the electromagnetic interaction. They are classified according to their quark content, total angular momentum, parity and various other properties, such as C-parity and G-parity. Although no meson is stable, those of lower mass are nonetheless more stable than the most massive mesons, and are easier to observe and study in particle accelerators or in cosmic ray experiments. They are also typically less massive than baryons, meaning that they are more easily produced in experiments, and thus exhibit certain higher energy phenomena more readily than baryons composed of the same quarks would. For example, the charm quark was first seen in the J/Psi meson (J/ψ) in 1974,[1][2] and the bottom quark in the upsilon meson (Υ) in 1977.[3]

22.1   History

From theoretical considerations, in 1934 Hideki Yukawa[4][5] predicted the existence and the approximate mass of the "meson" as the carrier of the nuclear force that holds atomic nuclei together. If there were no nuclear force, all nuclei with two or more protons would fly apart because of the electromagnetic repulsion. Yukawa called his carrier particle the meson, from μέσος *mesos*, the Greek word for "intermediate," because its predicted mass was between that of the electron and that of the proton, which has about 1,836 times the mass of the electron. Yukawa had originally named his particle the "mesotron", but he was corrected by the physicist Werner Heisenberg (whose father was a professor of Greek at the University of Munich). Heisenberg pointed out that there is no "tr" in the Greek word "mesos".[6]

The first candidate for Yukawa's meson, now known in modern terminology as the muon, was discovered in 1936 by Carl David Anderson and others in the decay products of cosmic ray interactions. The mu meson had about the right mass to be Yukawa's carrier of the strong nuclear force, but over the course of the next decade, it became evident that it was not the right particle. It was eventually found that the "mu meson" did not participate in the strong nuclear interaction at all, but rather behaved like a heavy version of the electron, and was eventually classed as a lepton like the electron, rather than a meson. Physicists in making this choice decided that properties other than particle mass should control their classification.

There were years of delays in the subatomic particle research during World War II in 1939–45, with most physicists working in applied projects for wartime necessities. When the war ended in August 1945, many physicists gradually returned to peacetime research. The first true meson to be discovered was what would later be called the "pi meson" (or pion). This discovery was made in 1947, by Cecil Powell, César Lattes, and Giuseppe Occhialini, who were investigating cosmic ray products at the University of Bristol in England, based on photographic films placed in the Andes mountains. Some mesons in these films had about the same mass as the already-known meson, yet seemed to decay into it, leading physicist Robert Marshak to hypothesize in 1947 that it was actually a new and different meson. Over the next few years, more experiments showed that the pion was indeed involved in strong interactions. The pion (as a virtual particle) is the primary force carrier for the nuclear force in atomic nuclei. Other mesons, such as the rho mesons are involved in mediating this force as well, but to lesser extents. Following the discovery of the pion, Yukawa was awarded the 1949 Nobel Prize in Physics for his predictions.

The word *meson* has at times been used to mean *any* force carrier, such as the "Z^0 meson", which is involved in mediating the weak interaction.[7] However, this spurious usage has fallen out of favor. Mesons are now defined as particles composed of pairs of quarks and antiquarks.

22.2   Overview

22.2.1   Spin, orbital angular momentum, and total angular momentum

Main articles: Spin (physics), angular momentum operator, Total angular momentum and Quantum numbers

Spin (quantum number S) is a vector quantity that represents the "intrinsic" angular momentum of a particle. It comes in increments of $\frac{1}{2}$ ℏ. The ℏ is often dropped because it is the "fundamental" unit of spin, and it is implied that "spin 1" means "spin 1 ℏ". (In some systems of natural units, ℏ is chosen to be 1, and therefore does not appear in equations).

Quarks are fermions—specifically in this case, particles having spin $\frac{1}{2}$ ($S = \frac{1}{2}$). Because spin projections vary in increments of 1 (that is 1 ℏ), a single quark has a spin vector of length $\frac{1}{2}$, and has two spin projections ($S_z = +\frac{1}{2}$ and $S_z = -\frac{1}{2}$). Two quarks can have their spins aligned, in which case the two spin vectors add to make a vector of length $S = 1$ and three spin projections ($S_z = +1$, $S_z = 0$, and $S_z = -1$), called the spin-1 triplet. If two quarks have unaligned spins, the spin vectors add up to make a vector of length S = 0 and only one spin projection ($S_z = 0$), called the spin-0 singlet. Because mesons are made of one quark and one antiquark, they can be found in triplet and singlet spin states.

There is another quantity of quantized angular momentum, called the orbital angular momentum (quantum number L), that comes in increments of 1 ℏ, which represent the angular momentum due to quarks orbiting around each other. The total angular momentum (quantum number J) of a particle is therefore the combination of intrinsic angular momentum (spin) and orbital angular momentum. It can take any value from $J = |L - S|$ to $J = |L + S|$, in increments of 1.

Particle physicists are most interested in mesons with no orbital angular momentum ($L = 0$), therefore the two groups of mesons most studied are the $S = 1$; $L = 0$ and $S = 0$; $L = 0$, which corresponds to $J = 1$ and $J = 0$, although they are not the only ones. It is also possible to obtain $J = 1$ particles from $S = 0$ and $L = 1$. How to distinguish between the $S = 1$, $L = 0$ and $S = 0$, $L = 1$ mesons is an active area of research in meson spectroscopy.

22.2.2 Parity

Main article: Parity (physics)

If the universe were reflected in a mirror, most of the laws of physics would be identical—things would behave the same way regardless of what we call "left" and what we call "right". This concept of mirror reflection is called parity (P). Gravity, the electromagnetic force, and the strong interaction all behave in the same way regardless of whether or not the universe is reflected in a mirror, and thus are said to conserve parity (P-symmetry). However, the weak interaction does distinguish "left" from "right", a phenomenon called parity violation (P-violation).

Based on this, one might think that, if the wavefunction for each particle (more precisely, the quantum field for each particle type) were simultaneously mirror-reversed, then the new set of wavefunctions would perfectly satisfy the laws of physics (apart from the weak interaction). It turns out that this is not quite true: In order for the equations to be satisfied, the wavefunctions of certain types of particles have to be multiplied by −1, in addition to being mirror-reversed. Such particle types are said to have *negative* or *odd* parity ($P = -1$, or alternatively $P = -$), whereas the other particles are said to have *positive* or *even* parity ($P = +1$, or alternatively $P = +$).

For mesons, the parity is related to the orbital angular momentum by the relation:[8]

$$P = (-1)^{L+1}$$

where the L is a result of the parity of the corresponding spherical harmonic of the wavefunction. The '+1' in the exponent comes from the fact that, according to the Dirac equation, a quark and an antiquark have opposite intrinsic parities. Therefore, the intrinsic parity of a meson is the product of the intrinsic parities of the quark (+1) and antiquark (−1). As these are different, their product is −1, and so it contributes a +1 in the exponent.

As a consequence, mesons with no orbital angular momentum ($L = 0$) all have odd parity ($P = -1$).

22.2.3 C-parity

Main article: C-parity

C-parity is only defined for mesons that are their own antiparticle (i.e. neutral mesons). It represents whether or not the wavefunction of the meson remains the same under the interchange of their quark with their antiquark.[9] If

$$|q\bar{q}\rangle = |\bar{q}q\rangle$$

then, the meson is "C even" (C = +1). On the other hand, if

$$|q\bar{q}\rangle = -|\bar{q}q\rangle$$

then the meson is "C odd" (C = −1).

C-parity rarely is studied on its own, but more commonly in combination with P-parity into CP-parity. CP-parity was thought to be conserved, but was later found to be violated in weak interactions.[10][11][12]

22.2.4 G-parity

Main article: G-parity

G parity is a generalization of the C-parity. Instead of simply comparing the wavefunction after exchanging quarks and antiquarks, it compares the wavefunction after exchanging the meson for the corresponding antimeson, regardless of quark content.[13] In the case of neutral meson, G-parity is equivalent to C-parity because neutral mesons are their own antiparticles.

If

$$|q_1 \bar{q}_2\rangle = |\bar{q}_1 q_2\rangle$$

then, the meson is "G even" (G = +1). On the other hand, if

$$|q_1 \bar{q}_2\rangle = -|\bar{q}_1 q_2\rangle$$

then the meson is "G odd" (G = −1).

22.2.5 Isospin and charge

Main article: Isospin

The concept of isospin was first proposed by Werner Heisenberg in 1932 to explain the similarities between protons and neutrons under the strong interaction.[14] Although they had different electric charges, their masses were so similar that physicists believed that they were actually the same particle. The different electric charges were explained as being the result of some unknown excitation similar to spin. This unknown excitation was later dubbed *isospin* by Eugene Wigner in 1937.[15] When the first mesons were discovered, they too were seen through the eyes of isospin and so the three pions were believed to be the same particle, but in different isospin states.

This belief lasted until Murray Gell-Mann proposed the quark model in 1964 (containing originally only the u, d, and s quarks).[16] The success of the isospin model is now understood to be the result of the similar masses of the u and d quarks. Because the u and d quarks have similar masses, particles made of the same number of them also have similar masses. The exact specific u and d quark composition determines the charge, because u quarks carry charge $+\frac{2}{3}$ whereas d quarks carry charge $-\frac{1}{3}$. For example the three pions all have different charges (π+ (ud), π0 (a quantum superposition of uu and dd states), π− (du)), but have similar masses (~140 MeV/c^2) as they are each made of a same number of total of up and down quarks and antiquarks. Under the isospin model, they were considered to be a single particle in different charged states.

The mathematics of isospin was modeled after that of spin. Isospin projections varied in increments of 1 just like those of spin, and to each projection was associated a "charged state". Because the "pion particle" had three "charged states", it was said to be of isospin I = 1. Its "charged states" π+, π0, and π−, corresponded to the isospin projections $I_3 = +1$, $I_3 = 0$, and $I_3 = -1$ respectively. Another example is the "rho particle", also with three charged states. Its "charged states" ρ+, ρ0, and ρ−, corresponded to the isospin projections $I_3 = +1$, $I_3 = 0$, and $I_3 = -1$ respectively. It was later noted that the isospin projections were related to the up and down quark content of particles by the relation

$$I_3 = \frac{1}{2}[(n_u - n_{\bar{u}}) - (n_d - n_{\bar{d}})],$$

where the n's are the number of up and down quarks and antiquarks.

In the "isospin picture", the three pions and three rhos were thought to be the different states of two particles. However, in the quark model, the rhos are excited states of pions. Isospin, although conveying an inaccurate picture of things, is still used to classify hadrons, leading to unnatural and often confusing nomenclature. Because mesons are hadrons, the isospin classification is also used, with $I_3 = +\frac{1}{2}$ for up quarks and down antiquarks, and $I_3 = -\frac{1}{2}$ for up antiquarks and down quarks.

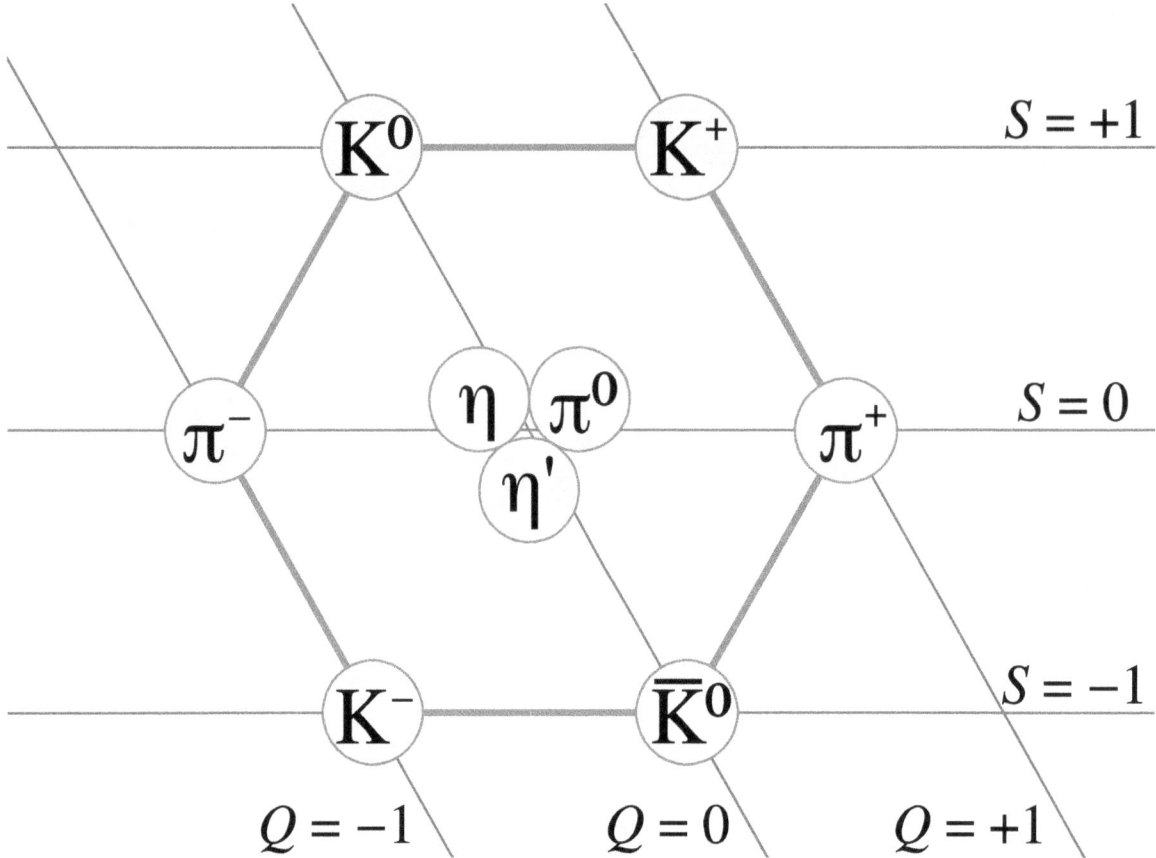

Combinations of one u, d or s quarks and one u, d, or s antiquark in $J^P = 0^-$ configuration form a nonet.

22.2.6 Flavour quantum numbers

Main article: Flavour (particle physics) § Flavour quantum numbers

The strangeness quantum number S (not to be confused with spin) was noticed to go up and down along with particle mass. The higher the mass, the lower the strangeness (the more s quarks). Particles could be described with isospin projections (related to charge) and strangeness (mass) (see the uds nonet figures). As other quarks were discovered, new quantum numbers were made to have similar description of udc and udb nonets. Because only the u and d mass are similar, this description of particle mass and charge in terms of isospin and flavour quantum numbers only works well for the nonets made of one u, one d and one other quark and breaks down for the other nonets (for example ucb nonet). If the quarks all had the same mass, their behaviour would be called *symmetric*, because they would all behave in exactly the same way with respect to the strong interaction. However, as quarks do not have the same mass, they do not interact in the same way (exactly like an electron placed in an electric field will accelerate more than a proton placed in the same field because of its lighter mass), and the symmetry is said to be broken.

It was noted that charge (Q) was related to the isospin projection (I_3), the baryon number (B) and flavour quantum numbers (S, C, B', T) by the Gell-Mann–Nishijima formula:[17]

$$Q = I_3 + \frac{1}{2}(B + S + C + B' + T),$$

where S, C, B', and T represent the strangeness, charm, bottomness and topness flavour quantum numbers respectively. They are related to the number of strange, charm, bottom, and top quarks and antiquark according to the relations:

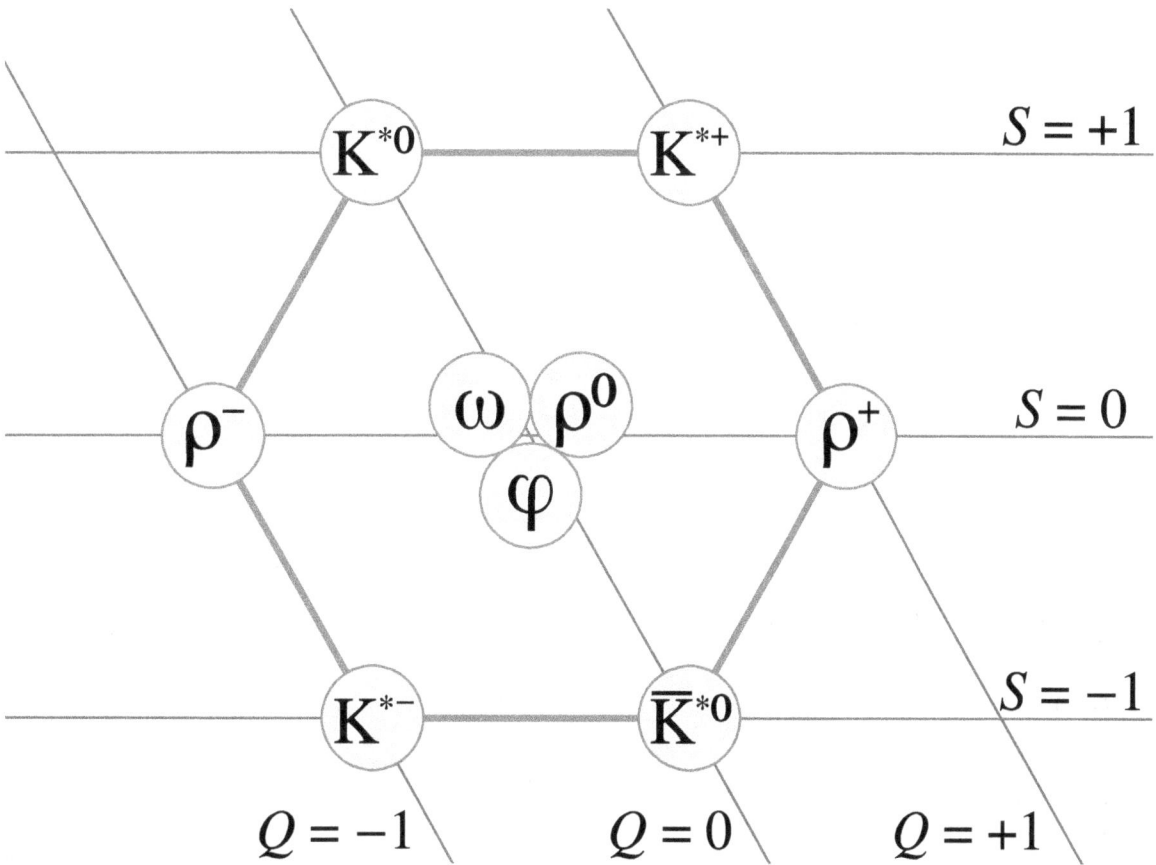

Combinations of one u, d or s quarks and one u, d, or s antiquark in $J^P = 1^-$ configuration also form a nonet.

$$S = -(n_s - n_{\bar{s}})$$
$$C = +(n_c - n_{\bar{c}})$$
$$B' = -(n_b - n_{\bar{b}})$$
$$T = +(n_t - n_{\bar{t}}),$$

meaning that the Gell-Mann–Nishijima formula is equivalent to the expression of charge in terms of quark content:

$$Q = \frac{2}{3}[(n_u - n_{\bar{u}}) + (n_c - n_{\bar{c}}) + (n_t - n_{\bar{t}})] - \frac{1}{3}[(n_d - n_{\bar{d}}) + (n_s - n_{\bar{s}}) + (n_b - n_{\bar{b}})].$$

22.3   Classification

Mesons are classified into groups according to their isospin (I), total angular momentum (J), parity (P), G-parity (G) or C-parity (C) when applicable, and quark (q) content. The rules for classification are defined by the Particle Data Group, and are rather convoluted.[18] The rules are presented below, in table form for simplicity.

22.3.1   Types of meson

Mesons are classified into types according to their spin configurations. Some specific configurations are given special names based on the mathematical properties of their spin configuration.

22.3.2 Nomenclature

Flavourless mesons

Flavourless mesons are mesons made of pair of quark and antiquarks of the same flavour (all their flavour quantum numbers are zero: $S = 0$, $C = 0$, $B' = 0$, $T = 0$).[20] The rules for flavourless mesons are:[18]

> [†] ^ The C parity is only relevant to neutral mesons.
> [††] ^ For $J^{PC}=1^{--}$, the ψ is called the J/ψ

In addition:

- When the spectroscopic state of the meson is known, it is added in parentheses.

- When the spectroscopic state is unknown, mass (in MeV/c^2) is added in parentheses.

- When the meson is in its ground state, nothing is added in parentheses.

Flavoured mesons

Flavoured mesons are mesons made of pair of quark and antiquarks of different flavours. The rules are simpler in this case: the main symbol depends on the heavier quark, the superscript depends on the charge, and the subscript (if any) depends on the lighter quark. In table form, they are:[18]

In addition:

- If J^P is in the "normal series" (i.e., $J^P = 0^+$, 1^-, 2^+, 3^-, ...), a superscript $*$ is added.

- If the meson is not pseudoscalar ($J^P = 0^-$) or vector ($J^P = 1^-$), J is added as a subscript.

- When the spectroscopic state of the meson is known, it is added in parentheses.

- When the spectroscopic state is unknown, mass (in MeV/c^2) is added in parentheses.

- When the meson is in its ground state, nothing is added in parentheses.

22.4 Exotic mesons

Main article: Exotic meson

There is experimental evidence for particles that are hadrons (i.e., are composed of quarks) and are color-neutral with zero baryon number, and thus by conventional definition are mesons. Yet, these particles do not consist of a single quark-antiquark pair, as all the other conventional mesons discussed above do. A tentative category for these particles is exotic mesons.

There are at least five exotic meson resonances that have been experimentally confirmed to exist by two or more independent experiments. The most statistically significant of these is the Z(4430), discovered by the Belle experiment in 2007 and confirmed by LHCb in 2014. It is a candidate for being a tetraquark: a particle composed of two quarks and two antiquarks.[21] See the main article above for other particle resonances that are candidates for being exotic mesons.

22.5   List

Main article: List of mesons

22.6   See also

- Standard Model

22.7   Notes

[1] J.J. Aubert *et al.* (1974)

[2] J.E. Augustin *et al.* (1974)

[3] S.W. Herb *et al.* (1977)

[4] The Noble Foundation (1949) Nobel Prize in Physics 1949 – Presentation Speech

[5] H. Yukawa (1935)

[6] G. Gamow (1961)

[7] J. Steinberger (1998)

[8] C. Amsler *et al.* (2008): Quark Model

[9] M.S. Sozzi (2008b)

[10] J.W. Cronin (1980)

[11] V.L. Fitch (1980)

[12] M.S. Sozzi (2008c)

[13] K. Gottfried, V.F. Weisskopf (1986)

[14] W. Heisenberg (1932)

[15] E. Wigner (1937)

[16] M. Gell-Mann (1964)

[17] S.S.M Wong (1998)

[18] C. Amsler *et al.* (2008): Naming scheme for hadrons

[19] W.E. Burcham, M. Jobes (1995)

[20] For the purpose of nomenclature, the isospin projection I_3 isn't considered a flavour quantum number. This means that the charged pion-like mesons ($\pi^\pm$, $a^\pm$, $b^\pm$, and $\rho^\pm$ mesons) follow the rules of flavourless mesons, even if they aren't truly "flavourless".

[21] LHCb collaborators (2014): Observation of the resonant character of the Z(4430)– state

22.8 References

- M.S. Sozzi (2008a). "Parity". *Discrete Symmetries and CP Violation: From Experiment to Theory*. Oxford University Press. pp. 15–87. ISBN 0-19-929666-9.

- M.S. Sozzi (2008b). "Charge Conjugation". *Discrete Symmetries and CP Violation: From Experiment to Theory*. Oxford University Press. pp. 88–120. ISBN 0-19-929666-9.

- M.S. Sozzi (2008c). "CP-Symmetry". *Discrete Symmetries and CP Violation: From Experiment to Theory*. Oxford University Press. pp. 231–275. ISBN 0-19-929666-9.

- C. Amsler *et al.* (Particle Data Group) (2008). "Review of Particle Physics". *Physics Letters B* **667** (1): 1–1340. Bibcode:2008PhLB..667....1P. doi:10.1016/j.physletb.2008.07.018.

- S.S.M. Wong (1998). "Nucleon Structure". *Introductory Nuclear Physics* (2nd ed.). New York (NY): John Wiley & Sons. pp. 21–56. ISBN 0-471-23973-9.

- W.E. Burcham, M. Jobes (1995). *Nuclear and Particle Physics* (2nd ed.). Longman Publishing. ISBN 0-582-45088-8.

- R. Shankar (1994). *Principles of Quantum Mechanics* (2nd ed.). New York (NY): Plenum Press. ISBN 0-306-44790-8.

- J. Steinberger (1989). "Experiments with high-energy neutrino beams". *Reviews of Modern Physics* **61** (3): 533–545. Bibcode:1989RvMP...61..533S. doi:10.1103/RevModPhys.61.533.

- K. Gottfried, V.F. Weisskopf (1986). "Hadronic Spectroscopy: G-parity". *Concepts of Particle Physics* **2**. Oxford University Press. pp. 303–311. ISBN 0-19-503393-0.

- J.W. Cronin (1980). "CP Symmetry Violation—The Search for its origin" (PDF). The Nobel Foundation.

- V.L. Fitch (1980). "The Discovery of Charge—Conjugation Parity Asymmetry" (PDF). The Nobel Foundation.

- S.W. Herb; Hom, D.; Lederman, L.; Sens, J.; Snyder, H.; Yoh, J.; Appel, J.; Brown, B.; et al. (1977). "Observation of a Dimuon Resonance at 9.5 Gev in 400-GeV Proton-Nucleus Collisions". *Physical Review Letters* **39** (5): 252–255. Bibcode:1977PhRvL..39..252H. doi:10.1103/PhysRevLett.39.252.

- J.J. Aubert; Becker, U.; Biggs, P.; Burger, J.; Chen, M.; Everhart, G.; Goldhagen, P.; Leong, J.; et al. (1974). "Experimental Observation of a Heavy Particle *J*". *Physical Review Letters* **33** (23): 1404–1406. Bibcode:1974 .doi:10.1103/PhysRevLett.33.1404.

- J.E. Augustin; Boyarski, A.; Breidenbach, M.; Bulos, F.; Dakin, J.; Feldman, G.; Fischer, G.; Fryberger, D.; et al. (1974). "Discovery of a Narrow Resonance in e$^+$e$^-$ Annihilation". *Physical Review Letters* **33** (23): 1406–1408. Bibcode:1974PhRvL..33.1406A. doi:10.1103/PhysRevLett.33.1406.

- M.Gell-Mann(1964). "A Schematic of Baryons and Mesons".*Physics Letters***8**(3): 214–215. Bibcode:1964PhL. doi:10.1016/S0031-9163(64)92001-3.

- Ishfaq Ahmad (1965). "the Interactions of 200 MeV $\pi\pm$ -Mesons with Complex Nuclei Proposal to Study the Interactions of 200 MeV $\pi\pm$ -Mesons with Complex Nuclei" (PDF). *CERN documents* **3** (5).

- G. Gamow (1988) [1961]. *The Great Physicists from Galileo to Einstein* (Reprint ed.). Dover Publications. p. 315. ISBN 978-0-486-25767-9.

- E. Wigner (1937). "On the Consequences of the Symmetry of the Nuclear Hamiltonian on the Spectroscopy of Nuclei". *Physical Review* **51** (2): 106–119. Bibcode:1937PhRv...51..106W. doi:10.1103/PhysRev.51.106.

- H. Yukawa (1935). "On the Interaction of Elementary Particles" (PDF). *Proc. Phys. Math. Soc. Jap.* **17** (48).

- W.Heisenberg(1932). "Über den Bau der Atomkerne I".*Zeitschrift für Physik*(in German)**77**: 1–11. Bibcode:1H. doi:10.1007/BF01342433.

- W. Heisenberg (1932). "Über den Bau der Atomkerne II". *Zeitschrift für Physik* (in German) **78** (3–4): 156–164. Bibcode:1932ZPhy...78..156H. doi:10.1007/BF01337585.

- W. Heisenberg (1932). "Über den Bau der Atomkerne III". *Zeitschrift für Physik* (in German) **80** (9–10): 587–596. Bibcode:1933ZPhy...80..587H. doi:10.1007/BF01335696.

22.9   External links

- A table of some mesons and their properties

- *Particle Data Group*—Compiles authoritative information on particle properties

- hep-ph/0211411: The light scalar mesons within quark models

- Naming scheme for hadrons (a PDF file)

- Mesons made thinkable, an interactive visualisation allowing physical properties to be compared

22.9.1   Recent findings

- What Happened to the Antimatter? Fermilab's DZero Experiment Finds Clues in Quick-Change Meson

- CDF experiment's definitive observation of matter-antimatter oscillations in the Bs meson

Chapter 23

Exotic meson

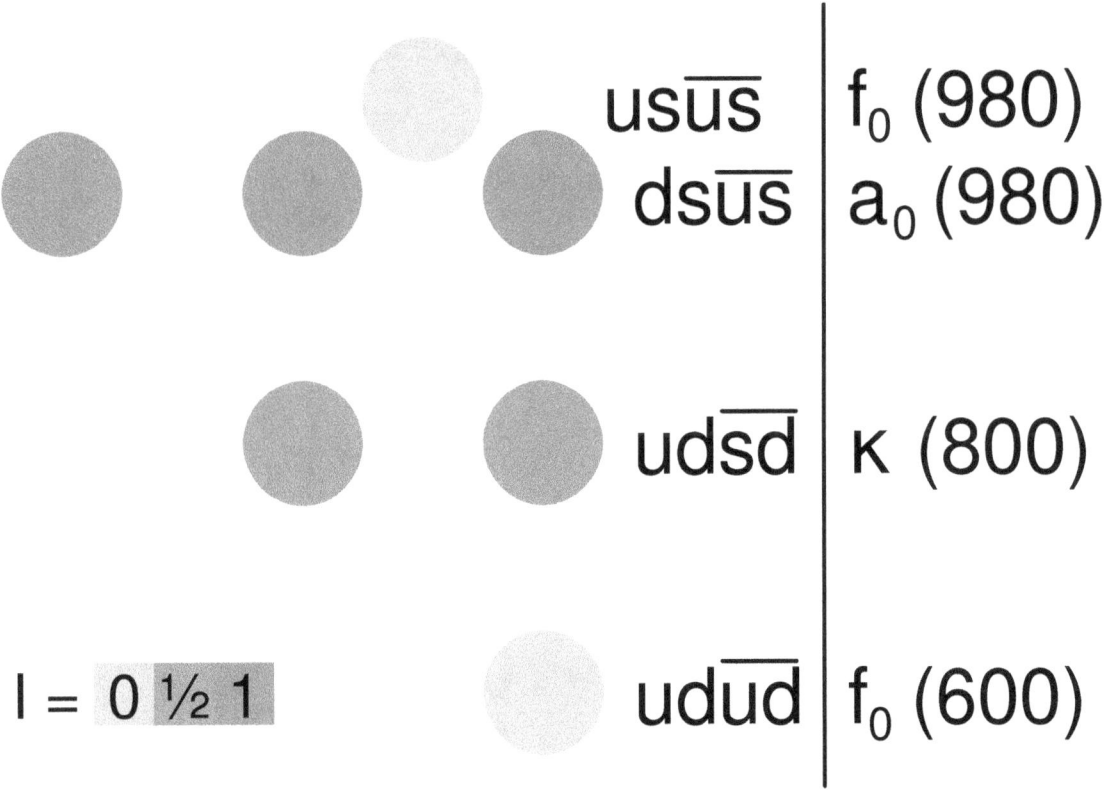

Identities and classification of possible tetraquark mesons. Green denotes I = 0 *states, blue,* I = 1/2 *and red,* I = 1. *The vertical axis is the mass.*

Non-quark model mesons include

1. **exotic mesons**, which have quantum numbers not possible for mesons in the quark model;

2. **glueballs** or **gluonium**, which have no valence quarks at all;

3. **tetraquarks**, which have two valence quark-antiquark pairs; and

4. **hybrid mesons**, which contain a valence quark-antiquark pair and one or more gluons.

All of these can be classed as mesons, because they are hadrons and carry zero baryon number. Of these, glueballs must be flavor singlets; that is, have zero isospin, strangeness, charm, bottomness, and topness. Like all particle states, they are specified by the quantum numbers which label representations of the Poincaré symmetry, q.e., J^{PC} (where J is the angular momentum, P is the intrinsic parity, and C is the charge conjugation parity) and by the mass. One also specifies the isospin I of the meson.

Typically, every quark model meson comes in SU(3) flavor nonet: an octet and a flavor singlet. A glueball shows up as an extra (*supernumerary*) particle outside the nonet. In spite of such seemingly simple counting, the assignment of any given state as a glueball, tetraquark, or hybrid remains tentative even today. Even when there is agreement that one of several states is one of these non-quark model mesons, the degree of mixing, and the precise assignment is fraught with uncertainties. There is also the considerable experimental labor of assigning quantum numbers to each state and cross-checking them in other experiments. As a result, all assignments outside the quark model are tentative. The remainder of this article outlines the situation as it stood at the end of 2004.

23.1   Lattice predictions

Lattice QCD predictions for glueballs are now fairly stable, at least when virtual quarks are neglected. The two lowest states are

0^{++} with mass of 1611 ± 163 MeV/c^2 and

2^{++} with mass of 2232 ± 310 MeV/c^2

The 0^{-+} and exotic glueballs such as 0^{--} are all expected to lie above 2 GeV/c^2. Glueballs are necessarily isoscalar, with isospin $I = 0$.

The ground state *hybrid mesons* 0^{-+}, 1^{-+}, 1^{--}, and 2^{-+} all lie a little below 2 GeV/c^2. The hybrid with exotic quantum numbers 1^{-+} is at 1.9 ± 0.2 GeV/c^2. The best lattice computations to date are made in the quenched approximation, which neglects virtual quarks loops. As a result, these computations miss mixing with meson states.

23.2   The 0^{++} states

The data show five isoscalar resonances: $f_0(500)$, $f_0(980)$, $f_0(1370)$, $f_0(1500)$, and $f_0(1710)$. Of these the $f_0(500)$ is usually identified with the σ of chiral models. The decays and production of $f_0(1710)$ give strong evidence that it is also a meson.

23.2.1   Glueball candidate

The $f_0(1370)$ and $f_0(1500)$ cannot both be a quark model meson, because one is supernumerary. The production of the higher mass state in two photon reactions such as $2\gamma \rightarrow 2\pi$ or $2\gamma \rightarrow 2K$ reactions is highly suppressed. The decays also give some evidence that one of these could be a glueball.

23.2.2   Tetraquark candidate

The $f_0(980)$ has been identified by some authors as a tetraquark meson, along with the $I = 1$ states $a_0(980)$ and $\kappa_0(800)$. Two long-lived (*narrow* in the jargon of particle spectroscopy) states: the scalar (0^{++}) state D*$\pm$
sJ(2317) and the vector (1^+) meson D*$\pm$
sJ(2460), observed at CLEO and BaBar, have also been tentatively identified as tetraquark states. However, for these, other explanations are possible.

23.3 The 2^{++} states

Two isoscalar states are definitely identified—f$_2$(1270) and the f$'_2$(1525). No other states have been consistently identified by all experiments. Hence it is difficult to say more about these states.

23.4 The 1^{-+} exotics and other states

The two isovector exotics π_1(1400) and π_1(1600) seem to be well established experimentally. They are clearly not glueballs, but could be either a tetraquark or a hybrid. The evidence for such assignments is weak.

The π(1800) (0^{-+}), ρ(1900) (1^{--}) and the η_2(1870) (2^{-+}) are fairly well identified states, which have been tentatively identified as hybrids by some authors. If this identification is correct, then it is a remarkable agreement with lattice computations, which place several hybrids in this range of masses.

23.5 See also

- Quark model, mesons, baryons, quarks, and gluons

- Exotic hadrons and exotic baryons

- Quantum chromodynamics, flavor, and the QCD vacuum

- GlueX, an experiment which will explore the spectrum of glueballs and exotic mesons

23.6 References and external links

- W.-M. Yao *et al.* (Particle Data Group) (2006). "Review of Particle Physics: Non-qq mesons" (PDF). *Journal of Physics G* **33**: 1. arXiv:astro-ph/0601168. Bibcode:2006JPhG...33....1Y. doi:10.1088/0954-3899/33/1/001.

Chapter 24

Proton

This article is about the proton as a subatomic particle. For other uses, see Proton (disambiguation).

The **proton** is a subatomic particle, symbol p or p+, with a positive electric charge of +1e elementary charge and mass slightly less than that of a neutron. Protons and neutrons, each with mass approximately one atomic mass unit, are collectively referred to as "nucleons". One or more protons are present in the nucleus of an atom. The number of protons in the nucleus is referred to as its atomic number. Since each element has a unique number of protons, each element has its own unique atomic number. The word *proton* is Greek for "first", and this name was given to the hydrogen nucleus by Ernest Rutherford in 1920. In previous years Rutherford had discovered that the hydrogen nucleus (known to be the lightest nucleus) could be extracted from the nuclei of nitrogen by collision. The proton was therefore a candidate to be a fundamental particle and a building block of nitrogen and all other heavier atomic nuclei.

In the modern Standard Model of particle physics, the proton is a hadron, and like the neutron, the other nucleon (particle present in atomic nuclei), is composed of three quarks. Although the proton was originally considered a fundamental or elementary particle, it is now known to be composed of three valence quarks: two up quarks and one down quark. The rest masses of the quarks contribute only about 1% of the proton's mass, however.[2] The remainder of the proton mass is due to the kinetic energy of the quarks and to the energy of the gluon fields that bind the quarks together. Because the proton is not a fundamental particle, it possesses a physical size; the radius of the proton is about 0.84–0.87 fm.[3]

At sufficiently low temperatures, free protons will bind to electrons. However, the character of such bound protons does not change, and they remain protons. A fast proton moving through matter will slow by interactions with electrons and nuclei, until it is captured by the electron cloud of an atom. The result is a protonated atom, which is a chemical compound of hydrogen. In vacuum, when free electrons are present, a sufficiently slow proton may pick up a single free electron, becoming a neutral hydrogen atom, which is chemically a free radical. Such "free hydrogen atoms" tend to react chemically with many other types of atoms at sufficiently low energies. When free hydrogen atoms react with each other, they form neutral hydrogen molecules (H_2), which are the most common molecular component of molecular clouds in interstellar space. Such molecules of hydrogen on Earth may then serve (among many other uses) as a convenient source of protons for accelerators (as used in proton therapy) and other hadron particle physics experiments that require protons to accelerate, with the most powerful and noted example being the Large Hadron Collider.

24.1 Description

Protons are spin-½ fermions and are composed of three valence quarks,[4] making them baryons (a sub-type of hadrons). The two up quarks and one down quark of the proton are held together by the strong force, mediated by gluons.[5]:21–22 A modern perspective has the proton composed of the valence quarks (up, up, down), the gluons, and transitory pairs of sea quarks. The proton has an approximately exponentially decaying positive charge distribution with a mean square radius of about 0.8 fm.[6]

Protons and neutrons are both nucleons, which may be bound together by the nuclear force to form atomic nuclei. The nucleus of the most common isotope of the hydrogen atom (with the chemical symbol "H") is a lone proton. The nuclei of the heavy hydrogen isotopes deuterium and tritium contain one proton bound to one and two neutrons, respectively. All other types of atomic nuclei are composed of two or more protons and various numbers of neutrons.

24.2 History

The concept of a hydrogen-like particle as a constituent of other atoms was developed over a long period. As early as 1815, William Prout proposed that all atoms are composed of hydrogen atoms (which he called "protyles"), based on a simplistic interpretation of early values of atomic weights (see Prout's hypothesis), which was disproved when more accurate values were measured.[7]:39–42

In 1886, Eugen Goldstein discovered canal rays (also known as anode rays) and showed that they were positively charged particles (ions) produced from gases. However, since particles from different gases had different values of charge-to-mass ratio (e/m), they could not be identified with a single particle, unlike the negative electrons discovered by J. J. Thomson.

Following the discovery of the atomic nucleus by Ernest Rutherford in 1911, Antonius van den Broek proposed that the place of each element in the periodic table (its atomic number) is equal to its nuclear charge. This was confirmed experimentally by Henry Moseley in 1913 using X-ray spectra.

In 1917 (in experiments reported in 1919), Rutherford proved that the hydrogen nucleus is present in other nuclei, a result usually described as the discovery of the proton.[8] Rutherford had earlier learned to produce hydrogen nuclei as a type of radiation produced as a product of the impact of alpha particles on nitrogen gas, and recognize them by their unique penetration signature in air and their appearance in scintillation detectors. These experiments were begun when Rutherford had noticed that, when alpha particles were shot into air (mostly nitrogen), his scintillation detectors showed the signatures of typical hydrogen nuclei as a product. After experimentation Rutherford traced the reaction to the nitrogen in air, and found that when alphas were produced into pure nitrogen gas, the effect was larger. Rutherford determined that this hydrogen could have come only from the nitrogen, and therefore nitrogen must contain hydrogen nuclei. One hydrogen nucleus was being knocked off by the impact of the alpha particle, producing oxygen-17 in the process. This was the first reported nuclear reaction, $^{14}N + \alpha \rightarrow {}^{17}O + p$. (This reaction would later be observed happening directly in a cloud chamber in 1925).

Rutherford knew hydrogen to be the simplest and lightest element and was influenced by Prout's hypothesis that hydrogen was the building block of all elements. Discovery that the hydrogen nucleus is present in all other nuclei as an elementary particle, led Rutherford to give the hydrogen nucleus a special name as a particle, since he suspected that hydrogen, the lightest element, contained only one of these particles. He named this new fundamental building block of the nucleus the *proton*, after the neuter singular of the Greek word for "first", πρῶτον. However, Rutherford also had in mind the word *protyle* as used by Prout. Rutherford spoke at the British Association for the Advancement of Science at its Cardiff meeting beginning 24 August 1920.[9] Rutherford was asked by Oliver Lodge for a new name for the positive hydrogen nucleus to avoid confusion with the neutral hydrogen atom. He initially suggested both *proton* and *prouton* (after Prout).[10] Rutherford later reported that the meeting had accepted his suggestion that the hydrogen nucleus be named the "proton", following Prout's word "protyle".[11] The first use of the word "proton" in the scientific literature appeared in 1920.[12]

24.3 Stability

Main article: Proton decay

The free proton (a proton not bound to nucleons or electrons) is a stable particle that has not been observed to break down spontaneously to other particles. Free protons are found naturally in a number of situations in which energies or temperatures are high enough to separate them from electrons, for which they have some affinity. Free protons exist in plasmas in which temperatures are too high to allow them to combine with electrons. Free protons of high energy and velocity make up 90% of cosmic rays, which propagate in vacuum for interstellar distances. Free protons are emitted directly from atomic nuclei in some rare types of radioactive decay. Protons also result (along with electrons and antineutrinos) from the radioactive decay of free neutrons, which are unstable.

The spontaneous decay of free protons has never been observed, and the proton is therefore considered a stable particle. However, some grand unified theories of particle physics predict that proton decay should take place with lifetimes of the order of 10^{36} years, and experimental searches have established lower bounds on the mean lifetime of the proton for various assumed decay products.[13][14][15]

Experiments at the Super-Kamiokande detector in Japan gave lower limits for proton mean lifetime of 6.6×10^{33} years for decay to an antimuon and a neutral pion, and 8.2×10^{33} years for decay to a positron and a neutral pion.[16] Another experiment at the Sudbury Neutrino Observatory in Canada searched for gamma rays resulting from residual nuclei resulting from the decay of a proton from oxygen-16. This experiment was designed to detect decay to any product, and established a lower limit to the proton lifetime of 2.1×10^{29} years.[17]

However, protons are known to transform into neutrons through the process of electron capture (also called inverse beta decay). For free protons, this process does not occur spontaneously but only when energy is supplied. The equation is:

$$p+ + e- \rightarrow n + \nu$$
$$e$$

The process is reversible; neutrons can convert back to protons through beta decay, a common form of radioactive decay. In fact, a free neutron decays this way, with a mean lifetime of about 15 minutes.

24.4   Quarks and the mass of the proton

In quantum chromodynamics, the modern theory of the nuclear force, most of the mass of the proton and the neutron is explained by special relativity. The mass of the proton is about 80–100 times greater than the sum of the rest masses of the quarks that make it up, while the gluons have zero rest mass. The extra energy of the quarks and gluons in a region within a proton, as compared to the rest energy of the quarks alone in the QCD vacuum, accounts for almost 99% of the mass. The rest mass of the proton is, thus, the invariant mass of the system of moving quarks and gluons that make up the particle, and, in such systems, even the energy of massless particles is still measured as part of the rest mass of the system.

Two terms are used in referring to the mass of the quarks that make up protons: *current quark mass* refers to the mass of a quark by itself, while *constituent quark mass* refers to the current quark mass plus the mass of the gluon particle field surrounding the quark.[18]:285–286 [19]:150–151 These masses typically have very different values. As noted, most of a proton's mass comes from the gluons that bind the current quarks together, rather than from the quarks themselves. While gluons are inherently massless, they possess energy—to be more specific, quantum chromodynamics binding energy (QCBE)— and it is this that contributes so greatly to the overall mass of the proton (see mass in special relativity). A proton has a mass of approximately $938 \, \text{MeV/c}^2$, of which the rest mass of its three valence quarks contributes only about $9.4 \, \text{MeV/c}^2$; much of the remainder can be attributed to the gluons' QCBE.[20][21][22]

The internal dynamics of the proton are complicated, because they are determined by the quarks' exchanging gluons, and interacting with various vacuum condensates. Lattice QCD provides a way of calculating the mass of the proton directly from the theory to any accuracy, in principle. The most recent calculations[23][24] claim that the mass is determined to better than 4% accuracy, even to 1% accuracy (see Figure S5 in Dürr *et al.*[24]). These claims are still controversial, because the calculations cannot yet be done with quarks as light as they are in the real world. This means that the predictions are found by a process of extrapolation, which can introduce systematic errors.[25] It is hard to tell whether these errors are controlled properly, because the quantities that are compared to experiment are the masses of the hadrons, which are known in advance.

These recent calculations are performed by massive supercomputers, and, as noted by Boffi and Pasquini: "a detailed description of the nucleon structure is still missing because ... long-distance behavior requires a nonperturbative and/or numerical treatment..."[26] More conceptual approaches to the structure of the proton are: the topological soliton approach originally due to Tony Skyrme and the more accurate AdS/QCD approach that extends it to include a string theory of gluons,[27] various QCD-inspired models like the bag model and the constituent quark model, which were popular in the 1980s, and the SVZ sum rules, which allow for rough approximate mass calculations.[28] These methods do not have the same accuracy as the more brute-force lattice QCD methods, at least not yet.

24.5 Charge radius

Main article: Charge radius

The internationally accepted value of the proton's charge radius is 0.8768 fm (see orders of magnitude for comparison to other sizes). This value is based on measurements involving a proton and an electron.

However, since 5 July 2010, an international research team has been able to make measurements involving an exotic atom made of a proton and a negatively charged muon. After a long and careful analysis of those measurements, the team concluded that the root-mean-square charge radius of a proton is "0.84184(67) fm, which differs by 5.0 standard deviations from the CODATA value of 0.8768(69) fm".[29] In January 2013, an updated value for the charge radius of a proton—0.84087(39) fm—was published. The precision was improved by 1.7 times, but the difference with CODATA value persisted at 7σ significance.[30]

The international research team that obtained this result at the Paul Scherrer Institut (PSI) in Villigen (Switzerland) includes scientists from the Max Planck Institute of Quantum Optics (MPQ) in Garching, the Ludwig-Maximilians-Universität (LMU) Munich and the Institut für Strahlwerkzeuge (IFWS) of the Universität Stuttgart (both from Germany), and the University of Coimbra, Portugal.[31][32] They are now attempting to explain the discrepancy, and re-examining the results of both previous high-precision measurements and complicated calculations. If no errors are found in the measurements or calculations, it could be necessary to re-examine the world's most precise and best-tested fundamental theory: quantum electrodynamics.[31] The proton radius remains a puzzle as of early 2015.[33]

24.6 Interaction of free protons with ordinary matter

Main article: Proton therapy

Although protons have affinity for oppositely charged electrons, free protons must lose sufficient velocity (and kinetic energy) in order to become closely associated and bound to electrons, since this is a relatively low-energy interaction. High energy protons, in traversing ordinary matter, lose energy by collisions with atomic nuclei, and by ionization of atoms (removing electrons) until they are slowed sufficiently to be captured by the electron cloud in a normal atom.

However, in such an association with an electron, the character of the bound proton is not changed, and it remains a proton. The attraction of low-energy free protons to any electrons present in normal matter (such as the electrons in normal atoms) causes free protons to stop and to form a new chemical bond with an atom. Such a bond happens at any sufficiently "cold" temperature (i.e., comparable to temperatures at the surface of the Sun) and with any type of atom. Thus, in interaction with any type of normal (non-plasma) matter, low-velocity free protons are attracted to electrons in any atom or molecule with which they come in contact, causing the proton and molecule to combine. Such molecules are then said to be "protonated", and chemically they often, as a result, become so-called Bronsted acids.

24.7 Proton in chemistry

24.7.1 Atomic number

In chemistry, the number of protons in the nucleus of an atom is known as the atomic number, which determines the chemical element to which the atom belongs. For example, the atomic number of chlorine is 17; this means that each chlorine atom has 17 protons and that all atoms with 17 protons are chlorine atoms. The chemical properties of each atom are determined by the number of (negatively charged) electrons, which for neutral atoms is equal to the number of (positive) protons so that the total charge is zero. For example, a neutral chlorine atom has 17 protons and 17 electrons, whereas a Cl⁻ anion has 17 protons and 18 electrons for a total charge of −1.

All atoms of a given element are not necessarily identical, however, as the number of neutrons may vary to form different isotopes, and energy levels may differ forming different nuclear isomers. For example, there are two stable isotopes of

chlorine: 35
17Cl with 35 − 17 = 18 neutrons and 37
17Cl with 37 − 17 = 20 neutrons.

24.7.2   Hydrogen ion

See also: Hydron (chemistry)
 In chemistry, the term proton refers to the hydrogen ion, H+
. Since the atomic number of hydrogen is 1, a hydrogen ion has no electrons and corresponds to a bare nucleus, consisting of a proton (and 0 neutrons for the most abundant isotope *protium* 1
1H). The proton is a "bare charge" with only about **1/64,000** of the radius of a hydrogen atom, and so is extremely reactive chemically. The free proton, thus, has an extremely short lifetime in chemical systems such as liquids and it reacts immediately with the electron cloud of any available molecule. In aqueous solution, it forms the hydronium ion, H_3O^+, which in turn is further solvated by water molecules in clusters such as $[H_5O_2]^+$ and $[H_9O_4]^+$.[34]

The transfer of H+
in an acid–base reaction is usually referred to as "proton transfer". The acid is referred to as a proton donor and the base as a proton acceptor. Likewise, biochemical terms such as proton pump and proton channel refer to the movement of hydrated H+
ions.

The ion produced by removing the electron from a deuterium atom is known as a deuteron, not a proton. Likewise, removing an electron from a tritium atom produces a triton.

24.7.3   Proton nuclear magnetic resonance (NMR)

Also in chemistry, the term "proton NMR" refers to the observation of hydrogen-1 nuclei in (mostly organic) molecules by nuclear magnetic resonance. This method uses the spin of the proton, which has the value one-half. The name refers to examination of protons as they occur in protium (hydrogen-1 atoms) in compounds, and does not imply that free protons exist in the compound being studied.

24.8   Human exposure

Main article: Effect of spaceflight on the human body

The Apollo Lunar Surface Experiments Packages (ALSEP) determined that more than 95% of the particles in the solar wind are electrons and protons, in approximately equal numbers.[35][36]

> Because the Solar Wind Spectrometer made continuous measurements, it was possible to measure how the Earth's magnetic field affects arriving solar wind particles. For about two-thirds of each orbit, the Moon is outside of the Earth's magnetic field. At these times, a typical proton density was 10 to 20 per cubic centimeter, with most protons having velocities between 400 and 650 kilometers per second. For about five days of each month, the Moon is inside the Earth's geomagnetic tail, and typically no solar wind particles were detectable. For the remainder of each lunar orbit, the Moon is in a transitional region known as the magnetosheath, where the Earth's magnetic field affects the solar wind but does not completely exclude it. In this region, the particle flux is reduced, with typical proton velocities of 250 to 450 kilometers per second. During the lunar night, the spectrometer was shielded from the solar wind by the Moon and no solar wind particles were measured.[35]

Protons also occur in from extrasolar origin in space, from galactic cosmic rays, where they make up about 90% of the total particle flux. These protons often have higher energy than solar wind protons, but their intensity is far more uniform and less variable than protons coming from the Sun, the production of which is heavily affected by solar proton events such as coronal mass ejections.

Research has been performed on the dose-rate effects of protons, as typically found in space travel, on human health.[36][37] To be more specific, there are hopes to identify what specific chromosomes are damaged, and to define the damage, during cancer development from proton exposure.[36] Another study looks into determining "the effects of exposure to proton irradiation on neurochemical and behavioral endpoints, including dopaminergic functioning, amphetamine-induced conditioned taste aversion learning, and spatial learning and memory as measured by the Morris water maze.[37] Electrical charging of a spacecraft due to interplanetary proton bombardment has also been proposed for study.[38] There are many more studies that pertain to space travel, including galactic cosmic rays and their possible health effects, and solar proton event exposure.

The American Biostack and Soviet Biorack space travel experiments have demonstrated the severity of molecular damage induced by heavy ions on micro organisms including Artemia cysts.[39]

24.9   Antiproton

Main article: Antiproton

CPT-symmetry puts strong constraints on the relative properties of particles and antiparticles and, therefore, is open to stringent tests. For example, the charges of the proton and antiproton must sum to exactly zero. This equality has been tested to one part in 10^8. The equality of their masses has also been tested to better than one part in 10^8. By holding antiprotons in a Penning trap, the equality of the charge to mass ratio of the proton and the antiproton has been tested to one part in 6×10^9.[40] The magnetic moment of the antiproton has been measured with error of 8×10^{-3} nuclear Bohr magnetons, and is found to be equal and opposite to that of the proton.

24.10   See also

- Fermion field
- Hydrogen
- Hydron (chemistry)
- List of particles
- Proton-proton chain reaction
- Quark model
- Proton spin crisis

24.11   References

[1] Mohr, P.J.; Taylor, B.N. and Newell, D.B. (2011), "The 2010 CODATA Recommended Values of the Fundamental Physical Constants", National Institute of Standards and Technology, Gaithersburg, MD, US.

[2] Cho, Adiran (2 April 2010). "Mass of the Common Quark Finally Nailed Down". *http://news.sciencemag.org*. American Association for the Advancement of Science. Retrieved 27 September 2014.

[3] "Proton size puzzle reinforced!". Paul Shearer Institute. 25 January 2013.

[4] Adair, R.K. (1989). *The Great Design: Particles, Fields, and Creation.* Oxford University Press. p. 214.

[5] Cottingham, W.N.; Greenwood, D.A. (1986). *An Introduction to Nuclear Physics.* Cambridge University Press. ISBN 9780521657334.

[6] Basdevant, J.-L.; Rich, J.; M. Spiro (2005). *Fundamentals in Nuclear Physics*. Springer. p. 155. ISBN 0-387-01672-4.

[7] Department of Chemistry and Biochemistry UCLA Eric R. Scerri Lecturer. *The Periodic Table : Its Story and Its Significance: Its Story and Its Significance*. Oxford University Press. ISBN 978-0-19-534567-4.

[8] Petrucci, R.H.; Harwood, W.S.; Herring, F.G. (2002). *General Chemistry* (8th ed.). p. 41.

[9] See meeting report and announcement

[10] Romer A (1997). "Proton or prouton? Rutherford and the depths of the atom". *Amer. J. Phys.* **65** (8): 707. Bibcode:1997AmJPh..65..707R. doi:10.1119/1.18640.

[11] Rutherford reported acceptance by the *British Association* in a footnote to Masson, O. (1921). "XXIV.The constitution of atoms". *Philosophical Magazine Series 6* **41** (242): 281. doi:10.1080/14786442108636219.

[12] Pais, A. (1986) *Inward Bound*, Oxford Press, ISBN 0198519974, p. 296. Pais believed the first science literature use of the word *proton* occurs in "Physics at the British Association". *Nature* **106** (2663): 357. 1920. doi:10.1038/106357a0.

[13] Buccella, F.; Miele, G.; Rosa, L.; Santorelli, P.; Tuzi, T. (1989). "An upper limit for the proton lifetime in SO(10)". *Physics Letters B* **233**: 178. doi:10.1016/0370-2693(89)90637-0.

[14] Lee, D. G.; Mohapatra, R.; Parida, M.; Rani, M. (1995). "Predictions for the proton lifetime in minimal nonsupersymmetric SO(10) models: An update". *Physical Review D* **51**: 229. arXiv:hep-ph/9404238. doi:10.1103/PhysRevD.51.229.

[15] "Proton lifetime is longer than 1034 years". Kamioka Observatory. November 2009.

[16] Nishino, H.; Clark, S.; Abe, K.; Hayato, Y.; Iida, T.; Ikeda, M.; Kameda, J.; Kobayashi, K.; Koshio, Y.; Miura, M.; Moriyama, S.; Nakahata, M.; Nakayama, S.; Obayashi, Y.; Ogawa, H.; Sekiya, H.; Shiozawa, M.; Suzuki, Y.; Takeda, A.; Takenaga, Y.; Takeuchi, Y.; Ueno, K.; Ueshima, K.; Watanabe, H.; Yamada, S.; Hazama, S.; Higuchi, I.; Ishihara, C.; Kajita, T.; et al. (2009). "Search for Proton Decay via $p \rightarrow e^+ \pi^0$ and $p \rightarrow \mu^+ \pi^0$ in a Large Water Cherenkov Detector". *Physical Review Letters* **102** (14): 141801. arXiv:0903.0676. Bibcode:2009PhRvL.102n1801N. doi:10.1103/PhysRevLett.102.141801. PMID 19392425.

[17] Ahmed, S.; Anthony, A.; Beier, E.; Bellerive, A.; Biller, S.; Boger, J.; Boulay, M.; Bowler, M.; Bowles, T.; Brice, S.; Bullard, T.; Chan, Y.; Chen, M.; Chen, X.; Cleveland, B.; Cox, G.; Dai, X.; Dalnoki-Veress, F.; Doe, P.; Dosanjh, R.; Doucas, G.; Dragowsky, M.; Duba, C.; Duncan, F.; Dunford, M.; Dunmore, J.; Earle, E.; Elliott, S.; Evans, H.; et al. (2004). "Constraints on Nucleon Decay via Invisible Modes from the Sudbury Neutrino Observatory". *Physical Review Letters* **92** (10): 102004. arXiv:hep-ex/0310030. Bibcode:2004PhRvL..92j2004A. doi:10.1103/PhysRevLett.92.102004. PMID 15089201.

[18] Watson, A. (2004). *The Quantum Quark*. Cambridge University Press. pp. 285–286. ISBN 0-521-82907-0.

[19] Timothy Paul Smith (2003). *Hidden Worlds: Hunting for Quarks in Ordinary Matter*. Princeton University Press. ISBN 0-691-05773-7.

[20] Weise, W.; Green, A.M. (1984). *Quarks and Nuclei*. World Scientific. pp. 65–66. ISBN 9971-966-61-1.

[21] Ball, Philip (Nov 20, 2008). "Nuclear masses calculated from scratch". Nature. doi:10.1038/news.2008.1246. Retrieved Aug 27, 2014.

[22] Reynolds, Mark (Apr 2009). "Calculating the Mass of a Proton". *CNRS international magazine* (CNRS) (13). ISSN 2270-5317. Retrieved Aug 27, 2014.

[23] See this news report and links

[24] Durr, S.; Fodor, Z.; Frison, J.; Hoelbling, C.; Hoffmann, R.; Katz, S. D.; Krieg, S.; Kurth, T.; Lellouch, L.; Lippert, T.; Szabo, K. K.; Vulvert, G. (2008). "Ab Initio Determination of Light Hadron Masses". *Science* **322** (5905): 1224–7. arXiv:0906.3599. doi:10.1126/science.1163233. PMID 19023076.

[25] Perdrisat, C. F.; Punjabi, V.; Vanderhaeghen, M. (2007). "Nucleon electromagnetic form factors". *Progress in Particle and Nuclear Physics* **59** (2): 694. arXiv:hep-ph/0612014. Bibcode:2007PrPNP..59..694P. doi:10.1016/j.ppnp.2007.05.001.

[26] Boffi, Sigfrido; Pasquini, Barbara (2007). "Generalized parton distributions and the structure of the nucleon". *Rivista del Nuovo Cimento* **30**: 387. arXiv:0711.2625. Bibcode:2007NCimR..30..387B. doi:10.1393/ncr/i2007-10025-7 (inactive 2015-10-20).

[27] Joshua, Erlich (December 2008). "Recent Results in AdS/QCD". *Proceedings, 8th Conference on Quark Confinement and the Hadron Spectrum, September 1–6, 2008, Mainz, Germany*. arXiv:0812.4976.

[28] Pietro, Colangelo; Alex, Khodjamirian (October 2000). "QCD Sum Rules, a Modern Perspective". In Shifman, M. *At the Frontier of Particle Physics / Handbook of QCD*. World Scientific. arXiv:hep-ph/0010175.

[29] Pohl, R.; Antognini, A.; Nez, F. O.; Amaro, F. D.; Biraben, F. O.; Cardoso, J. O. M. R.; Covita, D. S.; Dax, A.; Dhawan, S.; Fernandes, L. M. P.; Giesen, A.; Graf, T.; Hänsch, T. W.; Indelicato, P.; Julien, L.; Kao, C. Y.; Knowles, P.; Le Bigot, E. O.; Liu, Y. W.; Lopes, J. A. M.; Ludhova, L.; Monteiro, C. M. B.; Mulhauser, F. O.; Nebel, T.; Rabinowitz, P.; Dos Santos, J. M. F.; Schaller, L. A.; Schuhmann, K.; Schwob, C.; et al. (2010). "The size of the proton". *Nature* **466** (7303): 213–6. doi:10.1038/nature09250. PMID 20613837.

[30] Antognini, A.; Nez, F.; Schuhmann, K.; Amaro, F. D.; Biraben, F.; Cardoso, J. M. R.; Covita, D. S.; Dax, A.; Dhawan, S.; Diepold, M.; Fernandes, L. M. P.; Giesen, A.; Gouvea, A. L.; Graf, T.; Hänsch, T. W.; Indelicato, P.; Julien, L.; Kao, C. -Y.; Knowles, P.; Kottmann, F.; Le Bigot, E. -O.; Liu, Y. -W.; Lopes, J. A. M.; Ludhova, L.; Monteiro, C. M. B.; Mulhauser, F.; Nebel, T.; Rabinowitz, P.; Dos Santos, J. M. F.; Schaller, L. A. (2013). "Proton Structure from the Measurement of 2S-2P Transition Frequencies of Muonic Hydrogen". *Science* **339** (6118): 417–420. doi:10.1126/science.1230016. PMID 23349284.

[31] Researchers Observes Unexpectedly Small Proton Radius in a Precision Experiment. *Azonano*. July 9, 2010

[32] "The Proton Just Got Smaller". *Photonics.Com*. 12 July 2010. Retrieved 2010-07-19.

[33] Carlson, Carl E. (February 19, 2015), *The Proton Radius Puzzle*, arXiv:1502.05314

[34] Headrick, J.M.; Diken, E.G.; Walters, R. S.; Hammer, N. I.; Christie, R.A.; Cui, J.; Myshakin, E.M.; Duncan, M.A.; Johnson, M.A.; Jordan, K.D. (2005). "Spectral Signatures of Hydrated Proton Vibrations in Water Clusters". *Science* **308** (5729): 1765–69. Bibcode:2005Sci...308.1765H. doi:10.1126/science.1113094. PMID 15961665.

[35] "Apollo 11 Mission". Lunar and Planetary Institute. 2009. Retrieved 2009-06-12.

[36] "Space Travel and Cancer Linked? Stony Brook Researcher Secures NASA Grant to Study Effects of Space Radiation". Brookhaven National Laboratory. 12 December 2007. Retrieved 2009-06-12.

[37] Shukitt-Hale, B.; Szprengiel, A.; Pluhar, J.; Rabin, B.M.; Joseph, J.A. "The effects of proton exposure on neurochemistry and behavior". Elsevier/COSPAR. Retrieved 2009-06-12.

[38] Green, N.W.; Frederickson, A.R. "A Study of Spacecraft Charging due to Exposure to Interplanetary Protons" (PDF). Jet Propulsion Laboratory. Retrieved 2009-06-12.

[39] Planel, H. (2004). *Space and life: an introduction to space biology and medicine*. CRC Press. pp. 135–138. ISBN 0-415-31759-2.

[40] Gabrielse, G. (2006). "Antiproton mass measurements". *International Journal of Mass Spectrometry* **251** (2–3): 273–280. Bibcode:2006IJMSp.251..273G. doi:10.1016/j.ijms.2006.02.013.

24.12 External links

- Particle Data Group

- Large Hadron Collider

- Eaves, Laurence; Copeland, Ed; Padilla, Antonio (Tony) (2010). "The shrinking proton". *Sixty Symbols*. Brady Haran for the University of Nottingham.

Ernest Rutherford at the first Solvay Conference, 1911

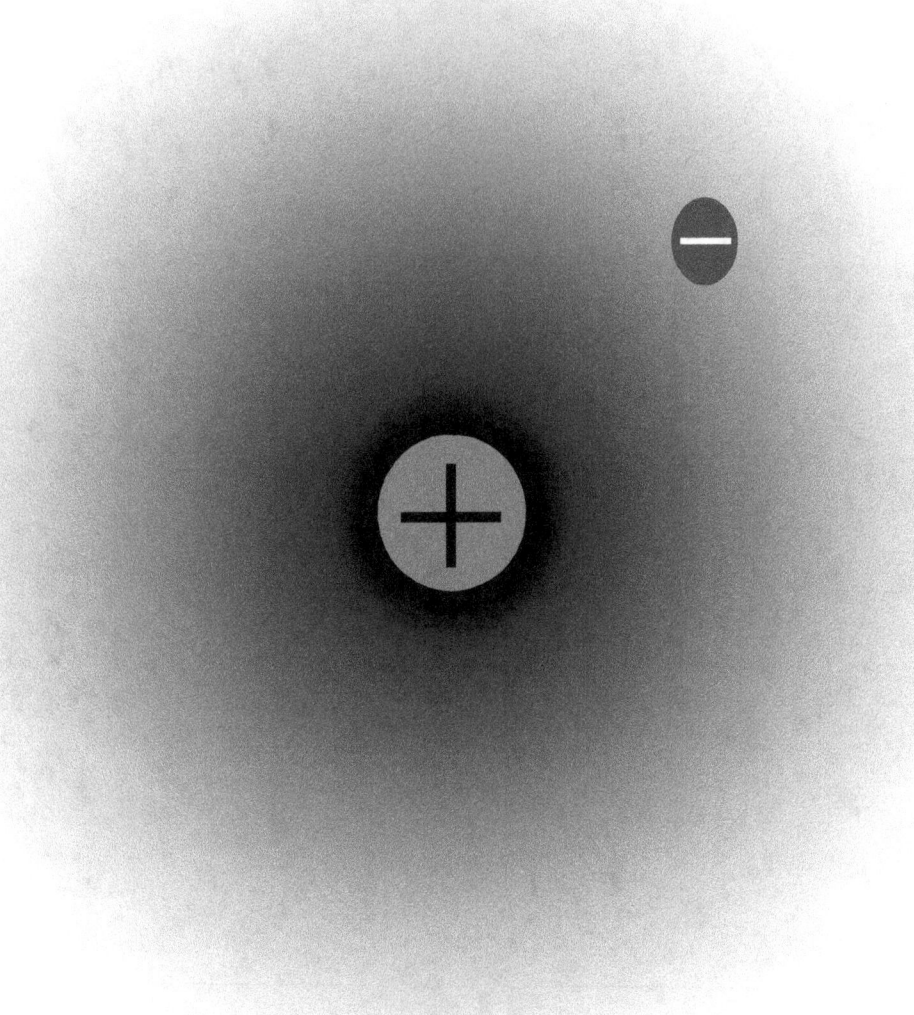

Protium, the most common isotope of hydrogen, consists of one proton and one electron (it has no neutrons). The term "hydrogen ion" (H+
) implies that that H-atom has lost its one electron, causing only a proton to remain. Thus, in chemistry, the terms "proton" and "hydrogen ion" (for the protium isotope) are used synonymously

Chapter 25

Neutron

This article is about the subatomic particle. For other uses, see Neutron (disambiguation).

The **neutron** is a subatomic particle, symbol n or n0, with no net electric charge and a mass slightly larger than that of a proton. Protons and neutrons, each with mass approximately one atomic mass unit, constitute the nucleus of an atom, and they are collectively referred to as nucleons.[4] Their properties and interactions are described by nuclear physics.

The nucleus consists of Z protons, where Z is called the atomic number, and N neutrons, where N is the neutron number. The atomic number defines the chemical properties of the atom, and the neutron number determines the isotope or nuclide.[5] The terms isotope and nuclide are often used synonymously, but they refer to chemical and nuclear properties, respectively. The atomic mass number, symbol A, equals Z+N. For example, carbon has atomic number 6, and its abundant carbon-12 isotope has 6 neutrons, whereas its rare carbon-13 isotope has 7 neutrons. Some elements occur in nature with only one stable isotope, such as fluorine (see stable nuclide). Other elements occur as many stable isotopes, such as tin with ten stable isotopes. Even though it is not a chemical element, the neutron is included in the table of nuclides.[6]

Within the nucleus, protons and neutrons are bound together through the nuclear force, and neutrons are required for the stability of nuclei. Neutrons are produced copiously in nuclear fission and fusion. They are a primary contributor to the nucleosynthesis of chemical elements within stars through fission, fusion, and neutron capture processes.

The neutron is essential to the production of nuclear power. In the decade after the neutron was discovered in 1932,[7] neutrons were used to effect many different types of nuclear transmutations. With the discovery of nuclear fission in 1938,[8] it was quickly realized that, if a fission event produced neutrons, each of these neutrons might cause further fission events, etc., in a cascade known as a nuclear chain reaction.[5] These events and findings led to the first self-sustaining nuclear reactor (Chicago Pile-1, 1942) and the first nuclear weapon (Trinity, 1945).

Free neutrons, or individual neutrons free of the nucleus, are effectively a form of ionizing radiation, and as such, are a biological hazard, depending upon dose.[5] A small natural "neutron background" flux of free neutrons exists on Earth, caused by cosmic ray muons, and by the natural radioactivity of spontaneously fissionable elements in the Earth's crust.[9] Dedicated neutron sources like neutron generators, research reactors and spallation sources produce free neutrons for use in irradiation and in neutron scattering experiments.

25.1 Description

Neutrons and protons are both nucleons, which are attracted and bound together by the nuclear force to form atomic nuclei. The nucleus of the most common isotope of the hydrogen atom (with the chemical symbol "H") is a lone proton. The nuclei of the heavy hydrogen isotopes deuterium and tritium contain one proton bound to one and two neutrons, respectively. All other types of atomic nuclei are composed of two or more protons and various numbers of neutrons. The most common nuclide of the common chemical element lead, ^{208}Pb has 82 protons and 126 neutrons, for example.

The free neutron has a mass of about 1.675×10^{-27} kg (equivalent to 939.6 MeV/c^2, or 1.0087 u).[3] The neutron has a mean square radius of about 0.8×10^{-15} m, or 0.8 fm,[10] and it is a spin-½ fermion.[11] The neutron has a magnetic moment with a negative value, because its orientation is opposite to the neutron's spin.[12] The neutron's magnetic moment causes its motion to be influenced by magnetic fields. Although the neutron has no net electric charge, it does have a slight distribution of charge within it. With its positive electric charge, the proton is directly influenced by electric fields, whereas the response of the neutron to this force is much weaker.

A free neutron is unstable, decaying to a proton, electron and antineutrino with a mean lifetime of just under 15 minutes (881.5±1.5 s). This radioactive decay, known as beta decay,[13] is possible since the mass of the neutron is slightly greater than the proton. The free proton is stable. Neutrons or protons bound in a nucleus can be stable or unstable, however, depending on the nuclide. Beta decay, in which neutrons decay to protons, or vice versa, is governed by the weak force, and it requires the emission or absorption of electrons and neutrinos, or their antiparticles.

Protons and neutrons behave almost identically under the influence of the nuclear force within the nucleus. The concept of isospin, in which the proton and neutron are viewed as two quantum states of the same particle, is used to model the interactions of nucleons by the nuclear or weak forces. Because of the strength of the nuclear force at short distances, the binding energy of nucleons is more than seven orders of magnitude larger than the electromagnetic energy binding electrons in atoms. Nuclear reactions (such as nuclear fission) therefore have an energy density that is more than ten million times that of chemical reactions. Because of the mass–energy equivalence, nuclear binding energies add or subtract from the mass of nuclei. Ultimately, the ability of the nuclear force to store energy arising from the electromagnetic repulsion of nuclear components is the basis for most of the energy that makes nuclear reactors or bombs possible. In nuclear fission, the absorption of a neutron by a heavy nuclide (e.g., uranium-235) causes the nuclide to become unstable and break into light nuclides and additional neutrons. The positively charged light nuclides then repel, releasing electromagnetic potential energy.

The neutron is classified as a hadron, since it is composed of quarks, and as a baryon, since it is composed of three quarks.[14] The finite size of the neutron and its magnetic moment indicate the neutron is a composite, rather than elementary, particle. The neutron consists of two down quarks with charge $-\frac{1}{3}e$ and one up quark with charge $+\frac{2}{3}e$, although this simple model belies the complexities of the Standard Model for nuclei.[15] The masses of the three quarks sum to only about 12 MeV/c^2, whereas the neutron's mass is about 940 MeV/c^2, for example.[15] Like the proton, the quarks of the neutron are held together by the strong force, mediated by gluons.[16] The nuclear force results from secondary effects of the more fundamental strong force.

25.2 Discovery

Main article: Discovery of the neutron

The story of the discovery of the neutron and its properties is central to the extraordinary developments in atomic physics that occurred in the first half of the 20th century, leading ultimately to the atomic bomb in 1945. In the 1911 Rutherford model, the atom consisted of a small positively charged massive nucleus surrounded by a much larger cloud of negatively charged electrons. In 1920 Rutherford suggested the nucleus consisted of positive protons and neutrally-charged particles, suggested to be a proton and an electron bound in some way.[17] Electrons were assumed to reside within the nucleus because it was known that beta radiation consisted of electrons emitted from the nucleus.[17] Rutherford called these uncharged particles *neutrons*, by the Latin root for *neutralis* (neuter) and the Greek suffix *-on* (a suffix used in the names of subatomic particles, i.e. *electron* and *proton*).[18][19] References to the word *neutron* in connection with the atom can be found in the literature as early as 1899, however.[20]

Throughout the 1920s, physicists assumed that the atomic nucleus was composed of protons and "nuclear electrons"[21][22] but there were obvious problems. It was difficult to reconcile the proton–electron model for nuclei with the Heisenberg uncertainty relation of quantum mechanics.[23][24] The Klein paradox,[25] discovered by Oskar Klein in 1928, presented further quantum mechanical objections to the notion of an electron confined within a nucleus.[23] Observed properties of atoms and molecules were inconsistent with the nuclear spin expected from proton–electron hypothesis. Since both protons and electrons carry an intrinsic spin of ½ $\hbar$, there is no way to arrange an odd number of spins ±½ $\hbar$ to give a spin integer multiple of $\hbar$. Nuclei with integer spin are common, e.g., ^{14}N.

In 1931, Walther Bothe and Herbert Becker found that if alpha particle radiation from polonium fell on beryllium, boron, or lithium, an unusually penetrating radiation was produced. The radiation was not influenced by an electric field, so Bothe and Becker assumed it was gamma radiation.[26][27] The following year Irène Joliot-Curie and Frédéric Joliot in Paris showed that if this "gamma" radiation fell on paraffin, or any other hydrogen-containing compound, it ejected protons of very high energy.[28] Neither Rutherford nor James Chadwick at the Cavendish Laboratory in Cambridge were convinced by the gamma ray interpretation.[21] Chadwick quickly performed a series of experiments that showed that the new radiation consisted of uncharged particles with about the same mass as the proton.[7][29][30] These particles were neutrons. Chadwick won the Nobel Prize in Physics for this discovery in 1935.[2]

Models for atomic nucleus consisting of protons and neutrons were quickly developed by Werner Heisenberg[31][32][33] and others.[34][35] The proton–neutron model explained the puzzle of nuclear spins. The origins of beta radiation were explained by Enrico Fermi in 1934 by the process of beta decay, in which the neutron decays to a proton by *creating* an electron and a (as yet undiscovered) neutrino.[36] In 1935 Chadwick and his doctoral student Maurice Goldhaber, reported the first accurate measurement of the mass of the neutron.[37][38]

By 1934, Fermi had bombarded heavier elements with neutrons to induce radioactivity in elements of high atomic number. In 1938, Fermi received the Nobel Prize in Physics *"for his demonstrations of the existence of new radioactive elements produced by neutron irradiation, and for his related discovery of nuclear reactions brought about by slow neutrons".*[39] In 1938 Otto Hahn, Lise Meitner, and Fritz Strassmann discovered nuclear fission, or the fractionation of uranium nuclei into light elements, induced by neutron bombardment.[40][41][42] In 1945 Hahn received the 1944 Nobel Prize in Chemistry *"for his discovery of the fission of heavy atomic nuclei."* [43][44][45] The discovery of nuclear fission would lead to the development of nuclear power and the atomic bomb by the end of World War II.

25.3   Beta decay and the stability of the nucleus

Under the Standard Model of particle physics, the only possible decay mode for the neutron that conserves baryon number is for one of the neutron's quarks to change flavour via the weak interaction. The decay of one of the neutron's down quarks into a lighter up quark can be achieved by the emission of a W boson. By this process, the Standard Model description of beta decay, the neutron decays into a proton (which contains one down and two up quarks), an electron, and an electron antineutrino.

Since interacting protons have a mutual electromagnetic repulsion that is stronger than their attractive nuclear interaction, neutrons are a necessary constituent of any atomic nucleus that contains more than one proton (see diproton and neutron–proton ratio).[46] Neutrons bind with protons and one another in the nucleus via the nuclear force, effectively moderating the repulsive forces between the protons and stabilizing the nucleus.

See also: Beta-decay stable isobars and Neutron emission

25.3.1   Free neutron decay

Outside the nucleus, free neutrons are unstable and have a mean lifetime of 881.5±1.5 s (about 14 minutes, 42 seconds); therefore the half-life for this process (which differs from the mean lifetime by a factor of $\ln(2) = 0.693$) is 611.0±1.0 s (about 10 minutes, 11 seconds).[13] Beta decay of the neutron, described above, can be denoted by the radioactive decay:[47]

n0 → p+ + e− + ν
e

where p+, e−, and ν

e denote the proton, electron and electron antineutrino, respectively. For the free neutron the decay energy for this process (based on the masses of the neutron, proton, and electron) is 0.782343 MeV. The maximal energy of the beta decay electron (in the process wherein the neutrino receives a vanishingly small amount of kinetic energy) has been measured at 0.782 ± .013 MeV.[48] The latter number is not well-enough measured to determine the comparatively tiny rest mass of the neutrino (which must in theory be subtracted from the maximal electron kinetic energy) as well as neutrino mass is constrained by many other methods.

A small fraction (about one in 1000) of free neutrons decay with the same products, but add an extra particle in the form of an emitted gamma ray:

n0 → p+ + e− + ν
e + γ

This gamma ray may be thought of as a sort of "internal bremsstrahlung" that arises as the emitted beta particle interacts with the charge of the proton in an electromagnetic way. Internal bremsstrahlung gamma ray production is also a minor feature of beta decays of bound neutrons (as discussed below).

A very small minority of neutron decays (about four per million) are so-called "two-body (neutron) decays", in which a proton, electron and antineutrino are produced as usual, but the electron fails to gain the 13.6 eV necessary energy to escape the proton, and therefore simply remains bound to it, as a neutral hydrogen atom (one of the "two bodies"). In this type of free neutron decay, in essence all of the neutron decay energy is carried off by the antineutrino (the other "body").

The transformation of a free proton to a neutron (plus a positron and a neutrino) is energetically impossible, since a free neutron has a greater mass than a free proton.

25.3.2 Bound neutron decay

Main article: Atomic nucleus

While a free neutron has a half life of about 10.2 min, most neutrons within nuclei are stable. According to the nuclear shell model, the protons and neutrons of a nuclide are a quantum mechanical system organized into discrete energy levels with unique quantum numbers. For a neutron to decay, the resulting proton requires an available state at lower energy than the initial neutron state. In stable nuclei the possible lower energy states are all filled, meaning they are each occupied by two protons with spin up and spin down. The Pauli exclusion principle therefore disallows the decay of a neutron to a proton within stable nuclei. The situation is similar to electrons of an atom, where electrons have distinct atomic orbitals and are prevented from decaying to lower energy states, with the emission of a photon, by the exclusion principle.

Neutrons in unstable nuclei can decay by beta decay as described above. In this case, an energetically allowed quantum state is available for the proton resulting from the decay. One example of this decay is carbon-14 (6 protons, 8 neutrons) that decays to nitrogen-14 (7 protons, 7 neutrons) with a half-life of about 5,730 years.

Inside a nucleus, a proton can transform into a neutron via inverse beta decay, if an energetically allowed quantum state is available for the neutron. This transformation occurs by emission of an antielectron (also called positron) and an electron neutrino:

p+ → n0 + e+ + ν
e

The transformation of a proton to a neutron inside of a nucleus is also possible through electron capture:

p+ + e− → n0 + ν
e

Positron capture by neutrons in nuclei that contain an excess of neutrons is also possible, but is hindered because positrons are repelled by the positive nucleus, and quickly annihilate when they encounter electrons.

25.3.3 Competition of beta decay types

Three types of beta decay in competition are illustrated by the single isotope copper-64 (29 protons, 35 neutrons), which has a half-life of about 12.7 hours. This isotope has one unpaired proton and one unpaired neutron, so either the proton or the neutron can decay. This particular nuclide (though not all nuclides in this situation) is almost equally likely to decay through proton decay by positron emission (18%) or electron capture (43%), as through neutron decay by electron emission (39%).

25.4 Intrinsic properties

25.4.1 Electric charge

The total electric charge of the neutron is $0\,e$. This zero value has been tested experimentally, and the present experimental limit for the charge of the neutron is $-2(8)\times10^{-22}\,e$,[49] or $-3(13)\times10^{-41}$ C. This value is consistent with zero, given the experimental uncertainties (indicated in parentheses). By comparison, the charge of the proton is, of course, $+1\,e$.

25.4.2 Electric dipole moment

Main article: Neutron electric dipole moment

The Standard Model of particle physics predicts a tiny separation of positive and negative charge within the neutron leading to a permanent electric dipole moment.[50] The predicted value is, however, well below the current sensitivity of experiments. From several unsolved puzzles in particle physics, it is clear that the Standard Model is not the final and full description of all particles and their interactions. New theories going beyond the Standard Model generally lead to much larger predictions for the electric dipole moment of the neutron. Currently, there are at least four experiments trying to measure for the first time a finite neutron electric dipole moment, including:

- Cryogenic neutron EDM experiment being set up at the Institut Laue–Langevin[51]

- nEDM experiment under construction at the new UCN source at the Paul Scherrer Institute[52]

- nEDM experiment being envisaged at the Spallation Neutron Source[53]

- nEDM experiment being built at the Institut Laue–Langevin[54]

25.4.3 Magnetic moment

Main article: Neutron magnetic moment

Even though the neutron is a neutral particle, the magnetic moment of a neutron is not zero. Since the neutron is a neutral particle, it is not affected by electric fields, but with its magnetic moment it is affected by magnetic fields. The magnetic moment of the neutron is an indication of its quark substructure and internal charge distribution.[55] The value for the neutron's magnetic moment was first directly measured by Luis Alvarez and Felix Bloch at Berkeley, California in 1940,[56] using an extension of the magnetic resonance methods developed by Rabi. Alvarez and Bloch determined the magnetic moment of the neutron to be $\mu_\mathrm{n} = -1.93(2)\,\mu$N, where μN is the nuclear magneton.

25.4.4 Structure and geometry of charge distribution

An article published in 2007 featuring a model-independent analysis concluded that the neutron has a negatively charged exterior, a positively charged middle, and a negative core.[57] In a simplified classical view, the negative "skin" of the neutron assists it to be attracted to the protons with which it interacts in the nucleus. (However, the main attraction between neutrons and protons is via the nuclear force, which does not involve charge.)

The simplified classical view of the neutron's charge distribution also "explains" the fact that the neutron magnetic dipole points in the opposite direction from its spin angular momentum vector (as compared to the proton). This gives the neutron, in effect, a magnetic moment which resembles a negatively charged particle. This can be reconciled classically with a neutral neutron composed of a charge distribution in which the negative sub-parts of the neutron have a larger average radius of distribution, and therefore contribute more to the particle's magnetic dipole moment, than do the positive parts that are, on average, nearer the core.

25.4.5 Mass

The mass of a neutron cannot be directly determined by mass spectrometry due to lack of electric charge. However, since the mass of protons and deuterons can be measured by mass spectrometry, the mass of a neutron can be deduced by subtracting proton mass from deuteron mass, with the difference being the mass of the neutron plus the binding energy of deuterium (expressed as a positive emitted energy). The latter can be directly measured by measuring the energy (B_d) of the single 0.7822 MeV gamma photon emitted when neutrons are captured by protons (this is exothermic and happens with zero-energy neutrons), plus the small recoil kinetic energy (E_{rd}) of the deuteron (about 0.06% of the total energy).

$$m_n = m_d - m_p + B_d - E_{rd}$$

The energy of the gamma ray can be measured to high precision by X-ray diffraction techniques, as was first done by Bell and Elliot in 1948. The best modern (1986) values for neutron mass by this technique are provided by Greene, et al.[58] These give a neutron mass of:

m_{neutron} = 1.008644904(14) u

The value for the neutron mass in MeV is less accurately known, due to less accuracy in the known conversion of u to MeV:[59]

m_{neutron} = 939.56563(28) MeV/c^2.

Another method to determine the mass of a neutron starts from the beta decay of the neutron, when the momenta of the resulting proton and electron are measured.

25.4.6 Anti-neutron

Main article: Antineutron

The antineutron is the antiparticle of the neutron. It was discovered by Bruce Cork in the year 1956, a year after the antiproton was discovered. CPT-symmetry puts strong constraints on the relative properties of particles and antiparticles, so studying antineutrons yields provide stringent tests on CPT-symmetry. The fractional difference in the masses of the neutron and antineutron is $(9\pm6)\times10^{-5}$. Since the difference is only about two standard deviations away from zero, this does not give any convincing evidence of CPT-violation.[13]

25.5 Neutron compounds

25.5.1 Dineutrons and tetraneutrons

Main articles: Dineutron and Tetraneutron

The existence of stable clusters of 4 neutrons, or tetraneutrons, has been hypothesised by a team led by Francisco-Miguel Marqués at the CNRS Laboratory for Nuclear Physics based on observations of the disintegration of beryllium−14 nuclei. This is particularly interesting because current theory suggests that these clusters should not be stable.

The dineutron is another hypothetical particle. In 2012, Artemis Spyrou from Michigan State University and coworkers reported that they observed, for the first time, the dineutron emission in the decay of ^{16}Be. The dineutron character is evidenced by a small emission angle between the two neutrons. The authors measured the two-neutron separation energy to be 1.35(10) MeV, in good agreement with shell model calculations, using standard interactions for this mass region.[60]

25.5.2   Neutronium and neutron stars

Main articles: Neutronium and Neutron star

At extremely high pressures and temperatures, nucleons and electrons are believed to collapse into bulk neutronic matter, called neutronium. This is presumed to happen in neutron stars.

The extreme pressure inside a neutron star may deform the neutrons into a cubic symmetry, allowing tighter packing of neutrons.[61]

25.6   Detection

Main article: Neutron detection

The common means of detecting a charged particle by looking for a track of ionization (such as in a cloud chamber) does not work for neutrons directly. Neutrons that elastically scatter off atoms can create an ionization track that is detectable, but the experiments are not as simple to carry out; other means for detecting neutrons, consisting of allowing them to interact with atomic nuclei, are more commonly used. The commonly used methods to detect neutrons can therefore be categorized according to the nuclear processes relied upon, mainly neutron capture or elastic scattering. A good discussion on neutron detection is found in chapter 14 of the book *Radiation Detection and Measurement* by Glenn F. Knoll (John Wiley & Sons, 1979).

25.6.1   Neutron detection by neutron capture

A common method for detecting neutrons involves converting the energy released from neutron capture reactions into electrical signals. Certain nuclides have a high neutron capture cross section, which is the probability of absorbing a neutron. Upon neutron capture, the compound nucleus emits more easily detectable radiation, for example an alpha particle, which is then detected. The nuclides 3He, 6Li, 10B, 233U, 235U, 237Np and 239Pu are useful for this purpose.

25.6.2   Neutron detection by elastic scattering

Neutrons can elastically scatter off nuclei, causing the struck nucleus to recoil. Kinematically, a neutron can transfer more energy to light nuclei such as hydrogen or helium than to heavier nuclei. Detectors relying on elastic scattering are called fast neutron detectors. Recoiling nuclei can ionize and excite further atoms through collisions. Charge and/or scintillation light produced in this way can be collected to produce a detected signal. A major challenge in fast neutron detection is discerning such signals from erroneous signals produced by gamma radiation in the same detector.

Fast neutron detectors have the advantage of not requiring a moderator, and therefore being capable of measuring the neutron's energy, time of arrival, and in certain cases direction of incidence.

25.7   Sources and production

Main articles: Neutron source, neutron generator and research reactor

Free neutrons are unstable, although they have the longest half-life of any unstable sub-atomic particle by several orders of magnitude. Their half-life is still only about 10 minutes, however, so they can be obtained only from sources that produce them freshly.

Natural neutron background. A small natural background flux of free neutrons exists everywhere on Earth. In the atmosphere and deep into the ocean, the "neutron background" is caused by muons produced by cosmic ray interaction with the atmosphere. These high energy muons are capable of penetration to considerable depths in water and soil. There, in striking atomic nuclei, among other reactions they induce spallation reactions in which a neutron is liberated from the nucleus. Within the Earth's crust a second source is neutrons produced primarily by spontaneous fission of uranium and thorium present in crustal minerals. The neutron background is not strong enough to be a biological hazard, but it is of importance to very high resolution particle detectors that are looking for very rare events, such as (hypothesized) interactions that might be caused by particles of dark matter.[9] Recent research has shown that even thunderstorms can produce neutrons with energies of up to several tens of MeV.[62]

Even stronger neutron background radiation is produced at the surface of Mars, where the atmosphere is thick enough to generate neutrons from cosmic ray muon production and neutron-spallation, but not thick enough to provide significant protection from the neutrons produced. These neutrons not only produce a Martian surface neutron radiation hazard from direct downward-going neutron radiation but may also produce a significant hazard from reflection of neutrons from the Martian surface, which will produce reflected neutron radiation penetrating upward into a Martian craft or habitat from the floor.[63]

Sources of neutrons for research. These include certain types of radioactive decay (spontaneous fission and neutron emission), and from certain nuclear reactions. Convenient nuclear reactions include tabletop reactions such as natural alpha and gamma bombardment of certain nuclides, often beryllium or deuterium, and induced nuclear fission, such as occurs in nuclear reactors. In addition, high-energy nuclear reactions (such as occur in cosmic radiation showers or accelerator collisions) also produce neutrons from disintigration of target nuclei. Small (tabletop) particle accelerators optimized to produce free neutrons in this way, are called neutron generators.

In practice, the most commonly used small laboratory sources of neutrons use radioactive decay to power neutron production. One noted neutron-producing radioisotope, californium−252 decays (half-life 2.65 years) by spontaneous fission 3% of the time with production of 3.7 neutrons per fission, and is used alone as a neutron source from this process. Nuclear reaction sources (that involve two materials) powered by radioisotopes use an alpha decay source plus a beryllium target, or else a source of high-energy gamma radiation from a source that undergoes beta decay followed by gamma decay, which produces photoneutrons on interaction of the high energy gamma ray with ordinary stable beryllium, or else with the deuterium in heavy water. A popular source of the latter type is radioactive antimony-124 plus beryllium, a system with a half-life of 60.9 days, which can be constructed from natural antimony (which is 42.8% stable antimony-123) by activating it with neutrons in a nuclear reactor, then transported to where the neutron source is needed.[64]

Nuclear fission reactors naturally produce free neutrons; their role is to sustain the energy-producing chain reaction. The intense neutron radiation can also be used to produce various radioisotopes through the process of neutron activation, which is a type of neutron capture.

Experimental nuclear fusion reactors produce free neutrons as a waste product. However, it is these neutrons that possess most of the energy, and converting that energy to a useful form has proved a difficult engineering challenge. Fusion reactors that generate neutrons are likely to create radioactive waste, but the waste is composed of neutron-activated lighter isotopes, which have relatively short (50–100 years) decay periods as compared to typical half-lives of 10,000 years for fission waste, which is long due primarily to the long half-life of alpha-emitting transuranic actinides.[65]

25.7.1 Neutron beams and modification of beams after production

Free neutron beams are obtained from neutron sources by neutron transport. For access to intense neutron sources, researchers must go to a specialist neutron facility that operates a research reactor or a spallation source.

The neutron's lack of total electric charge makes it difficult to steer or accelerate them. Charged particles can be accelerated, decelerated, or deflected by electric or magnetic fields. These methods have little effect on neutrons. However, some effects may be attained by use of inhomogeneous magnetic fields because of the neutron's magnetic moment. Neutrons can be controlled by methods that include moderation, reflection, and velocity selection. Thermal neutrons can be polarized by transmission through magnetic materials in a method analogous to the Faraday effect for photons. Cold neutrons of wavelengths of 6–7 angstroms can be produced in beams of a high degree of polarization, by use of magnetic mirrors and magnetized interference filters.[66]

25.8   Applications

The neutron plays an important role in many nuclear reactions. For example, neutron capture often results in neutron activation, inducing radioactivity. In particular, knowledge of neutrons and their behavior has been important in the development of nuclear reactors and nuclear weapons. The fissioning of elements like uranium-235 and plutonium-239 is caused by their absorption of neutrons.

Cold, *thermal* and *hot* neutron radiation is commonly employed in neutron scattering facilities, where the radiation is used in a similar way one uses X-rays for the analysis of condensed matter. Neutrons are complementary to the latter in terms of atomic contrasts by different scattering cross sections; sensitivity to magnetism; energy range for inelastic neutron spectroscopy; and deep penetration into matter.

The development of "neutron lenses" based on total internal reflection within hollow glass capillary tubes or by reflection from dimpled aluminum plates has driven ongoing research into neutron microscopy and neutron/gamma ray tomography.[67][68][69]

A major use of neutrons is to excite delayed and prompt gamma rays from elements in materials. This forms the basis of neutron activation analysis (NAA) and prompt gamma neutron activation analysis (PGNAA). NAA is most often used to analyze small samples of materials in a nuclear reactor whilst PGNAA is most often used to analyze subterranean rocks around bore holes and industrial bulk materials on conveyor belts.

Another use of neutron emitters is the detection of light nuclei, in particular the hydrogen found in water molecules. When a fast neutron collides with a light nucleus, it loses a large fraction of its energy. By measuring the rate at which slow neutrons return to the probe after reflecting off of hydrogen nuclei, a neutron probe may determine the water content in soil.

25.9   Medical therapies

Main articles: Fast neutron therapy and Neutron capture therapy of cancer

Because neutron radiation is both penetrating and ionizing, it can be exploited for medical treatments. Neutron radiation can have the unfortunate side-effect of leaving the affected area radioactive, however. Neutron tomography is therefore not a viable medical application.

Fast neutron therapy utilizes high energy neutrons typically greater than 20 MeV to treat cancer. Radiation therapy of cancers is based upon the biological response of cells to ionizing radiation. If radiation is delivered in small sessions to damage cancerous areas, normal tissue will have time to repair itself, while tumor cells often cannot.[70] Neutron radiation can deliver energy to a cancerous region at a rate an order of magnitude larger than gamma radiation[71]

Beams of low energy neutrons are used in boron capture therapy to treat cancer. In boron capture therapy, the patient is given a drug that contains boron and that preferentially accumulates in the tumor to be targeted. The tumor is then bombarded with very low energy neutrons (although often higher than thermal energy) which are captured by the boron-10 isotope in the boron, which produces an excited state of boron-11 that then decays to produce lithium-7 and an alpha particle that have sufficient energy to kill the malignant cell, but insufficient range to damage nearby cells. For such a therapy to be applied to the treatment of cancer, a neutron source having an intensity of the order of billion (10^9) neutrons per second per cm^2 is preferred. Such fluxes require a research nuclear reactor.

25.10 Protection

Exposure to free neutrons can be hazardous, since the interaction of neutrons with molecules in the body can cause disruption to molecules and atoms, and can also cause reactions that give rise to other forms of radiation (such as protons). The normal precautions of radiation protection apply: Avoid exposure, stay as far from the source as possible, and keep exposure time to a minimum. Some particular thought must be given to how to protect from neutron exposure, however. For other types of radiation, e.g. alpha particles, beta particles, or gamma rays, material of a high atomic number and with high density make for good shielding; frequently, lead is used. However, this approach will not work with neutrons, since the absorption of neutrons does not increase straightforwardly with atomic number, as it does with alpha, beta, and gamma radiation. Instead one needs to look at the particular interactions neutrons have with matter (see the section on detection above). For example, hydrogen-rich materials are often used to shield against neutrons, since ordinary hydrogen both scatters and slows neutrons. This often means that simple concrete blocks or even paraffin-loaded plastic blocks afford better protection from neutrons than do far more dense materials. After slowing, neutrons may then be absorbed with an isotope that has high affinity for slow neutrons without causing secondary capture radiation, such as lithium-6.

Hydrogen-rich ordinary water affects neutron absorption in nuclear fission reactors: Usually, neutrons are so strongly absorbed by normal water that fuel enrichment with fissionable isotope is required. The deuterium in heavy water has a very much lower absorption affinity for neutrons than does protium (normal light hydrogen). Deuterium is, therefore, used in CANDU-type reactors, in order to slow (moderate) neutron velocity, to increase the probability of nuclear fission compared to neutron capture.

25.11 Neutron temperature

Main article: Neutron temperature

25.11.1 Thermal neutrons

A *thermal neutron* is a free neutron that is Boltzmann distributed with $kT = 0.0253$ eV (4.0×10^{-21} J) at room temperature. This gives characteristic (not average, or median) speed of 2.2 km/s. The name 'thermal' comes from their energy being that of the room temperature gas or material they are permeating. (see *kinetic theory* for energies and speeds of molecules). After a number of collisions (often in the range of 10–20) with nuclei, neutrons arrive at this energy level, provided that they are not absorbed.

In many substances, thermal neutron reactions show a much larger effective cross-section than reactions involving faster neutrons, and thermal neutrons can therefore be absorbed more readily (i.e., with higher probability) by any atomic nuclei that they collide with, creating a heavier — and often unstable — isotope of the chemical element as a result.

Most fission reactors use a neutron moderator to slow down, or *thermalize* the neutrons that are emitted by nuclear fission so that they are more easily captured, causing further fission. Others, called fast breeder reactors, use fission energy neutrons directly.

25.11.2 Cold neutrons

Cold neutrons are thermal neutrons that have been equilibrated in a very cold substance such as liquid deuterium. Such a *cold source* is placed in the moderator of a research reactor or spallation source. Cold neutrons are particularly valuable for neutron scattering experiments.

25.11.3 Ultracold neutrons

Ultracold neutrons are produced by inelastically scattering cold neutrons in substances with a temperature of a few kelvins, such as solid deuterium or superfluid helium. An alternative production method is the mechanical deceleration of cold neutrons.

25.11.4 Fission energy neutrons

Main article: nuclear fission

A *fast neutron* is a free neutron with a kinetic energy level close to 1 MeV (1.6×10^{-13} J), hence a speed of ~14000 km/s (~ 5% of the speed of light). They are named *fission energy* or *fast* neutrons to distinguish them from lower-energy thermal neutrons, and high-energy neutrons produced in cosmic showers or accelerators. Fast neutrons are produced by nuclear processes such as nuclear fission. Neutrons produced in fission, as noted above, have a Maxwell–Boltzmann distribution of kinetic energies from 0 to ~14 MeV, a mean energy of 2 MeV (for U-235 fission neutrons), and a mode of only 0.75 MeV, which means that more than half of them do not qualify as fast (and thus have almost no chance of initiating fission in fertile materials, such as U-238 and Th-232).

Fast neutrons can be made into thermal neutrons via a process called moderation. This is done with a neutron moderator. In reactors, typically heavy water, light water, or graphite are used to moderate neutrons.

25.11.5 Fusion neutrons

For more details on this topic, see Nuclear fusion § Criteria and candidates for terrestrial reactions.

D–T (deuterium–tritium) fusion is the fusion reaction that produces the most energetic neutrons, with 14.1 MeV of kinetic energy and traveling at 17% of the speed of light. D–T fusion is also the easiest fusion reaction to ignite, reaching near-peak rates even when the deuterium and tritium nuclei have only a thousandth as much kinetic energy as the 14.1 MeV that will be produced.

14.1 MeV neutrons have about 10 times as much energy as fission neutrons, and are very effective at fissioning even non-fissile heavy nuclei, and these high-energy fissions produce more neutrons on average than fissions by lower-energy neutrons. This makes D–T fusion neutron sources such as proposed tokamak power reactors useful for transmutation of transuranic waste. 14.1 MeV neutrons can also produce neutrons by knocking them loose from nuclei.

On the other hand, these very high energy neutrons are less likely to simply be captured without causing fission or spallation. For these reasons, nuclear weapon design extensively utilizes D–T fusion 14.1 MeV neutrons to cause more fission. Fusion neutrons are able to cause fission in ordinarily non-fissile materials, such as depleted uranium (uranium-238), and these materials have been used in the jackets of thermonuclear weapons. Fusion neutrons also can cause fission in substances that are unsuitable or difficult to make into primary fission bombs, such as reactor grade plutonium. This physical fact thus causes ordinary non-weapons grade materials to become of concern in certain nuclear proliferation discussions and treaties.

Other fusion reactions produce much less energetic neutrons. D–D fusion produces a 2.45 MeV neutron and helium-3 half of the time, and produces tritium and a proton but no neutron the other half of the time. D–^{3}He fusion produces no neutron.

25.11.6 Intermediate-energy neutrons

A fission energy neutron that has slowed down but not yet reached thermal energies is called an epithermal neutron.

Cross sections for both capture and fission reactions often have multiple resonance peaks at specific energies in the epithermal energy range. These are of less significance in a fast neutron reactor, where most neutrons are absorbed before slowing down to this range, or in a well-moderated thermal reactor, where epithermal neutrons interact mostly with moderator nuclei, not with either fissile or fertile actinide nuclides. However, in a partially moderated reactor with more interactions of epithermal neutrons with heavy metal nuclei, there are greater possibilities for transient changes in reactivity that might make reactor control more difficult.

Ratios of capture reactions to fission reactions are also worse (more captures without fission) in most nuclear fuels such as plutonium-239, making epithermal-spectrum reactors using these fuels less desirable, as captures not only waste the one neutron captured but also usually result in a nuclide that is not fissile with thermal or epithermal neutrons, though still fissionable with fast neutrons. The exception is uranium-233 of the thorium cycle, which has good capture-fission ratios at all neutron energies.

25.11.7 High-energy neutrons

These neutrons have much more energy than fission energy neutrons and are generated as secondary particles by particle accelerators or in the atmosphere from cosmic rays. They can have energies as high as tens of joules per neutron. These neutrons are extremely efficient at ionization and far more likely to cause cell death than X-rays or protons.[72][73]

25.12 See also

- Ionizing radiation
- Isotope
- List of particles
- Neutronium
- Neutron magnetic moment
- Neutron radiation and the Sievert radiation scale
- Nuclear reaction
- Thermal reactor
- Nucleosynthesis
    - Neutron capture nucleosynthesis
    - R-process
    - S-process

25.12.1 Neutron sources

- Neutron generator
- Neutron sources

25.12.2 Processes involving neutrons

- Neutron bomb
- Neutron diffraction
- Neutron flux
- Neutron transport

25.13   References

[1] Ernest Rutherford. Chemed.chem.purdue.edu. Retrieved on 2012-08-16.

[2] 1935 Nobel Prize in Physics. Nobelprize.org. Retrieved on 2012-08-16.

[3] Mohr, P.J.; Taylor, B.N. and Newell, D.B. (2011), "The 2010 CODATA Recommended Values of the Fundamental Physical Constants" (Web Version 6.0). The database was developed by J. Baker, M. Douma, and S. Kotochigova. (2011-06-02). National Institute of Standards and Technology, Gaithersburg, Maryland 20899.

[4] Thomas, A.W.; Weise, W. (2001), *The Structure of the Nucleon*, Wiley-WCH, Berlin, ISBN 3-527-40297-7

[5] Glasstone, Samuel; Dolan, Philip J., eds. (1977), *The Effects of Nuclear Weapons, Third Edition*, U.S. Dept. of Defense and Energy Research and Development Administration, U.S. Government Printing Office, ISBN 1-60322-016-X

[6] Nudat 2. Nndc.bnl.gov. Retrieved on 2010-12-04.

[7] Chadwick, James (1932). "Possible Existence of a Neutron". *Nature* **129** (3252): 312. Bibcode:1932Natur.129Q.312C. doi:10.1038/129312a0.

[8] O. Hahn and F. Strassmann (1939). "Über den Nachweis und das Verhalten der bei der Bestrahlung des Urans mittels Neutronen entstehenden Erdalkalimetalle ("On the detection and characteristics of the alkaline earth metals formed by irradiation of uranium with neutrons")". *Naturwissenschaften* **27** (1): 11–15. Bibcode:1939NW.....27...11H. doi:10.1007/BF01488241.. The authors were identified as being at the Kaiser-Wilhelm-Institut für Chemie, Berlin-Dahlem. Received 22 December 1938.

[9] M. J. Carson; et al. (2004). "Neutron background in large-scale xenon detectors for dark matter searches". *Astroparticle Physics* **21** (6): 667–687. doi:10.1016/j.astropartphys.2004.05.001.

[10] Povh, B.; Rith, K.; Scholz, C.; Zetsche, F. (2002). *Particles and Nuclei: An Introduction to the Physical Concepts*. Berlin: Springer-Verlag. p. 73. ISBN 978-3-540-43823-6.

[11] J.-L. Basdevant, J. Rich, M. Spiro (2005). *Fundamentals in Nuclear Physics*. Springer. p. 155. ISBN 0-387-01672-4.

[12] Paul Allen Tipler, Ralph A. Llewellyn (2002). *Modern Physics* (4 ed.). Macmillan. p. 310. ISBN 0-7167-4345-0.

[13] Nakamura, K (2010). "Review of Particle Physics". *Journal of Physics G: Nuclear and Particle Physics* **37** (7A): 075021. Bibcode:2010JPhG...37g5021N. doi:10.1088/0954-3899/37/7A/075021. PDF with 2011 partial update for the 2012 edition The exact value of the mean lifetime is still uncertain, due to conflicting results from experiments. The Particle Data Group reports values up to six seconds apart (more than four standard deviations), commenting that "our 2006, 2008, and 2010 Reviews stayed with 885.7±0.8 s; but we noted that in light of SEREBROV 05 our value should be regarded as suspect until further experiments clarified matters. Since our 2010 Review, PICHLMAIER 10 has obtained a mean life of 880.7±1.8 s, closer to the value of SEREBROV 05 than to our average. And SEREBROV 10B[...] claims their values should be lowered by about 6 s, which would bring them into line with the two lower values. However, those reevaluations have not received an enthusiastic response from the experimenters in question; and in any case the Particle Data Group would have to await published changes (by those experimenters) of published values. At this point, we can think of nothing better to do than to average the seven best but discordant measurements, getting 881.5±1.5s. Note that the error includes a scale factor of 2.7. This is a jump of 4.2 old (and 2.8 new) standard deviations. This state of affairs is a particularly unhappy one, because the value is so important. We again call upon the experimenters to clear this up."

[14] R.K. Adair (1989). *The Great Design: Particles, Fields, and Creation*. Oxford University Press. p. 214.

[15] Cho, Adiran (2 April 2010). "Mass of the Common Quark Finally Nailed Down". *http://news.sciencemag.org*. American Association for the Advancement of Science. Retrieved 27 September 2014.

[16] W.N.Cottingham,D.A.Greenwood(1986).*An Introduction to Nuclear Physics*. Cambridge University Press. ISBN97805216.

[17] E.Rutherford(1920). "Nuclear Constitution of Atoms".*Proceedings of the Royal Society A***97**(686): 374–400. Bibcode:1920R. doi:10.1098/rspa.1920.0040.

[18] "Wolfgang Pauli". *Sources in the History of Mathematics and Physical Sciences*. Sources in the History of Mathematics and Physical Sciences **6**: 105–144. 1985. doi:10.1007/978-3-540-78801-0_3. ISBN 978-3-540-13609-5. |chapter= ignored (help)

[19] Hendry, John, ed. (1984), *Cambridge Physics in the Thirties*, Adam Hilger Ltd, Bristol, ISBN 0852747616

[20] N.Feather(1960). "A history of neutrons and nuclei. Part1".*Contemporary Physics***1**(3): 191–203. doi:10.1080/0010751.

[21] Brown,Laurie M. (1978). "The idea of the neutrino".*Physics Today***31**(9): 23. Bibcode:1978PhT....31i..23B.doi:10.1063/1.

[22] Friedlander G., Kennedy J.W. and Miller J.M. (1964) *Nuclear and Radiochemistry* (2nd edition), Wiley, pp. 22–23 and 38–39

[23] Stuewer, Roger H. (1985). "Niels Bohr and Nuclear Physics". In French, A. P.; Kennedy, P. J. *Niels Bohr: A Centenary Volume.* Harvard University Press. pp. 197–220. ISBN 0674624165.

[24] Pais, Abraham (1986). *Inward Bound.* Oxford: Oxford University Press. p. 299. ISBN 0198519974.

[25] Klein, O. (1929). "Die Reflexion von Elektronen an einem Potentialsprung nach der relativistischen Dynamik von Dirac". *Zeitschrift für Physik* **53** (3–4): 157–165. Bibcode:1929ZPhy...53..157K. doi:10.1007/BF01339716.

[26] Bothe, W.; Becker, H. (1930). "Künstliche Erregung von Kern-γ-Strahlen" [Artificial excitation of nuclear γ-radiation]. *Zeitschrift für Physik* **66** (5–6): 289–306. Bibcode:1930ZPhy...66..289B. doi:10.1007/BF01390908.

[27] Becker, H.; Bothe, W. (1932). "Die in Bor und Beryllium erregten γ-Strahlen" [Γ-rays excited in boron and beryllium]. *Zeitschrift für Physik* **76** (7–8): 421–438. Bibcode:1932ZPhy...76..421B. doi:10.1007/BF01336726.

[28] Joliot-Curie, Irène and Joliot, Frédéric (1932). "Émission de protons de grande vitesse par les substances hydrogénées sous l'influence des rayons γ très pénétrants" [Emission of high-speed protons by hydrogenated substances under the influence of very penetrating γ-rays]. *Comptes Rendus* **194**: 273.

[29] "Atop the Physics Wave: Rutherford Back in Cambridge, 1919–1937". *Rutherford's Nuclear World.* American Institute of Physics. 2011–2014. Retrieved 19 August 2014.

[30] Chadwick, J. (1933). "Bakerian Lecture. The Neutron". *Proceedings of the Royal Society A: Mathematical, Physical and Engineering Sciences* **142** (846): 1–25. Bibcode:1933RSPSA.142....1C. doi:10.1098/rspa.1933.0152.

[31] Heisenberg, W. (1932). "Über den Bau der Atomkerne. I". *Z. Phys.* **77**: 1–11. doi:10.1007/BF01342433.

[32] Heisenberg, W. (1932). "Über den Bau der Atomkerne. II". *Z. Phys.* **78** (3–4): 156–164. doi:10.1007/BF01337585.

[33] Heisenberg, W. (1933). "Über den Bau der Atomkerne. III". *Z. Phys.* **80** (9–10): 587–596. doi:10.1007/BF01335696.

[34] Iwanenko, D.D., The neutron hypothesis, Nature **129** (1932) 798.

[35] Miller A. I. *Early Quantum Electrodynamics: A Sourcebook*, Cambridge University Press, Cambridge, 1995, ISBN 0521568919, pp. 84–88.

[36] Wilson, Fred L. (1968). "Fermi's Theory of Beta Decay". *Am. J. Phys.* **36** (12): 1150–1160. Bibcode:1968AmJPh..36.1150W. doi:10.1119/1.1974382.

[37] Chadwick, J.; Goldhaber, M. (1934). "A nuclear photo-effect: disintegration of the diplon by gamma rays". *Nature* **134**: 237–238. doi:10.1038/134237a0.

[38] Chadwick,J.;Goldhaber,M. (1935). "A nuclear photoelectric effect"(PDF).*Proc.R.Soc.Lond***151**: 479–493. doi:10.1098/rspa.

[39] Cooper, Dan (1999). *Enrico Fermi: And the Revolutions in Modern physics.* New York: Oxford University Press. ISBN 0-19-511762-X. OCLC 39508200.

[40] Hahn, O. (1958). "The Discovery of Fission". *Scientific American* **198** (2): 76. doi:10.1038/scientificamerican0258-76.

[41] Rife, Patricia (1999). *Lise Meitner and the dawn of the nuclear age.* Basel, Switzerland: Birkhäuser. ISBN 0-8176-3732-X.

[42] Hahn, O.; Strassmann, F. (10 February 1939). "Proof of the Formation of Active Isotopes of Barium from Uranium and Thorium Irradiated with Neutrons; Proof of the Existence of More Active Fragments Produced by Uranium Fission". *Die Naturwissenschaften* **27**: 89–95.

[43] "The Nobel Prize in Chemistry 1944". Nobel Foundation. Retrieved 2007-12-17.

[44] Bernstein, Jeremy (2001). *Hitler's uranium club: the secret recordings at Farm Hall.* New York: Copernicus. p. 281. ISBN 0-387-95089-3.

[45] "The Nobel Prize in Chemistry 1944: Presentation Speech". Nobel Foundation. Retrieved 2008-01-03.

[46] Sir James Chadwick's Discovery of Neutrons. ANS Nuclear Cafe. Retrieved on 2012-08-16.

[47] Particle Data Group Summary Data Table on Baryons. lbl.gov (2007). Retrieved on 2012-08-16.

[48] Basic Ideas and Concepts in Nuclear Physics: An Introductory Approach, Third Edition K. Heyde Taylor & Francis 2004. Print ISBN 978-0-7503-0980-6. eBook ISBN 978-1-4200-5494-1. DOI: 10.1201/9781420054941.ch5. full text

[49] Olive, K.A.; et al. (2014). "Review of Particle Physics". *Chin. Phys. C* **38**: 090001. doi:10.1088/1674-1137/38/9/090001. |first2= missing |last2= in Authors list (help)

[50] "Pear-shaped particles probe big-bang mystery" (Press release). University of Sussex. 20 February 2006. Retrieved 2009-12-14.

[51] A cryogenic experiment to search for the EDM of the neutron. Hepwww.rl.ac.uk. Retrieved on 2012-08-16.

[52] Search for the neutron electric dipole moment: nEDM. Nedm.web.psi.ch (2001-09-12). Retrieved on 2012-08-16.

[53] SNS Neutron EDM Experiment. P25ext.lanl.gov. Retrieved on 2012-08-16.

[54] Measurement of the Neutron Electric Dipole Moment. Nrd.pnpi.spb.ru. Retrieved on 2012-08-16.

[55] Gell, Y.; Lichtenberg, D. B. (1969). "Quark model and the magnetic moments of proton and neutron". *Il Nuovo Cimento A*. Series 10 **61**: 27–40. Bibcode:1969NCimA..61...27G. doi:10.1007/BF02760010.

[56] Alvarez, L. W; Bloch, F. (1940). "A quantitative determination of the neutron magnetic moment in absolute nuclear magnetons". *Physical Review* **57**: 111–122. doi:10.1103/physrev.57.111.

[57] Miller,G.A. (2007). "Charge Densities of the Neutron and Proton".*Physical Review Letters***99**(11): 112001. Bibcode:2007Ph. doi:10.1103/PhysRevLett.99.112001.

[58] Greene, GL; et al. (1986). "New determination of the deuteron binding energy and the neutron mass". *Phys. Rev. Lett.* **56**: 819–822. Bibcode:1986PhRvL..56..819G. doi:10.1103/PhysRevLett.56.819.

[59] Byrne, J. *Neutrons, Nuclei, and Matter*, Dover Publications, Mineola, New York, 2011, ISBN 0486482383, pp. 18–19

[60] Spyrou, A.; et al. (2012). "First Observation of Ground State Dineutron Decay: 16Be". *Physical Review Letters* **108** (10): 102501. Bibcode:2012PhRvL.108j2501S. doi:10.1103/PhysRevLett.108.102501. PMID 22463404.

[61] Llanes-Estrada, Felipe J.; Moreno Navarro, Gaspar (2011). "Cubic neutrons". arXiv:1108.1859v1 [nucl-th].

[62] Köhn, C., Ebert, U., Calculation of beams of positrons, neutrons and protons associated with terrestrial gamma-ray flashes, J. Geophys. Res. Atmos. (2015), vol. 23, doi:10.1002/2014JD022229

[63] Clowdsley, MS; Wilson, JW; Kim, MH; Singleterry, RC; Tripathi, RK; Heinbockel, JH; Badavi, FF; Shinn, JL (2001). "Neutron Environments on the Martian Surface" (PDF). *Physica Medica* **17** (Suppl 1): 94–6. PMID 11770546.

[64] Byrne, J. *Neutrons, Nuclei, and Matter*, Dover Publications, Mineola, New York, 2011, ISBN 0486482383, pp. 32–33.

[65] Science/Nature | Q&A: Nuclear fusion reactor. BBC News (2006-02-06). Retrieved on 2010-12-04.

[66] Byrne, J. *Neutrons, Nuclei, and Matter*, Dover Publications, Mineola, New York, 2011, ISBN 0486482383, p. 453.

[67] Kumakhov, M. A.; Sharov, V. A. (1992). "A neutron lens". *Nature* **357** (6377): 390–391. Bibcode:1992Natur.357..390K. doi:10.1038/357390a0.

[68] Physorg.com, "New Way of 'Seeing': A 'Neutron Microscope'". Physorg.com (2004-07-30). Retrieved on 2012-08-16.

[69] "NASA Develops a Nugget to Search for Life in Space". NASA.gov (2007-11-30). Retrieved on 2012-08-16.

[70] Hall EJ. Radiobiology for the Radiologist. Lippincott Williams & Wilkins; 5th edition (2000)

[71] Johns HE and Cunningham JR. The Physics of Radiology. Charles C Thomas 3rd edition 1978

[72] Tami Freeman (May 23, 2008). "Facing up to secondary neutrons". Medical Physics Web. Retrieved 2011-02-08.

[73] Heilbronn, L.; Nakamura, T; Iwata, Y; Kurosawa, T; Iwase, H; Townsend, LW (2005). "Expand+Overview of secondary neutron production relevant to shielding in space". *Radiation Protection Dosimetry* **116** (1–4): 140–143. doi:10.1093/rpd/nci033. PMID 16604615.

25.14   Further reading

- Annotated bibliography for neutrons from the Alsos Digital Library for Nuclear Issues

- Abraham Pais, *Inward Bound*, Oxford: Oxford University Press, 1986. ISBN 0198519974.

- Sin-Itiro Tomonaga, *The Story of Spin*, The University of Chicago Press, 1997

- Herwig Schopper, *Weak interactions and nuclear beta decay*, Publisher, North-Holland Pub. Co., 1966.

25.15   External links

- neutron properties at Particle Data Group, Lawrence Berkeley National Laboratory in Berkeley, CA. (pdgLive)

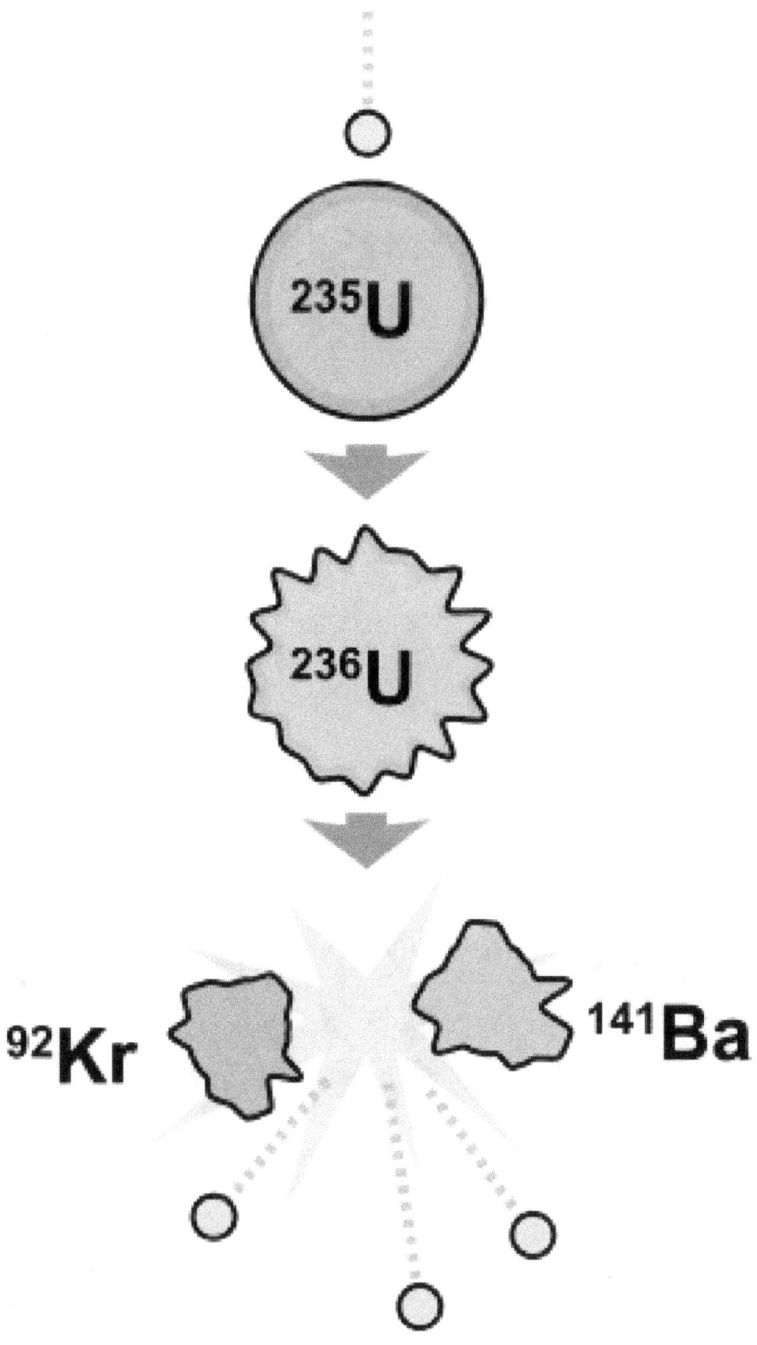

Nuclear fission caused by absorption of a neutron by uranium-235.
The heavy nuclide fragments into lighter components and additional neutrons.

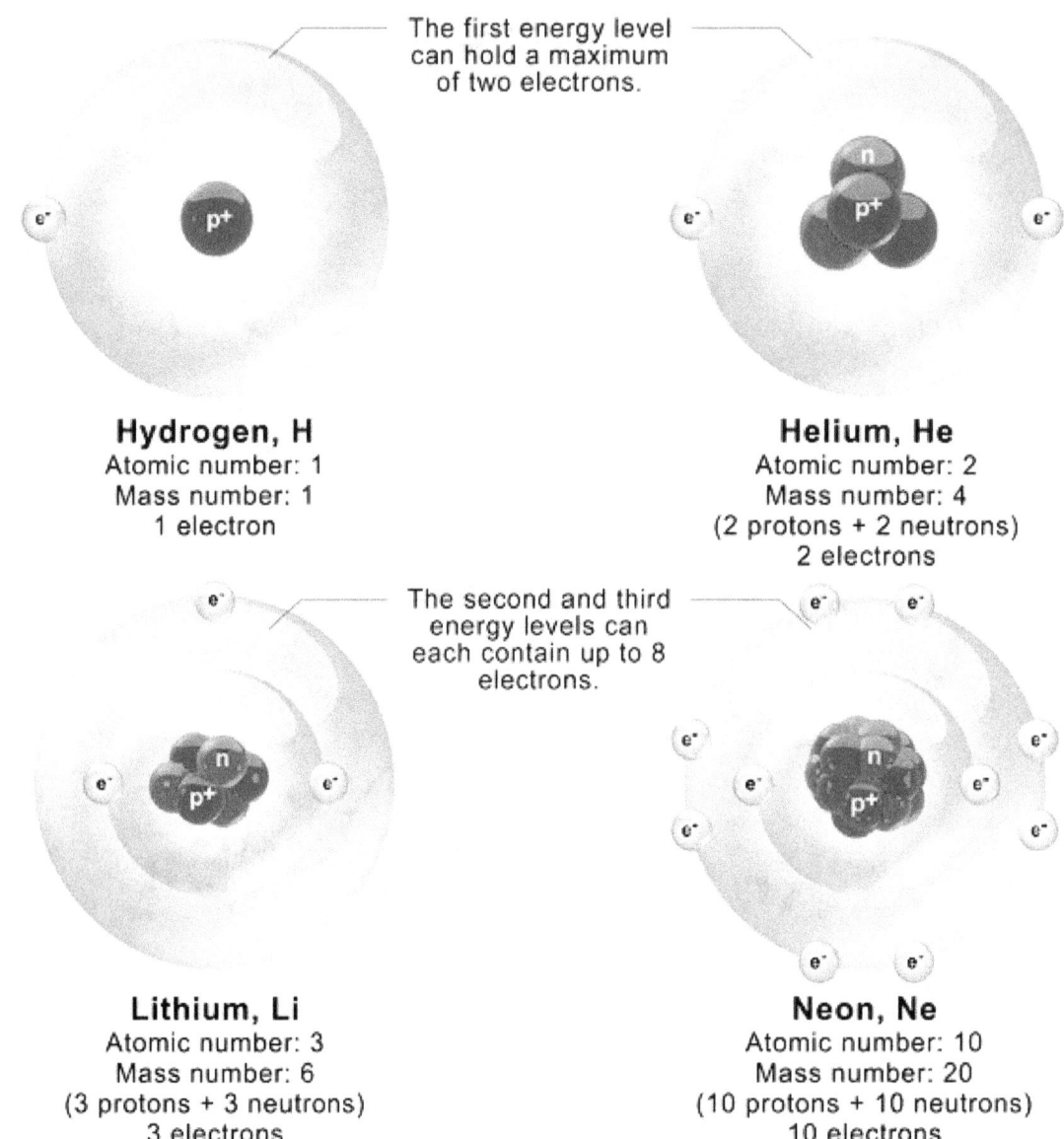

Models depicting the nucleus and electron energy levels in hydrogen, helium, lithium, and neon atoms. In reality, the diameter of the nucleus is about 100,000 times smaller than the diameter of the atom.

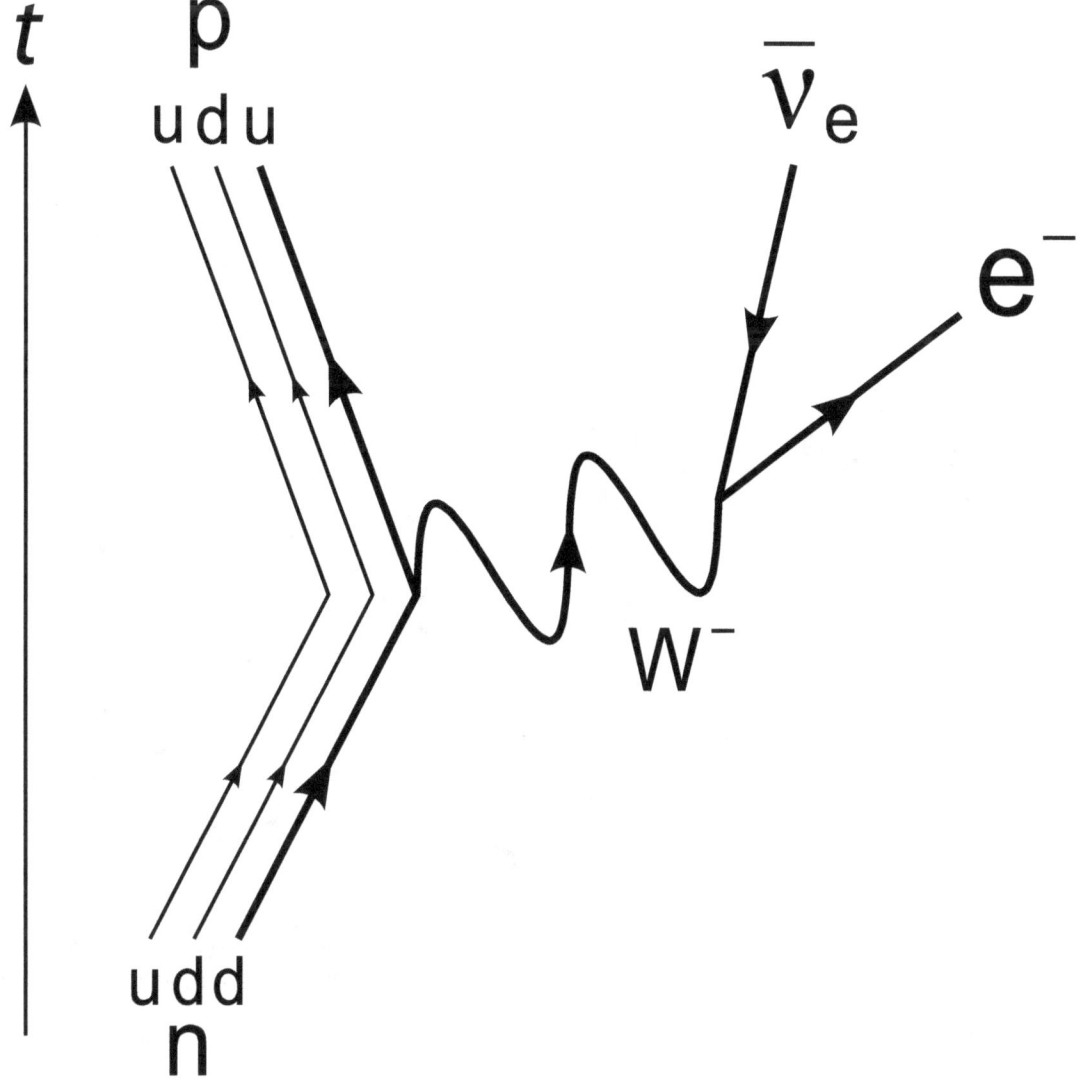

The Feynman diagram for beta decay of a neutron into a proton, electron, and electron antineutrino via an intermediate heavy W boson

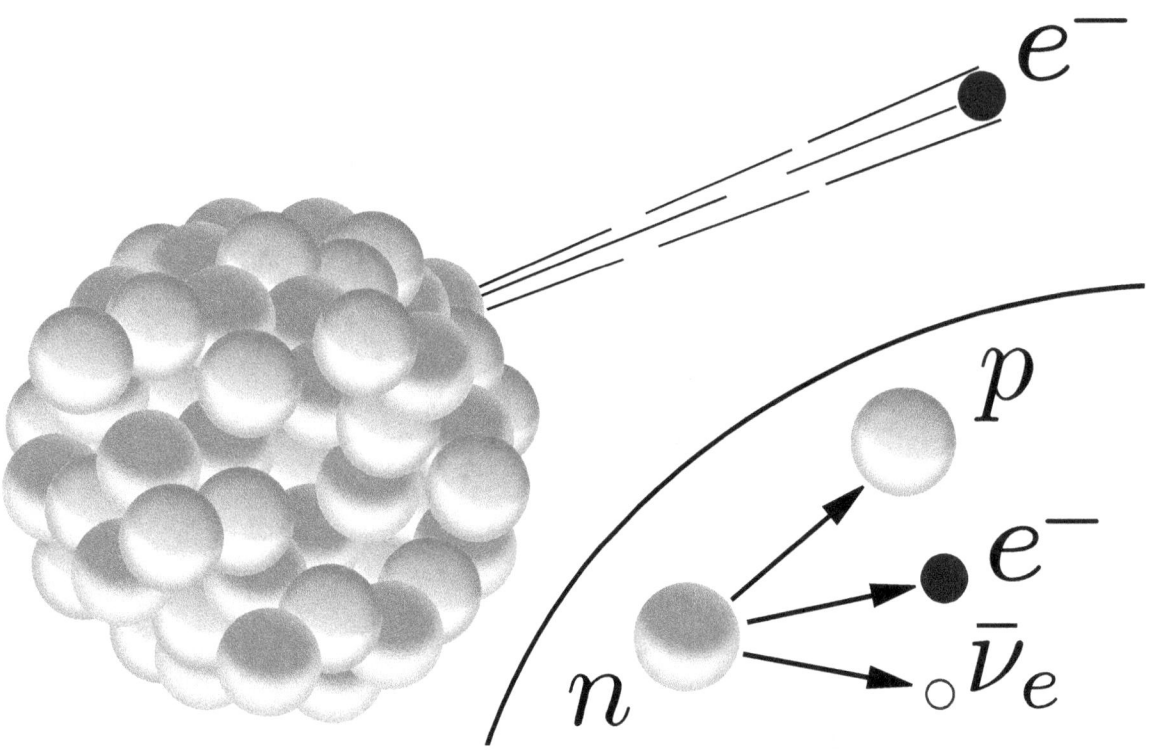

A schematic of the nucleus of an atom indicating β− radiation, the emission of a fast electron from the nucleus (the accompanying antineutrino is omitted). In the Rutherford model for the nucleus, red spheres were protons with positive charge and blue spheres were protons tightly bound to an electron with no net charge.
*The **inset** shows beta decay of a free neutron as it is understood today; an electron and antineutrino are created in this process.*

Institut Laue–Langevin (ILL) in Grenoble, France – a major neutron research facility.

Example of Cold Neutron Source

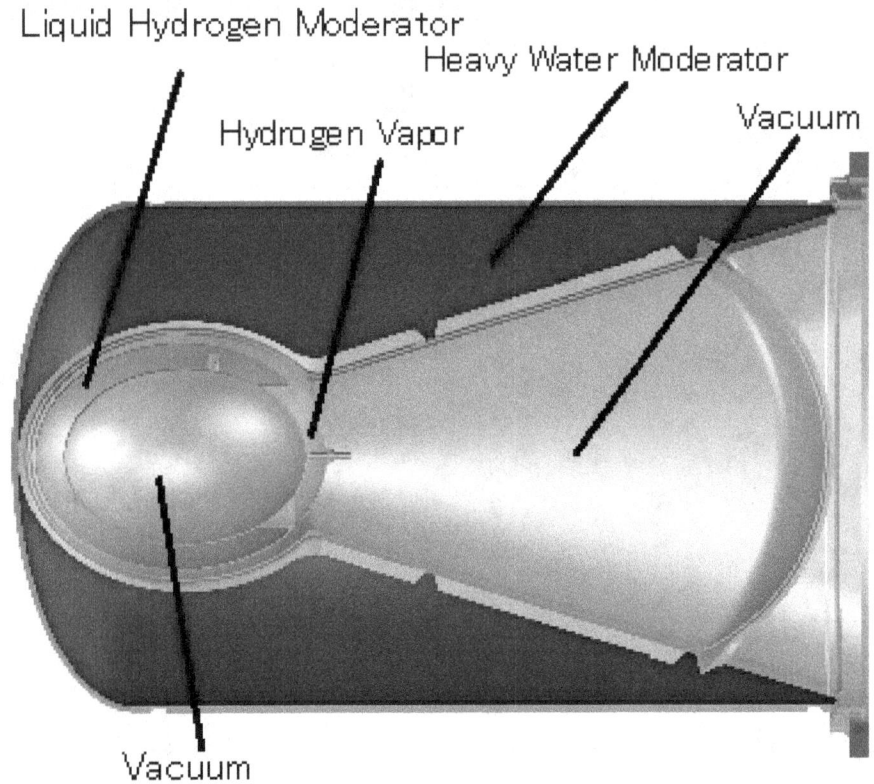

Cold neutron source providing neutrons at about the temperature of liquid hydrogen

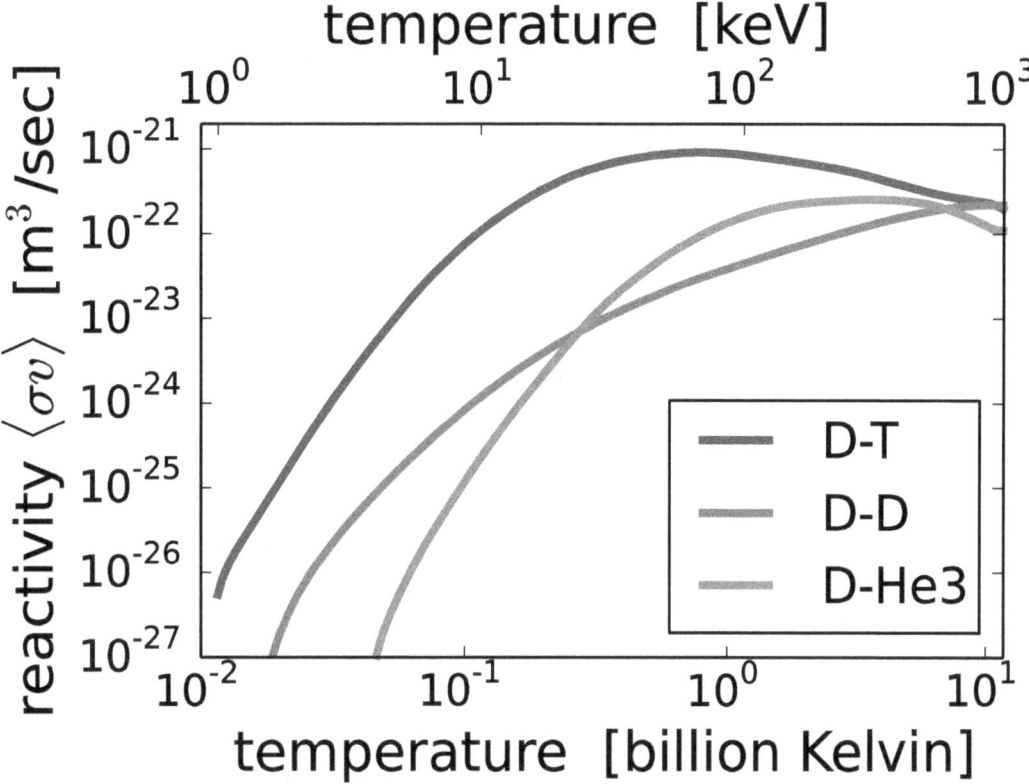

The fusion reaction rate increases rapidly with temperature until it maximizes and then gradually drops off. The DT rate peaks at a lower temperature (about 70 keV, or 800 million kelvins) and at a higher value than other reactions commonly considered for fusion energy.

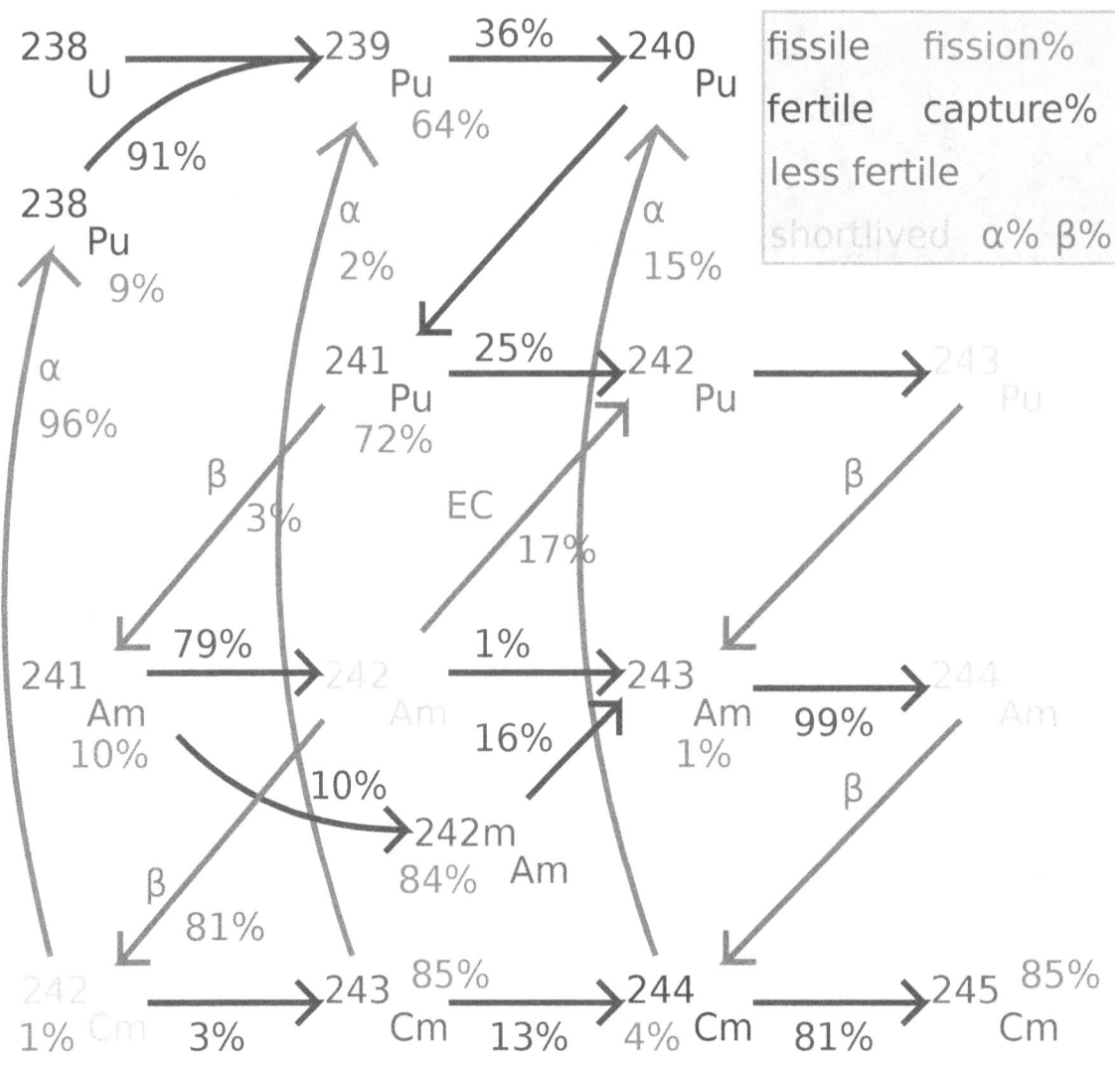

Transmutation flow in light water reactor, which is a thermal-spectrum reactor

Chapter 26

Antiparticle

Corresponding to most kinds of particles, there is an associated antimatter **antiparticle** with the same mass and opposite charge (including electric charge). For example, the antiparticle of the electron is the positively charged positron, which is produced naturally in certain types of radioactive decay.

The laws of nature are very nearly symmetrical with respect to particles and antiparticles. For example, an antiproton and a positron can form an antihydrogen atom, which is believed to have the same properties as a hydrogen atom. This leads to the question of why the formation of matter after the Big Bang resulted in a universe consisting almost entirely of matter, rather than being a half-and-half mixture of matter and antimatter. The discovery of Charge Parity violation helped to shed light on this problem by showing that this symmetry, originally thought to be perfect, was only approximate.

Particle-antiparticle pairs can annihilate each other, producing photons; since the charges of the particle and antiparticle are opposite, total charge is conserved. For example, the positrons produced in natural radioactive decay quickly annihilate themselves with electrons, producing pairs of gamma rays, a process exploited in positron emission tomography.

Antiparticles are produced naturally in beta decay, and in the interaction of cosmic rays in the Earth's atmosphere.

Because charge is conserved, it is not possible to create an antiparticle without either destroying a particle of the same charge (as in beta decay, when a proton (positive charge) is destroyed, a neutron created and a positron (positive charge, antiparticle) is also created and emitted) or by creating a particle of the opposite charge. The latter is seen in many processes in which both a particle and its antiparticle are created simultaneously, as in particle accelerators. This is the inverse of the particle-antiparticle annihilation process.

Although particles and their antiparticles have opposite charges, electrically neutral particles need not be identical to their antiparticles. The neutron, for example, is made out of quarks, the antineutron from antiquarks, and they are distinguishable from one another because neutrons and antineutrons annihilate each other upon contact. However, other neutral particles are their own antiparticles, such as photons, hypothetical gravitons, and some WIMPs.

26.1 History

26.1.1 Experiment

In 1932, soon after the prediction of positrons by Paul Dirac, Carl D. Anderson found that cosmic-ray collisions produced these particles in a cloud chamber— a particle detector in which moving electrons (or positrons) leave behind trails as they move through the gas. The electric charge-to-mass ratio of a particle can be measured by observing the radius of curling of its cloud-chamber track in a magnetic field. Positrons, because of the direction that their paths curled, were at first mistaken for electrons travelling in the opposite direction. Positron paths in a cloud-chamber trace the same helical path as an electron but rotate in the opposite direction with respect to the magnetic field direction due to their having the same magnitude of charge-to-mass ratio but with opposite charge and, therefore, opposite signed charge-to-mass ratios.

The antiproton and antineutron were found by Emilio Segrè and Owen Chamberlain in 1955 at the University of Califor-

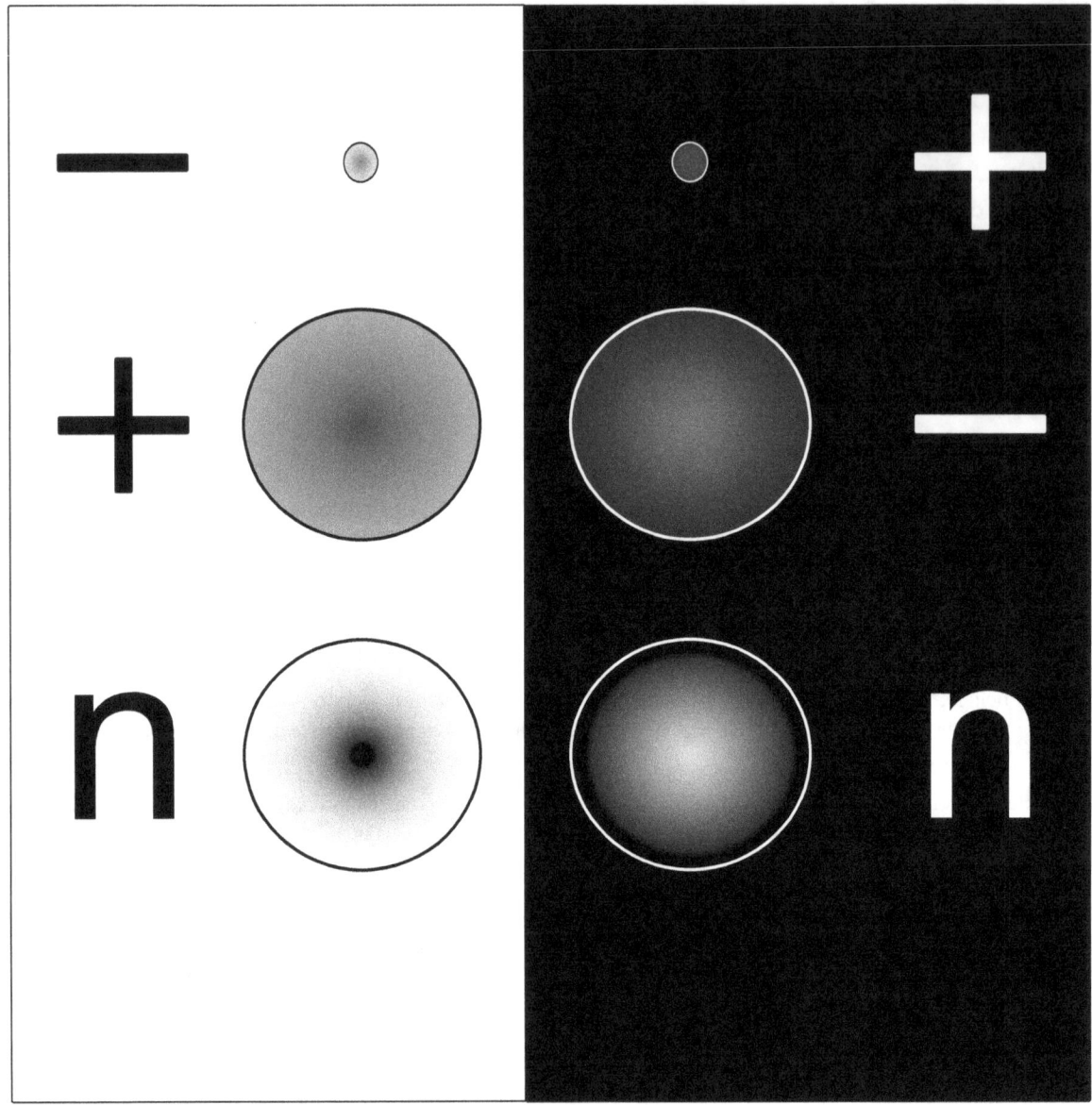

Illustration of electric charge of particles (left) and antiparticles (right). From top to bottom; electron/positron, proton/antiproton, neutron/antineutron.

nia, Berkeley. Since then, the antiparticles of many other subatomic particles have been created in particle accelerator experiments. In recent years, complete atoms of antimatter have been assembled out of antiprotons and positrons, collected in electromagnetic traps.[1]

26.1.2 Dirac's Hole theory

... the development of quantum field theory made the interpretation of antiparticles as holes unnecessary, even though it lingers on in many textbooks.

Steven Weinberg[2]

Solutions of the Dirac equation contained negative energy quantum states. As a result, an electron could always radiate energy and fall into a negative energy state. Even worse, it could keep radiating infinite amounts of energy because there were infinitely many negative energy states available. To prevent this unphysical situation from happening, Dirac proposed that a "sea" of negative-energy electrons fills the universe, already occupying all of the lower-energy states so that, due to the Pauli exclusion principle, no other electron could fall into them. Sometimes, however, one of these negative-energy particles could be lifted out of this Dirac sea to become a positive-energy particle. But, when lifted out, it would leave behind a *hole* in the sea that would act exactly like a positive-energy electron with a reversed charge. These he interpreted as "negative-energy electrons" and attempted to identify them with protons in his 1930 paper *A Theory of Electrons and Protons*[3] However, these "negative-energy electrons" turned out to be positrons, and not protons.

This picture implied an infinite negative charge for the universe--a problem of which Dirac was aware. Dirac tried to argue that we would perceive this as the normal state of zero charge. Another difficulty was the difference in masses of the electron and the proton. Dirac tried to argue that this was due to the electromagnetic interactions with the sea, until Hermann Weyl proved that hole theory was completely symmetric between negative and positive charges. Dirac also predicted a reaction $e- + p+ \rightarrow \gamma + \gamma$, where an electron and a proton annihilate to give two photons. Robert Oppenheimer and Igor Tamm proved that this would cause ordinary matter to disappear too fast. A year later, in 1931, Dirac modified his theory and postulated the positron, a new particle of the same mass as the electron. The discovery of this particle the next year removed the last two objections to his theory.

However, the problem of infinite charge of the universe remains. Also, as we now know, bosons also have antiparticles, but since bosons do not obey the Pauli exclusion principle (only fermions do), hole theory does not work for them. A unified interpretation of antiparticles is now available in quantum field theory, which solves both these problems.

26.2   Particle-antiparticle annihilation

Main article: Annihilation

If a particle and antiparticle are in the appropriate quantum states, then they can annihilate each other and produce

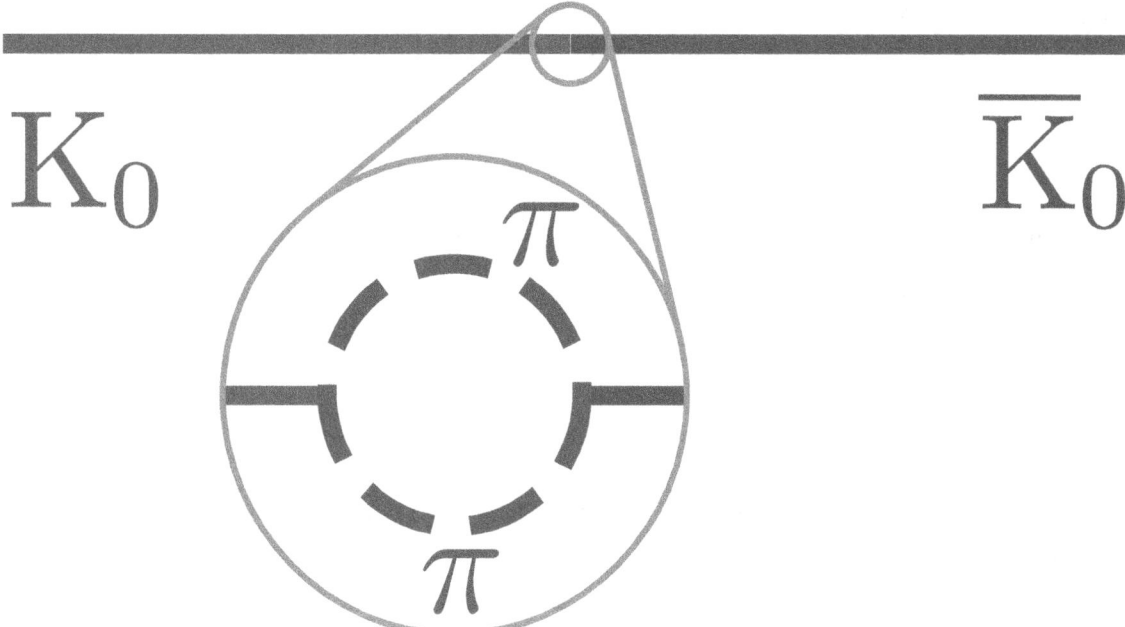

An example of a virtual pion pair that influences the propagation of a kaon, causing a neutral kaon to mix with the antikaon. This is an example of renormalization in quantum field theory— the field theory being necessary because of the change in particle number.

other particles. Reactions such as $e- + e+ \rightarrow \gamma + \gamma$ (the two-photon annihilation of an electron-positron pair) are an example. The single-photon annihilation of an electron-positron pair, $e- + e+ \rightarrow \gamma$, cannot occur in free space because it is impossible to conserve energy and momentum together in this process. However, in the Coulomb field of a nucleus the translational invariance is broken and single-photon annihilation may occur.[4] The reverse reaction (in free space, without an atomic nucleus) is also impossible for this reason. In quantum field theory, this process is allowed only as an intermediate quantum state for times short enough that the violation of energy conservation can be accommodated by the uncertainty principle. This opens the way for virtual pair production or annihilation in which a one particle quantum state may *fluctuate* into a two particle state and back. These processes are important in the vacuum state and renormalization of a quantum field theory. It also opens the way for neutral particle mixing through processes such as the one pictured here, which is a complicated example of mass renormalization.

26.3 Properties of antiparticles

Quantum states of a particle and an antiparticle can be interchanged by applying the charge conjugation (**C**), parity (**P**), and time reversal (**T**) operators. If $|p, \sigma, n\rangle$ denotes the quantum state of a particle (**n**) with momentum **p**, spin **J** whose component in the z-direction is σ, then one has

$$CPT\, |p, \sigma, n\rangle \;=\; (-1)^{J-\sigma}\, |p, -\sigma, n^c\rangle,$$

where n^c denotes the charge conjugate state, *i.e.*, the antiparticle. This behaviour under **CPT** is the same as the statement that the particle and its antiparticle lie in the same irreducible representation of the Poincaré group. Properties of antiparticles can be related to those of particles through this. If **T** is a good symmetry of the dynamics, then

$$T\, |p, \sigma, n\rangle \;\propto\; |-p, -\sigma, n\rangle,$$
$$CP\, |p, \sigma, n\rangle \;\propto\; |-p, \sigma, n^c\rangle,$$
$$C\, |p, \sigma, n\rangle \;\propto\; |p, \sigma, n^c\rangle,$$

where the proportionality sign indicates that there might be a phase on the right hand side. In other words, particle and antiparticle must have

- the same mass **m**
- the same spin state **J**
- opposite electric charges **q** and **-q**.

26.4 Quantum field theory

This section draws upon the ideas, language and notation of canonical quantization of a quantum field theory.

One may try to quantize an electron field without mixing the annihilation and creation operators by writing

$$\psi(x) = \sum_k u_k(x) a_k e^{-iE(k)t},$$

where we use the symbol k to denote the quantum numbers p and σ of the previous section and the sign of the energy, $E(k)$, and a_k denotes the corresponding annihilation operators. Of course, since we are dealing with fermions, we have to have the operators satisfy canonical anti-commutation relations. However, if one now writes down the Hamiltonian

$$H = \sum_k E(k) a_k^\dagger a_k,$$

then one sees immediately that the expectation value of H need not be positive. This is because $E(k)$ can have any sign whatsoever, and the combination of creation and annihilation operators has expectation value 1 or 0.

So one has to introduce the charge conjugate *antiparticle* field, with its own creation and annihilation operators satisfying the relations

$$b_{k\prime} = a_k^\dagger \text{ and } b_{k\prime}^\dagger = a_k,$$

where k has the same p, and opposite σ and sign of the energy. Then one can rewrite the field in the form

$$\psi(x) = \sum_{k_+} u_k(x) a_k e^{-iE(k)t} + \sum_{k_-} u_k(x) b_k^\dagger e^{-iE(k)t},$$

where the first sum is over positive energy states and the second over those of negative energy. The energy becomes

$$H = \sum_{k_+} E_k a_k^\dagger a_k + \sum_{k_-} |E(k)| b_k^\dagger b_k + E_0,$$

where E_0 is an infinite negative constant. The vacuum state is defined as the state with no particle or antiparticle, *i.e.*, $a_k|0\rangle = 0$ and $b_k|0\rangle = 0$. Then the energy of the vacuum is exactly E_0. Since all energies are measured relative to the vacuum, **H** is positive definite. Analysis of the properties of *ak* and *bk* shows that one is the annihilation operator for particles and the other for antiparticles. This is the case of a fermion.

This approach is due to Vladimir Fock, Wendell Furry and Robert Oppenheimer. If one quantizes a real scalar field, then one finds that there is only one kind of annihilation operator; therefore, real scalar fields describe neutral bosons. Since complex scalar fields admit two different kinds of annihilation operators, which are related by conjugation, such fields describe charged bosons.

26.4.1   Feynman–Stueckelberg interpretation

By considering the propagation of the negative energy modes of the electron field backward in time, Ernst Stueckelberg reached a pictorial understanding of the fact that the particle and antiparticle have equal mass **m** and spin **J** but opposite charges **q**. This allowed him to rewrite perturbation theory precisely in the form of diagrams. Richard Feynman later gave an independent systematic derivation of these diagrams from a particle formalism, and they are now called Feynman diagrams. Each line of a diagram represents a particle propagating either backward or forward in time. This technique is the most widespread method of computing amplitudes in quantum field theory today.

Since this picture was first developed by Ernst Stueckelberg, and acquired its modern form in Feynman's work, it is called the *Feynman-Stueckelberg interpretation* of antiparticles to honor both scientists.

As a consequence of this interpretation, Villata argued that the assumption of antimatter as CPT-transformed matter would imply that the gravitational interaction between matter and antimatter is repulsive.[5]

26.5   See also

- Gravitational interaction of antimatter

- Parity, charge conjugation and time reversal symmetry.

- CP violations and the baryon asymmetry of the universe.

- Quantum field theory and the list of particles

- Baryogenesis

26.6   References

[1] http://news.nationalgeographic.com/news/2010/11/101118-antimatter-trapped-engines-bombs-nature-science-cern/

[2] Weinberg, Steve. *The quantum theory of fields, Volume 1 : Foundations.* p. 14. ISBN 0-521-55001-7.

[3] Dirac,Paul(1930). "A Theory of Electrons and Protons".*Proceedings of the Royal Society A***126**(801): 360–365. Bibcode:. doi:10.1098/rspa.1930.0013.

[4] Sodickson, L.; W. Bowman; J. Stephenson (1961). "Single-Quantum Annihilation of Positrons". *Physical Review* **124** (6): 1851–1861. Bibcode:1961PhRv..124.1851S. doi:10.1103/PhysRev.124.1851.

[5] M. Villata, CPT symmetry and antimatter gravity in general relativity, 2011, EPL (Europhysics Letters) 94, 20001

- Feynman, R. P. (1987). "The reason for antiparticles". In R. P. Feynman and S. Weinberg. *The 1986 Dirac memorial lectures.* Cambridge University Press. ISBN 0-521-34000-4.

- Weinberg, S. (1995). *The Quantum Theory of Fields, Volume 1: Foundations.* Cambridge University Press. ISBN 0-521-55001-7.

Chapter 27

Exotic hadron

Exotic hadrons are subatomic particles composed of quarks and gluons, but which do not fit into the usual scheme of hadrons. While bound by the strong interaction they are not predicted by the simple quark model. That is, exotic hadrons do not have the same quark content as ordinary hadrons: **exotic baryons** have more than just the three quarks of ordinary baryons and **exotic mesons** do not have one quark and one antiquark like ordinary mesons. Exotic hadrons can be searched for by looking for S-matrix poles with quantum numbers forbidden to ordinary hadrons. Experimental signatures for such exotic hadrons have been seen recently[1] but remain a topic of controversy in particle physics.

Jaffe and Low [2] suggested that the exotic hadrons manifest themselves as poles of the P matrix, and not of the S matrix. Experimental P-matrix poles are determined reliably in both the meson-meson channels and nucleon-nucleon channels.

27.1 History

When the quark model was first postulated by Murray Gell-Mann and others in the 1960s, it was to organize the states known then to be in existence in a meaningful way. As Quantum Chromodynamics (QCD) developed over the next decade, it became apparent that there was no reason why only 3-quark and quark-antiquark combinations could exist. In addition, it seemed that gluons, the mediator particles of the strong interaction, could also form bound states by themselves (glueballs) and with quarks (hybrid hadrons). Several decades have passed without conclusive evidence of an exotic hadron that could be associated with the S-matrix pole.

In April 2014, The LHCb collaboration confirmed the existence of the Z(4430)$^-$. Examinations of the character of the particle suggest that it may be exotic.[3]

27.2 Candidates

There are several exotic hadron candidates:

- X(3872) – Discovered by the Belle detector at KEK in Japan, this particle has been variously hypothesized to be diquark or a mesonic molecule.

- Y(3940) – This particle fails to fit into the Charmonium spectrum predicted by theorists.

- Y(4140) – Discovered at Fermilab in March 2009 .

- Y(4260) – Discovered by the BaBar detector at SLAC in Menlo Park, California this particle is hypothesized to be made up of a gluon bound to a quark and antiquark.

- Zc(3900) – Discovered by Belle and BES III

- Z(4430) – Discovered by Belle and later confirmed by LHCb with 13.9σ significance

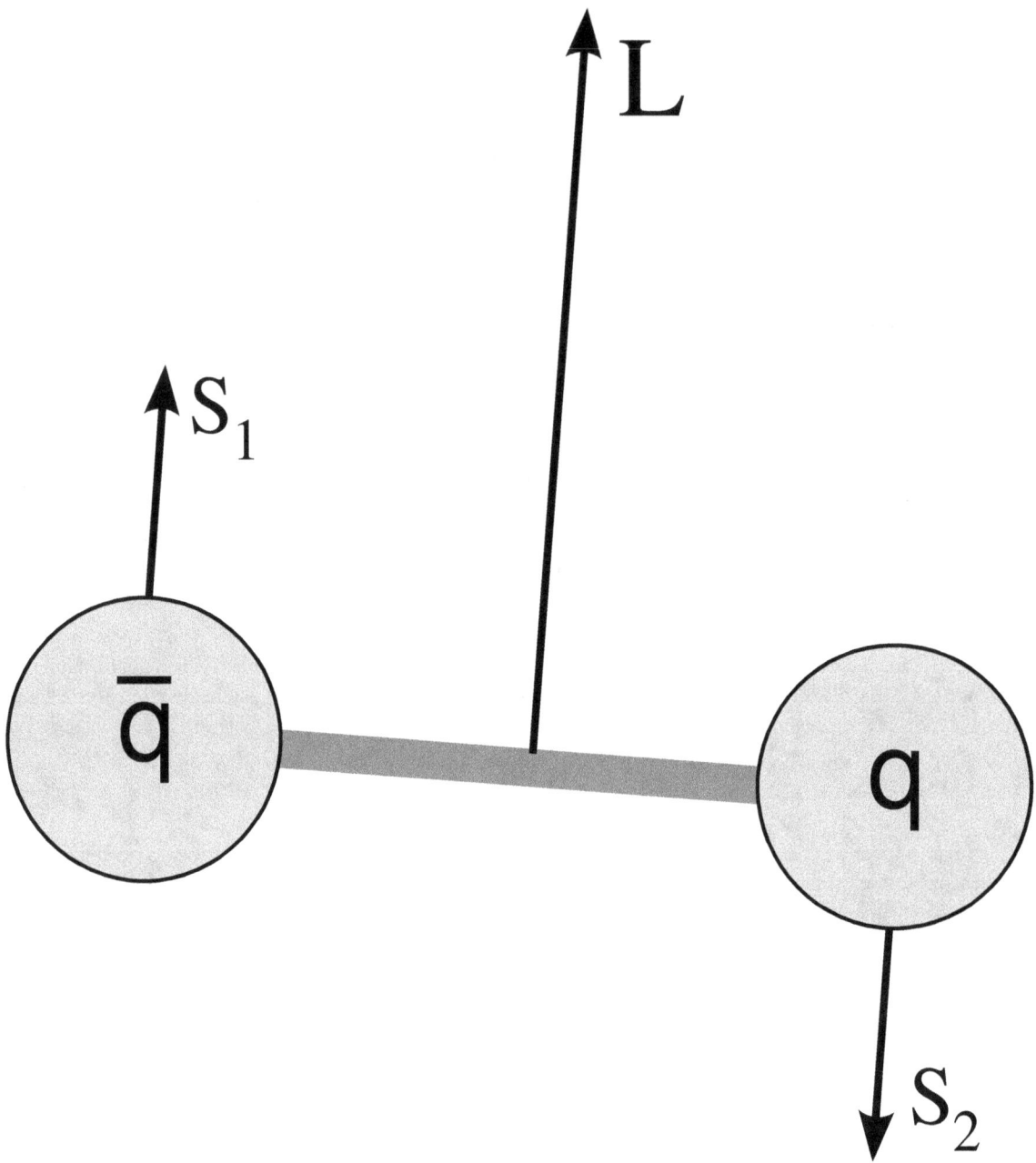

A regular meson made from a quark (q) and an antiquark (q̄) with spins s_2 and s_1 respectively and having an overall angular momentum L

27.3 See also

- Pentaquark

- Tetraquark

27.4   Notes

[1] See "note on non-q qbar mesons" in PDG 2006, Journal of Physics, G 33 (2006) 1.

[2] R. L. Jaffe and F. E. Low, Phys. Rev. D 19, 2105 (1979). doi:10.1103/PhysRevD.19.2105

[3] LHCb collaboration (7 April 2014). "Observation of the resonant character of the Z(4430)⁻ state". arXiv:1404.1903.

Chapter 28

Tetraquark

A **tetraquark**, in particle physics, is an exotic meson composed of four valence quarks. In principle, a tetraquark state may be allowed in quantum chromodynamics, the modern theory of strong interactions. Any established tetraquark state would be an example of an exotic hadron which lies outside the quark model classification.

28.1 History

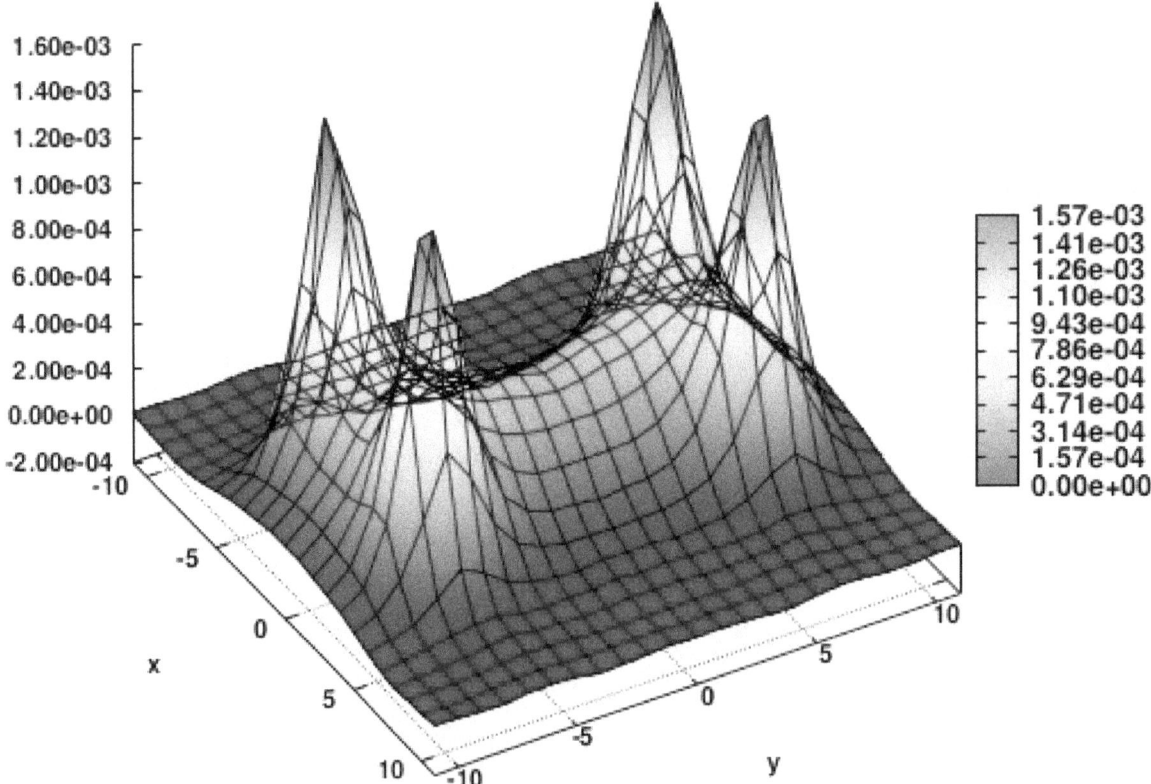

Colour flux tubes produced by four static quark and antiquark charges, computed in lattice QCD.[1] Confinement in Quantum Chromo Dynamics leads to the production of flux tubes connecting colour charges. The flux tubes act as attractive QCD string-like potentials.

In 2003 a particle temporarily called X(3872), by the Belle experiment in Japan, was proposed to be a tetraquark candidate,[2] as originally theorized.[3] The name X is a temporary name, indicating that there are still some questions

219

about its properties to be tested. The number following is the mass of the particle in 10^0 MeV/c^2.

In 2004, the D_sJ(2632) state seen in Fermilab's SELEX was suggested as a possible tetraquark candidate.

In 2007, Belle announced the observation of the Z(4430) state, a ccdu tetraquark candidate. In 2014, the Large Hadron Collider experiment LHCb confirmed this resonance with a significance of over 13.9σ.[4][5] There are also indications that the Y(4660), also discovered by Belle in 2007, could be a tetraquark state.[6]

In 2009, Fermilab announced that they have discovered a particle temporarily called Y(4140), which may also be a tetraquark.[7]

In 2010, two physicists from DESY and a physicist from Quaid-i-Azam University re-analyzed former experimental data and announced that, in connection with the Υ(5S) meson (a form of bottomonium), a well-defined tetraquark resonance exists.[8][9]

In June 2013, two independent groups reported on Z_c(3900).[10] [11]

28.2   See also

- Color confinement

- Hadron

- Pentaquark

28.3   References

[1] N. Cardoso, M. Cardoso, and P. Bicudo (2011). "Colour Fields Computed in SU(3) Lattice QCD for the Static Tetraquark System". *Physical Review D* **84** (5): 054508. arXiv:1107.1355. doi:10.1103/PhysRevD.84.054508.

[2] D. Harris (13 April 2008). "The charming case of X(3872)". *Symmetry Magazine*. Retrieved 2009-12-17.

[3] L. Maiani, F. Piccinini, V. Riquer and A.D. Polosa (2005). "Diquark-antidiquarks with hidden or open charm and the nature of X(3872)".*Physical Review D***71**: 014028. arXiv:hep-ph/0412098. Bibcode:2005PhRvD..71a4028M.doi:10.1103/PhysRevD..

[4] "LHCb confirms existence of exotic hadrons".

[5] LHCb collaboration, LHCb; Aaij, R.; Adeva, B.; Adinolfi, M.; Affolder, A.; Ajaltouni, Z.; Albrecht, J.; Alessio, F.; Alexander, M.; Ali, S.; Alkhazov, G.; Alvarez Cartelle, P.; Alves Jr, A. A.; Amato, S.; Amerio, S.; Amhis, Y.; An, L.; Anderlini, L.; Anderson, J.; Andreassen, R.; Andreotti, M.; Andrews, J. E.; Appleby, R. B.; Aquines Gutierrez, O.; Archilli, F.; Artamonov, A.; Artuso, M.; Aslanides, E.; Auriemma, G.; et al. (2014). "Observation of the resonant character of the Z(4430)– state". arXiv:1404.1903v1 [hep-ex].

[6] G. Cotugno, R. Faccini, A.D. Polosa and C. Sabelli (2010). "Charmed Baryonium". *Physical Review Letters* **104** (13): 132005. arXiv:0911.2178. Bibcode:2010PhRvL.104m2005C. doi:10.1103/PhysRevLett.104.132005.

[7] Anne Minard (2009-03-18). "New Particle Throws Monkeywrench in Particle Physics". Universetoday.com. Retrieved 2014-04-12.

[8] "Evidence grows for tetraquarks". physicsworld.com. Retrieved 2014-04-12.

[9] A. Ali, C. Hambrock, M.J. Aslam; Hambrock; Aslam (2010). "Tetraquark Interpretation of the BELLE Data on the Anomalous Υ(1S)π+π- and Υ(2S)π+π- Production near the Υ(5S) Resonance". *Physical Review Letters* **104** (16): 162001. arXiv:0912.5016. Bibcode:2010PhRvL.104p2001A. doi:10.1103/PhysRevLett.104.162001.

[10] "Physics - New Particle Hints at Four-Quark Matter". Physics.aps.org. 2013-06-17. Retrieved 2014-04-12.

[11] Eric Swanson (2013). "Viewpoint: New Particle Hints at Four-Quark Matter". *Physics* **69** (6). Bibcode:2013PhyOJ...6...69S. doi:10.1103/Physics.6.69.

28.4 External links

- The Belle experiment (press release)

- O'Luanaigh, Cian. "LHCb confirms existence of exotic hadrons". *cern.ch*. Geneva, Switzerland: CERN. Retrieved 2014-04-12.

Chapter 29

Pentaquark

Two models of a generic pentaquark

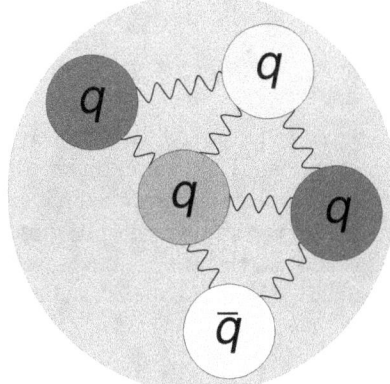

A five-quark "bag"

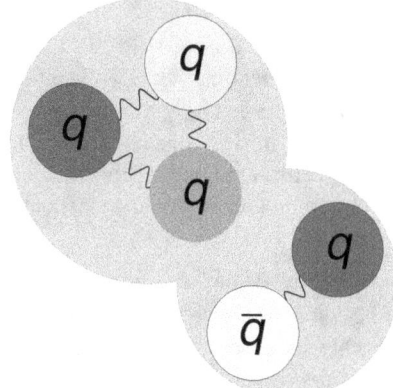

A "meson-baryon molecule"

q indicates a quark, whereas $\bar{q}$ indicates an antiquark. The wavy lines are gluons, which mediate the strong interaction between the quarks. The colours correspond to the various colour charges of quarks. The colours red, green, and blue must each be present and the remaining quark and antiquark must share corresponding colour and anticolour, here chosen to be blue and antiblue (shown as yellow).

A **pentaquark** is a subatomic particle consisting of four quarks and one antiquark bound together.

As quarks have a baryon number of $+\frac{1}{3}$, and antiquarks of $-\frac{1}{3}$, the pentaquark would have a total baryon number of 1, and thus would be a baryon. Further, because it has five quarks instead of the usual three found in regular baryons (aka 'triquarks'), it would be classified as an exotic baryon. The name pentaquark was coined by Harry J. Lipkin in 1987,[1] however, the possibility of five-quark particles was identified as early as 1964 when Murray Gell-Mann first postulated the existence of quarks.[2] Although predicted for decades, pentaquarks have proved surprisingly difficult to discover and some physicists were beginning to suspect that an unknown law of nature prevented their production.[3]

The first claim of pentaquark discovery was recorded at LEPS in Japan in 2003, and several experiments in the mid-2000s also reported discoveries of other pentaquark states.[4] Others were not able to replicate the LEPS results, however, and the other pentaquark discoveries were not accepted because of poor data and statistical analysis.[5] On 13 July 2015, the LHCb collaboration at CERN reported results consistent with pentaquark states in the decay of bottom Lambda baryons (Λ^0_b).[6]

Outside of particle physics laboratories pentaquarks also could be produced naturally by supernovae as part of the process of forming a neutron star.[7] The scientific study of pentaquarks might offer insights into how these stars form, as well as, allowing more thorough study of particle interactions and the strong force.

29.1 Background

Main article: Quark

A quark is a type of elementary particle that has mass, electric charge, and colour charge, as well as an additional property called flavour, which describes what type of quark it is (up, down, strange, charm, top, or bottom). Due to an effect known as colour confinement, quarks are never seen on their own. Instead, they form composite particles known as hadrons so that their colour charges cancel out. Hadrons made of one quark and one antiquark are known as mesons, while those made of three quarks are known as baryons. These 'regular' hadrons are well documented and characterized, however, there is nothing in theory to prevent quarks from forming 'exotic' hadrons such as tetraquarks with two quarks and two antiquarks, or pentaquarks with four quarks and one antiquark.[3]

29.2 Structure

A wide variety of pentaquarks are possible, with different quark combinations producing different particles. To identify which quarks compose a given pentaquark, physicists use the notation $qqqq\bar{q}$, where q and $\bar{q}$ respectively refer to any of the six flavours of quarks and antiquarks. The symbols u, d, s, c, b, and t stand for the up, down, strange, charm, bottom, and top quarks respectively, with the symbols of $\bar{u}$, $\bar{d}$, $\bar{s}$, $\bar{c}$, $\bar{b}$, $\bar{t}$ corresponding to the respective antiquarks. For instance a pentaquark made of two up quarks, one down quark, one charm quark, and one charm antiquark would be denoted uudc$\bar{c}$.

The quarks are bound together by the strong force, which acts in such a way as to cancel the colour charges within the particle. In a meson, this means a quark is partnered with an antiquark with an opposite colour charge – blue and antiblue, for example – while in a baryon, the three quarks have between them all three colour charges – red, blue, and green.[nb 1] In a pentaquark, the colours also need to cancel out, and the only feasible combination is to have one quark with one colour (e.g. red), one quark with a second colour (e.g. green), two quarks with the third colour (e.g. blue), and one antiquark to counteract the surplus colour (e.g. antiblue).[8]

The binding mechanism for pentaquarks is not yet clear. They may consist of five quarks tightly bound together, but it is also possible that they are more loosely bound and consist of a three-quark baryon and a two-quark meson interacting relatively weakly with each other via pion exchange (the same force that binds atomic nuclei) in a "meson-baryon molecule".[9][2][10]

29.3 History

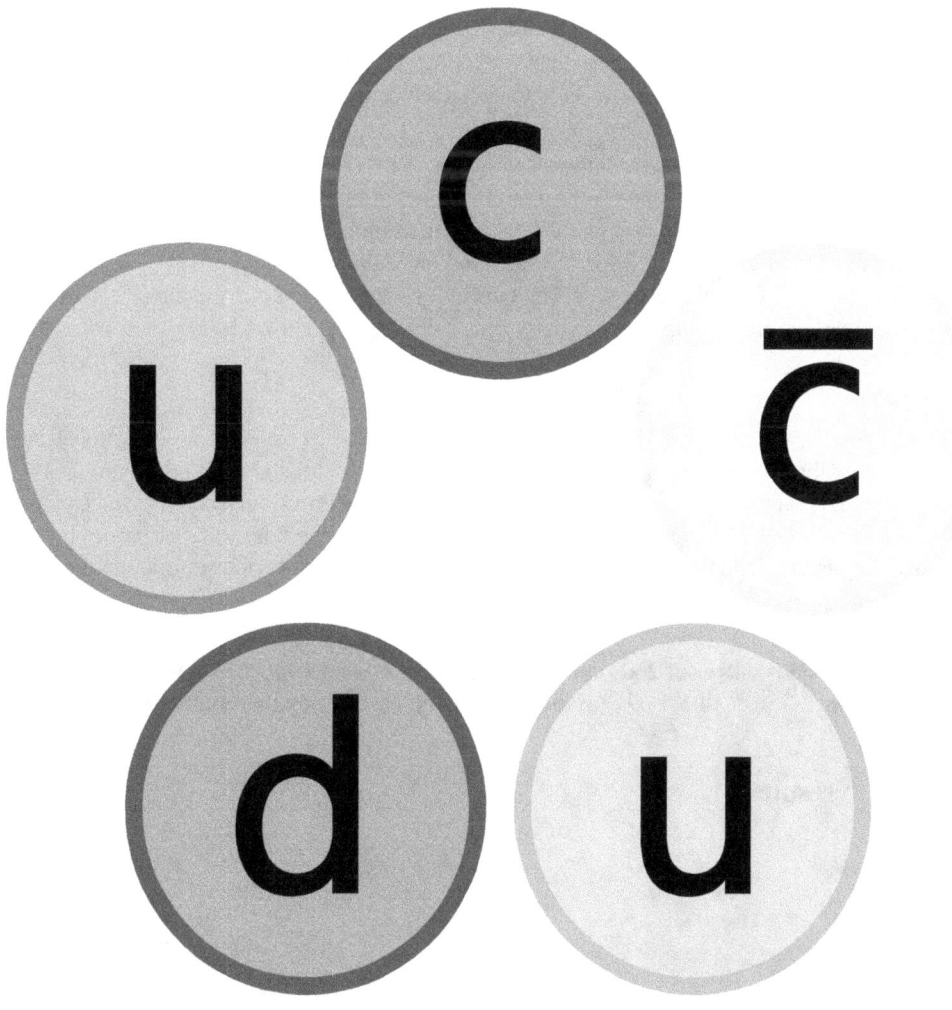

A diagram of the P+
c type pentaquark possibly discovered in July 2015, showing the flavours of each quark and one possible colour configuration.

29.3.1   Mid-2000s

The requirement to include an antiquark means that many classes of pentaquark are hard to identify experimentally – if the flavour of the antiquark matches the flavour of any other quark in the quintuplet, it will cancel out and the particle will resemble its three-quark hadron cousin. For this reason, early pentaquark searches looked for particles where the antiquark did not cancel.[8] In the mid-2000s, several experiments claimed to reveal pentaquark states. In particular, a resonance with a mass of 1540 MeV/c^2 (4.6 σ) was reported by LEPS in 2003, the Θ+.[11] This coincided with a pentaquark state with a mass of 1530 MeV/c^2 predicted in 1997.[12]

The proposed state was composed of two up quarks, two down quarks, and one strange antiquark (uudds). Following this announcement, nine other independent experiments reported seeing narrow peaks from nK+ and pK0, with masses between 1522 MeV/c^2 and 1555 MeV/c^2, all above 4 σ.[11] While concerns existed about the validity of these states, the

Particle Data Group gave the Θ+ a 3-star rating (out of 4) in the 2004 *Review of Particle Physics*.[11] Two other pentaquark states were reported albeit with low statistical significance—the Φ−− (ddssu), with a mass of 1860 MeV/c^2 and the Θ0 c (uuddc), with a mass of 3099 MeV/c^2. Both were later found to be statistical effects rather than true resonances.[11]

Ten experiments then looked for the Θ+, but came out empty-handed.[11] Two in particular (one at BELLE, and the other at CLAS) had nearly the same conditions as other experiments which claimed to have detected the Θ+ (DIANA and SAPHIR respectively).[11] The 2006 *Review of Particle Physics* concluded:[11]

> [T]here has not been a high-statistics confirmation of any of the original experiments that claimed to see the Θ+; there have been two high-statistics repeats from Jefferson Lab that have clearly shown the original positive claims in those two cases to be wrong; there have been a number of other high-statistics experiments, none of which have found any evidence for the Θ+; and all attempts to confirm the two other claimed pentaquark states have led to negative results. The conclusion that pentaquarks in general, and the Θ+, in particular, do not exist, appears compelling.

The 2008 *Review of Particle Physics* went even further:[5]

> There are two or three recent experiments that find weak evidence for signals near the nominal masses, but there is simply no point in tabulating them in view of the overwhelming evidence that the claimed pentaquarks do not exist... The whole story—the discoveries themselves, the tidal wave of papers by theorists and phenomenologists that followed, and the eventual "undiscovery"—is a curious episode in the history of science.

Despite these null results, LEPS results as of 2009 continue to show the existence of a narrow state with a mass of 1524±4 MeV/c^2, with a statistical significance of 5.1 σ.[13] Experiments continue to study this controversy.

29.3.2 2015 LHCb results

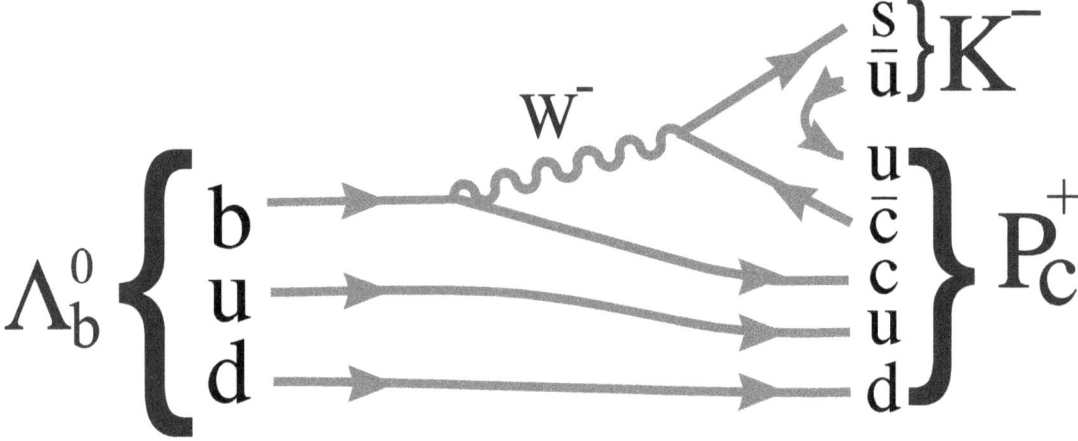

Feynman diagram representing the decay of a lambda baryon Λ0 b into a kaon K− and a pentaquark P+ c.

In July 2015, the LHCb collaboration identified pentaquarks in the Λ0 b→J/ψK− p channel, which represents the decay of the bottom lambda baryon (Λ0

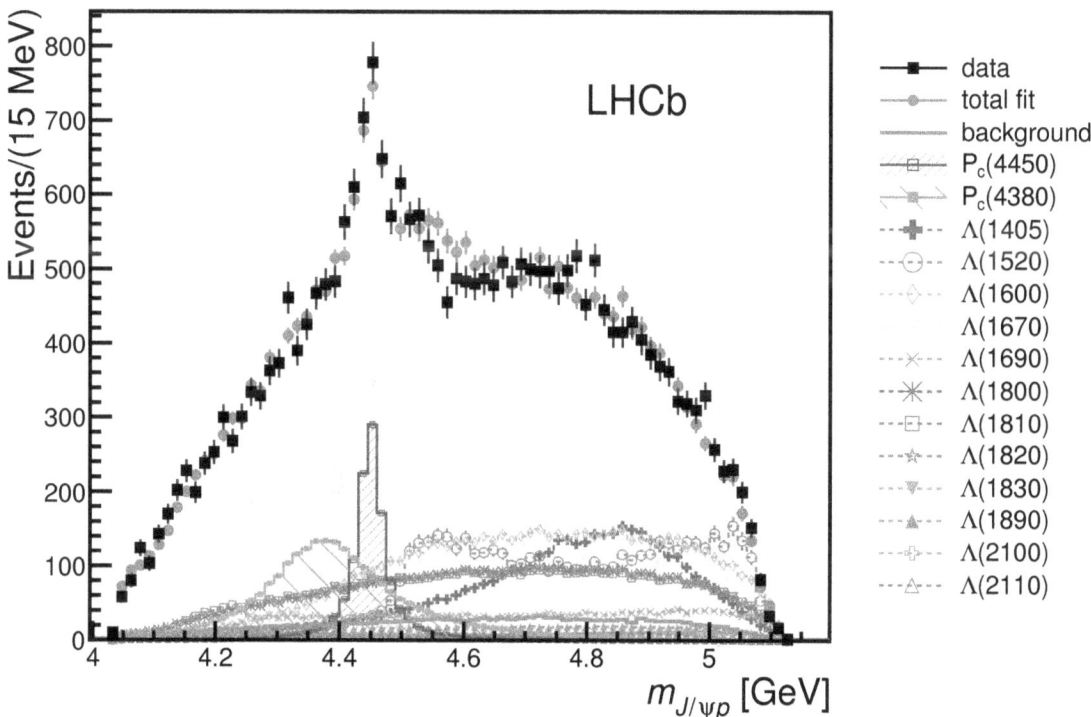

A fit to the J/ψp invariant mass spectrum for the Λ0
b→J/ψK−
p decay, with each fit component shown individually. The contribution of the pentaquarks are shown by hatched histograms.

b) into a J/ψ meson (J/ψ), a kaon (K−
) and a proton (p). The results showed that sometimes, instead of decaying directly into mesons and baryons, the Λ0
b decayed via intermediate pentaquark states. The two states, named P+
c(4380) and P+
c(4450), had individual statistical significances of 9 σ and 12 σ, respectively, and a combined significance of 15 σ —
enough to claim a formal discovery. The analysis ruled out the possibility that the effect was caused by conventional
particles.[2] The two pentaquark states were both observed decaying strongly to J/ψp, hence must have a valence quark
content of two up quarks, a down quark, a charm quark, and an anti-charm quark (uudcc), making them charmonium-
pentaquarks.[6][7][14]

The search for pentaquarks was not an objective of the LHCb experiment (which is primarily designed to investigate
matter-antimatter asymmetry)[15] and the apparent discovery of pentaquarks was described as an "accident" and "something we've stumbled across" by a CERN spokesperson.[9]

29.4 Applications

The discovery of pentaquarks will allow physicists to study the strong force in greater detail and aid understanding of
quantum chromodynamics. In addition, current theories suggest that some very large stars produce pentaquarks as they
collapse. The study of pentaquarks might help shed light on the physics of neutron stars.[7]

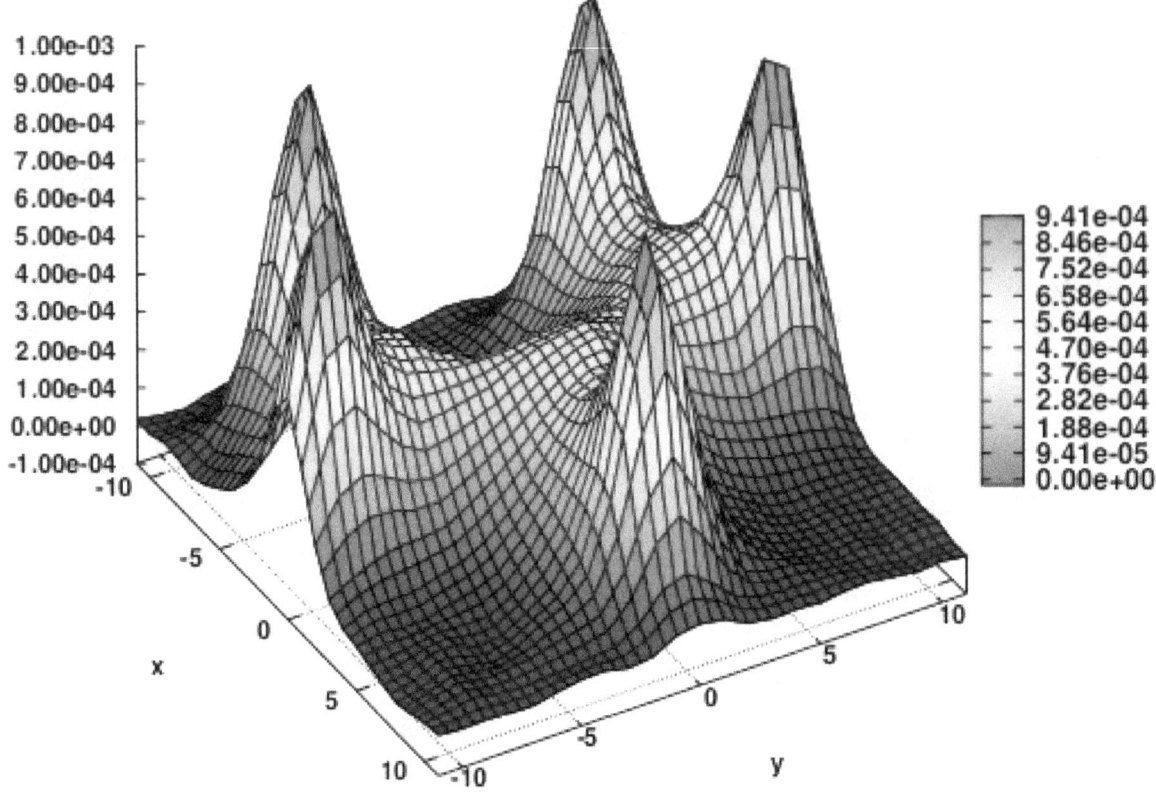

Colour flux tubes produced by five static quark and antiquark charges, computed in lattice QCD.[16] Confinement in Quantum Chromo Dynamics leads to the production of flux tubes connecting colour charges. The flux tubes act as attractive QCD string-like potentials.

29.5   See also

- Exotic matter

- List of particles

- Quark model

- Tetraquark

- Triquark

29.6   Footnotes

[1] The colour charges do not correspond to physical visible colours. They are arbitrary labels used to help scientists describe and visualise the charges of quarks.

29.7   References

[1] H. J. Lipkin (1987). "New possibilities for exotic hadrons — anticharmed strange baryons". *Physics Letters B* **195** (3): 484–488. Bibcode:1987PhLB..195..484L. doi:10.1016/0370-2693(87)90055-4.

[2] "Observation of particles composed of five quarks, pentaquark-charmonium states, seen in Λ0 b→J/ψpK⁻ decays". CERN/LHCb. 14 July 2015. Retrieved 2015-07-14.

[3] H. Muir (2 July 2003). "Pentaquark discovery confounds sceptics". *New Scientist*. Retrieved 2010-01-08.

[4] K. Hicks (23 July 2003). "Physicists find evidence for an exotic baryon". Ohio University. Retrieved 2010-01-08.

[5] See p. 1124 in C. Amsler et al. (Particle Data Group) (2008). "Review of particle physics" (PDF). *Physics Letters B* **667** (1-5): 1. Bibcode:2008PhLB..667....1A. doi:10.1016/j.physletb.2008.07.018.

[6] R. Aaij et al. (LHCb collaboration) (2015). "Observation of J/ψp resonances consistent with pentaquark states in Λ0 b→J/ψK− p decays". *Physical Review Letters* **115** (7). doi:10.1103/PhysRevLett.115.072001.

[7] I. Sample (14 July 2015). "Large Hadron Collider scientists discover new particles: pentaquarks". *The Guardian*. Retrieved 2015-07-14.

[8] J. Pochodzalla (2005). "Duets of strange quarks". *Hadron Physics*. p. 268. ISBN 161499014X.

[9] G. Amit (14 July 2015). "Pentaquark discovery at LHC shows long-sought new form of matter". *New Scientist*. Retrieved 2015-07-14.

[10] T. D. Cohen, P. M. Hohler, R. F. Lebed (2005). "On the Existence of Heavy Pentaquarks: The large N_c and Heavy Quark Limits and Beyond".*Physical Review D***72**(7): 074010. arXiv:hep-ph/0508199. Bibcode:2005PhRvD..72g4010C.doi:10.1103/Phys.

[11] W.-M. Yao et al. (Particle Data Group) (2006). "Review of particle physics: Θ+" (PDF). *Journal of Physics G* **33**: 1. arXiv:astro-ph/0601168. Bibcode:2006JPhG...33....1Y. doi:10.1088/0954-3899/33/1/001.

[12] D. Diakonov, V. Petrov, and M. Polyakov (1997). "Exotic anti-decuplet of baryons: prediction from chiral solitons". *Zeitschrift für Physik A* **359** (3): 305. arXiv:hep-ph/9703373. Bibcode:1997ZPhyA.359..305D. doi:10.1007/s002180050406.

[13] T. Nakano et al. (LEPS Collaboration) (2009). "Evidence of the Θ^+ in the γd→K⁺K⁻pn reaction". *Physical Review C* **79** (2): 025210. arXiv:0812.1035. Bibcode:2009PhRvC..79b5210N. doi:10.1103/PhysRevC.79.025210.

[14] P. Rincon (14 July 2015). "Large Hadron Collider discovers new pentaquark particle". *BBC News*. Retrieved 2015-07-14.

[15] "Where has all the antimatter gone?". CERN/LHCb. 2008. Retrieved 2015-07-15.

[16] N. Cardoso, M. Cardoso, and P. Bicudo (2013). "Color fields of the static pentaquark system computed in SU(3) lattice QCD". *Physical Review D* **87** (3): 034504. arXiv:1209.1532. doi:10.1103/PhysRevD.87.034504.

29.8   Further reading

- David Whitehouse (1 July 2003). "Behold the Pentaquark (BBC News)". BBC News. Retrieved 2010-01-08.

- Thomas E. Browder, Igor R. Klebanov, Daniel R. Marlow (2004). "Prospects for Pentaquark Production at Meson Factories".*Physics Letters B***587**: 62. arXiv:hep-ph/0401115. Bibcode:2004PhLB..587...62B.doi:10.1016/j.

- Akio Sugamoto (2004). "An Attempt to Study Pentaquark Baryons in String Theory". arXiv:hep-ph/0404019 [hep-ph].

- Kenneth Hicks (2005). "An Experimental Review of the Θ+ Pentaquark". *Journal of Physics: Conference Series* **9**: 183. arXiv:hep-ex/0412048. Bibcode:2005JPhCS...9..183H. doi:10.1088/1742-6596/9/1/035.

- Mark Peplow (18 April 2005). "Doubt is Cast on Pentaquarks". *Nature*. doi:10.1038/news050418-1.

- Maggie McKie (20 April 2005). "Pentaquark hunt draws blanks". *New Scientist*. Retrieved 2010-01-08.

- Thomas Jefferson National Accelerator Facility (21 April 2005). "Is It Or Isn't It? Pentaquark Debate Heats Up". *Space Daily*. Retrieved 2010-01-08.

- Dmitri Diakonov (2005). "Relativistic Mean Field Approximation to Baryons". *European Physical Journal A* **24**: 3. Bibcode:2005EPJAS..24a...3D. doi:10.1140/epjad/s2005-05-001-3.

- Schumacher, R. A. (2006). "The Rise and Fall of Pentaquarks in Experiments". *AIP Conference Proceedings* **842**: 409. arXiv:nucl-ex/0512042. doi:10.1063/1.2220285.

- Kandice Carter (2006). "The Rise and Fall of the Pentaquark". *Symmetry Magazine* **3** (7): 16.

29.9 External links

- "Pentaquark on arxiv.org".

Chapter 30

Color confinement

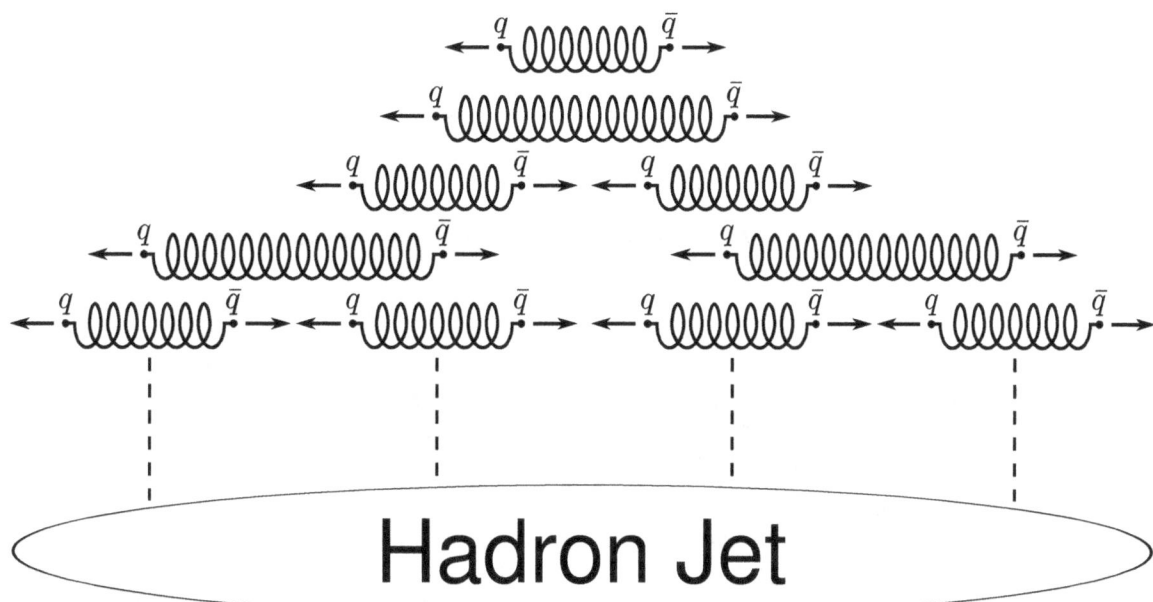

The color force favors confinement because at a certain range it is more energetically favorable to create a quark-antiquark pair than to continue to elongate the color flux tube. This is analoguous to the behavior of an elongated rubber-band.

Color confinement, often simply called **confinement**, is the phenomenon that color charged particles (such as quarks) cannot be isolated singularly, and therefore cannot be directly observed.[1] Quarks, by default, clump together to form groups, or hadrons. The two types of hadrons are the mesons (one quark, one antiquark) and the baryons (three quarks).

The constituent quarks in a group cannot be separated from their parent hadron, and this is why quarks currently cannot be studied or observed in any more direct way than at a hadron level.[2]

30.1   Origin

The reasons for quark confinement are somewhat complicated; no analytic proof exists that quantum chromodynamics should be confining. The current theory is that confinement is due to the force-carrying gluons having color charge. As any two electrically charged particles separate, the electric fields between them diminish quickly, allowing (for example) electrons to become unbound from atomic nuclei. However, as a quark-antiquark pair separates, the gluon field forms a narrow tube (or string) of color field between them. This is quite different from the behavior of the electric field of

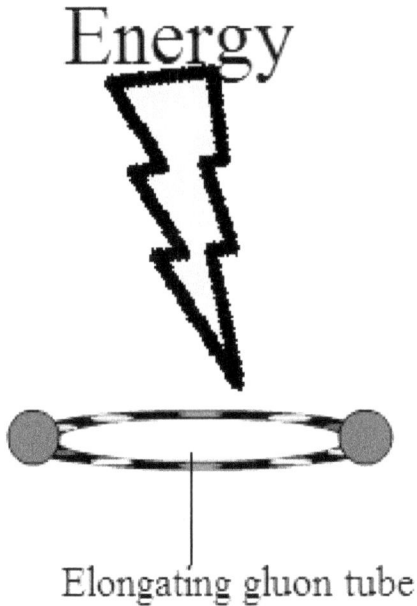

An animation of color confinement. Energy is supplied to the quarks, and the gluon tube elongates until it reaches a point where it "snaps" and forms a quark-antiquark pair.

a pair of positive and negative electric charges, which extends into the whole surrounding space and diminishes at large distances. Because of this behavior of the gluon field, a strong force between the quark pair acts constantly—regardless of their distance[3][4]—with a force of around 10,000 newtons. [5]

When two quarks become separated, as happens in particle accelerator collisions, at some point it is more energetically favorable for a new quark–antiquark pair to spontaneously appear, than to allow the tube to extend further. As a result of this, when quarks are produced in particle accelerators, instead of seeing the individual quarks in detectors, scientists see "jets" of many color-neutral particles (mesons and baryons), clustered together. This process is called *hadronization*, *fragmentation*, or *string breaking*, and is one of the least understood processes in particle physics.

The confining phase is usually defined by the behavior of the action of the Wilson loop, which is simply the path in spacetime traced out by a quark–antiquark pair created at one point and annihilated at another point. In a non-confining theory, the action of such a loop is proportional to its perimeter. However, in a confining theory, the action of the loop is instead proportional to its area. Since the area will be proportional to the separation of the quark–antiquark pair, free quarks are suppressed. Mesons are allowed in such a picture, since a loop containing another loop in the opposite direction will have only a small area between the two loops.

30.2 Models exhibiting confinement

Besides QCD in four spacetime dimensions, another model which exhibits confinement is the Schwinger model.[6] Compact Abelian gauge theories also exhibit confinement in 2 and 3 spacetime dimensions.[7] Confinement has recently been found in elementary excitations of magnetic systems called spinons.[8]

30.3   Models of fully screened quarks

Besides the quark confinement idea, there is a potential possibility, that color charge of quarks gets fully screened by the gluonic color, surrounding the quark. Exact solutions of SU(3) classical Yang–Mills theory, which provide full screening (by gluon fields) of the color charge of a quark have been found.[9] However, such classical solutions do not take into account non-trivial properties of QCD vacuum. Therefore, a significance of such full gluonic screening solutions for a separated quark is not clear.

30.4   See also

- Gluon field strength tensor

- Asymptotic freedom

- Center vortices

- Deconfining phase

- Quantum mechanics

- Particle physics

- Fundamental force

- Dual superconducting model

- Beta-function

- Infrared safety

30.5   References

[1] V. Barger, R. Phillips (1997). *Collider Physics*. Addison–Wesley. ISBN 0-201-14945-1.

[2] T.-Y. Wu, W.-Y. Pauchy Hwang (1991). *Relativistic quantum mechanics and quantum fields*. World Scientific. p. 321. ISBN 981-02-0608-9.

[3] T. Muta (2009). *Foundations of quantum chromodynamics: an introduction to perturbative methods in gauge theories* (3rd ed.). World Scientific. ISBN 978-981-279-353-9.

[4] A. Smilga (2001). *Lectures on quantum chromodynamics*. World Scientific. ISBN 978-981-02-4331-9.

[5] Fritzsch, op. cite, p. 164. The author states that the force between differently coloured quarks remains constant at any distance after they travel only a tiny distance from each other, and is equal to that need to raise one ton, which is 1000 kg x 9.8 m/s^2 = ~10,000 N.

[6] Wilson, Kenneth G. (1974-10-15). "Confinement of Quarks". *Physical Review D* (College Park, MD, USA: American Physical Society) **10**: 2445–2459. Bibcode:1974PhRvD..10.2445W. doi:10.1103/PhysRevD.10.2445. ISSN 1550-2368. OCLC 55589778. Retrieved 2014-04-12.

[7] Schön, Verena; Michael, Thies (2000-08-22). "2d Model Field Theories at Finite Temperature and Density (Section 2.5)". arXiv:hep-th/0008175v1 [hep-th].

[8] Lake, Bella; Tsvelik, Alexei M.; Notbohm, Susanne; Tennant, D. Alan; Perring, Toby G.; Reehuis, Manfred; Sekar, Chinnathambi; Krabbes, Gernot; Büchner, Bernd (2009-11-29). "Confinement of fractional quantum number particles in a condensed-matter system". *Nature Physics* (London, UK: Nature Publishing Group) **6** (1): 50–55. arXiv:0908.1038. Bibcode:2010NatPh... .doi:10.1038/nphys1462. ISSN 1745-2481. OCLC 150143123. Retrieved 2014-04-12. (subscription required(help)).

[9] Cahill, Kevin (1978-08-28). "Example of Color Screening". *Physical Review Letters* (American Physical Society) **41** (9): 599–601. Bibcode:1978PhRvL..41..599C. doi:10.1103/PhysRevLett.41.599. ISSN 1079-7114. OCLC 31492939. Retrieved 2014-04-12. (subscription required (help)).

30.6 External links

- Quarks

Chapter 31

Fermion

In particle physics, a **fermion** (a name coined by Paul Dirac[1] from the surname of Enrico Fermi) is any particle characterized by Fermi–Dirac statistics. These particles obey the Pauli exclusion principle. Fermions include all quarks and leptons, as well as any composite particle made of an odd number of these, such as all baryons and many atoms and nuclei. Fermions differ from bosons, which obey Bose–Einstein statistics.

A fermion can be an elementary particle, such as the electron, or it can be a composite particle, such as the proton. According to the spin-statistics theorem in any reasonable relativistic quantum field theory, particles with integer spin are bosons, while particles with half-integer spin are fermions.

Besides this spin characteristic, fermions have another specific property: they possess conserved baryon or lepton quantum numbers. Therefore what is usually referred as the spin statistics relation is in fact a spin statistics-quantum number relation.[2]

As a consequence of the Pauli exclusion principle, only one fermion can occupy a particular quantum state at any given time. If multiple fermions have the same spatial probability distribution, then at least one property of each fermion, such as its spin, must be different. Fermions are usually associated with matter, whereas bosons are generally force carrier particles, although in the current state of particle physics the distinction between the two concepts is unclear. At low temperature fermions show superfluidity for uncharged particles and superconductivity for charged particles. Composite fermions, such as protons and neutrons, are the key building blocks of everyday matter. Weakly interacting fermions can also display bosonic behavior under extreme conditions, such as superconductivity.

31.1 Elementary fermions

The Standard Model recognizes two types of elementary fermions, quarks and leptons. In all, the model distinguishes 24 different fermions. There are six quarks (up, down, strange, charm, bottom and top quarks), and six leptons (electron, electron neutrino, muon, muon neutrino, tau particle and tau neutrino), along with the corresponding antiparticle of each of these.

Mathematically, fermions come in three types - Weyl fermions (massless), Dirac fermions (massive), and Majorana fermions (each its own antiparticle). Most Standard Model fermions are believed to be Dirac fermions, although it is unknown at this time whether the neutrinos are Dirac or Majorana fermions. Dirac fermions can be treated as a combination of two Weyl fermions.[3]:106 So far there is no known example of Weyl fermion in particle physics. In July 2015, Weyl fermions have been experimentally realized in Weyl semimetals.

Enrico Fermi

31.2 Composite fermions

See also: List of particles § Composite particles

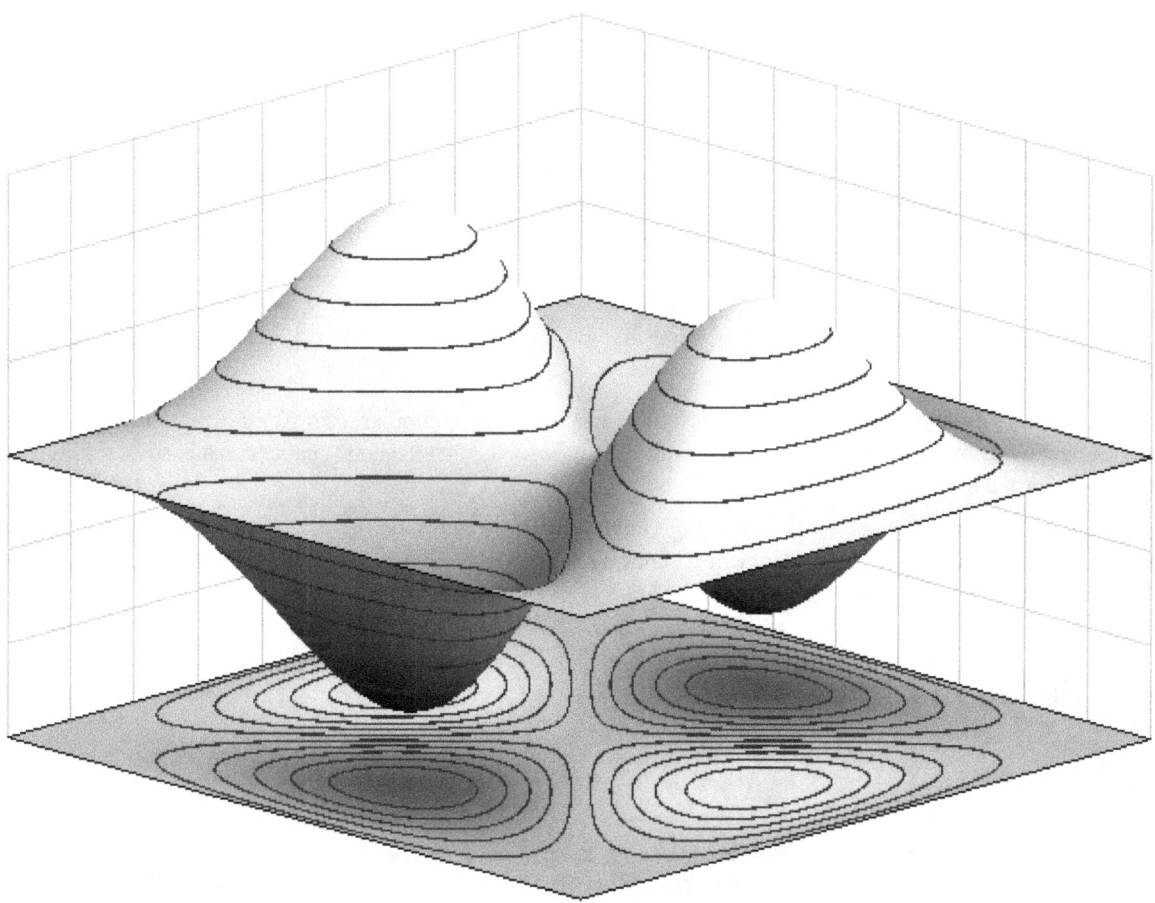

Antisymmetric wavefunction for a (fermionic) 2-particle state in an infinite square well potential.

Composite particles (such as hadrons, nuclei, and atoms) can be bosons or fermions depending on their constituents. More precisely, because of the relation between spin and statistics, a particle containing an odd number of fermions is itself a fermion. It will have half-integer spin.

Examples include the following:

- A baryon, such as the proton or neutron, contains three fermionic quarks and thus it is a fermion.

- The nucleus of a carbon-13 atom contains six protons and seven neutrons and is therefore a fermion.

- The atom helium-3 (^{3}He) is made of two protons, one neutron, and two electrons, and therefore it is a fermion.

The number of bosons within a composite particle made up of simple particles bound with a potential has no effect on whether it is a boson or a fermion.

Fermionic or bosonic behavior of a composite particle (or system) is only seen at large (compared to size of the system) distances. At proximity, where spatial structure begins to be important, a composite particle (or system) behaves according to its constituent makeup.

Fermions can exhibit bosonic behavior when they become loosely bound in pairs. This is the origin of superconductivity and the superfluidity of helium-3: in superconducting materials, electrons interact through the exchange of phonons, forming Cooper pairs, while in helium-3, Cooper pairs are formed via spin fluctuations.

The quasiparticles of the fractional quantum Hall effect are also known as composite fermions, which are electrons with an even number of quantized vortices attached to them.

31.2.1 Skyrmions

Main article: Skyrmion

In a quantum field theory, there can be field configurations of bosons which are topologically twisted. These are coherent states (or solitons) which behave like a particle, and they can be fermionic even if all the constituent particles are bosons. This was discovered by Tony Skyrme in the early 1960s, so fermions made of bosons are named skyrmions after him.

Skyrme's original example involved fields which take values on a three-dimensional sphere, the original nonlinear sigma model which describes the large distance behavior of pions. In Skyrme's model, reproduced in the large N or string approximation to quantum chromodynamics (QCD), the proton and neutron are fermionic topological solitons of the pion field.

Whereas Skyrme's example involved pion physics, there is a much more familiar example in quantum electrodynamics with a magnetic monopole. A bosonic monopole with the smallest possible magnetic charge and a bosonic version of the electron will form a fermionic dyon.

The analogy between the Skyrme field and the Higgs field of the electroweak sector has been used[4] to postulate that all fermions are skyrmions. This could explain why all known fermions have baryon or lepton quantum numbers and provide a physical mechanism for the Pauli exclusion principle.

31.3 See also

31.4 Notes

[1] Notes on Dirac's lecture *Developments in Atomic Theory* at Le Palais de la Découverte, 6 December 1945, UKNATARCHI Dirac Papers BW83/2/257889. See note 64 on page 331 in "The Strangest Man: The Hidden Life of Paul Dirac, Mystic of the Atom" by Graham Farmelo

[2] Physical Review D volume 87, page 0550003, year 2013, author Weiner, Richard M., title "Spin-statistics-quantum number connection and supersymmetry" arxiv:1302.0969

[3] T. Morii; C. S. Lim; S. N. Mukherjee (1 January 2004). *The Physics of the Standard Model and Beyond*. World Scientific. ISBN 978-981-279-560-1.

[4] Weiner, Richard M. (2010). "The Mysteries of Fermions". *International Journal of Theoretical Physics* **49** (5): 1174–1180. arXiv:0901.3816. Bibcode:2010IJTP...49.1174W. doi:10.1007/s10773-010-0292-7.

Chapter 32

Pion

In particle physics, a **pion** (or a **pi meson**, denoted with the Greek letter pi: π) is any of three subatomic particles: π0, π+, and π−. Each pion consists of a quark and an antiquark and is therefore a meson. Pions are the lightest mesons (and, more generally, the lightest hadrons), because they are composed of the lightest quarks (the u and d quarks). They are unstable, with the charged pions π+ and π− decaying with a mean lifetime of 26 nanoseconds (2.6×10^{-8} seconds), and the neutral pion π0 decaying with a much shorter lifetime of 8.4×10^{-17} seconds. Charged pions most often decay into muons and muon neutrinos, and neutral pions into gamma rays.

The exchange of virtual pions, along with the vector, rho and omega mesons, provides an explanation for the residual strong force between nucleons. Pions are not produced in radioactive decay, but are produced commonly in high energy accelerators in collisions between hadrons. All types of pions are also produced in natural processes when high energy cosmic ray protons and other hadronic cosmic ray components interact with matter in the Earth's atmosphere. Recently, detection of characteristic gamma rays originating from decay of neutral pions in two supernova remnant stars has shown that pions are produced copiously in supernovas, most probably in conjunction with production of high energy protons that are detected on Earth as cosmic rays.[1]

The concept of mesons as the carrier particles of the nuclear force was first proposed in 1935 by Hideki Yukawa. While the muon was first proposed to be this particle after its discovery in 1936, later work found that it did not participate in the strong nuclear interaction. The pions, which turned out to be examples of Yukawa's proposed mesons, were discovered later: the charged pions in 1947, and the neutral pion in 1950.

32.1   History

Theoretical work by Hideki Yukawa in 1935 had predicted the existence of mesons as the carrier particles of the strong nuclear force. From the range of the strong nuclear force (inferred from the radius of the atomic nucleus), Yukawa predicted the existence of a particle having a mass of about 100 MeV. Initially after its discovery in 1936, the muon (initially called the "mu meson") was thought to be this particle, since it has a mass of 106 MeV. However, later particle physics experiments showed that the muon did not participate in the strong nuclear interaction. In modern terminology, this makes the muon a lepton, and not a true meson. However, some communities of nuclear physicists, continue to call the muon a "mu-meson."

In 1947, the first true mesons, the charged pions, were found by the collaboration of Cecil Powell, César Lattes, Giuseppe Occhialini, *et al.*, at the University of Bristol, in England. Since the advent of particle accelerators had not yet come, high-energy subatomic particles were only obtainable from atmospheric cosmic rays. Photographic emulsions, which used the gelatin-silver process, were placed for long periods of time in sites located at high altitude mountains, first at Pic du Midi de Bigorre in the Pyrenees, and later at Chacaltaya in the Andes Mountains, where they were impacted by cosmic rays.

After the development of the photographic plates, microscopic inspection of the emulsions revealed the tracks of charged subatomic particles. Pions were first identified by their unusual "double meson" tracks, which were left by their decay into another "meson". (It was actually the muon, which is not classified as a meson in modern particle physics.) In

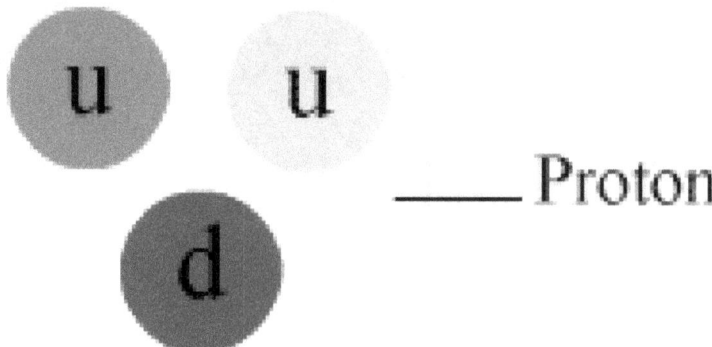

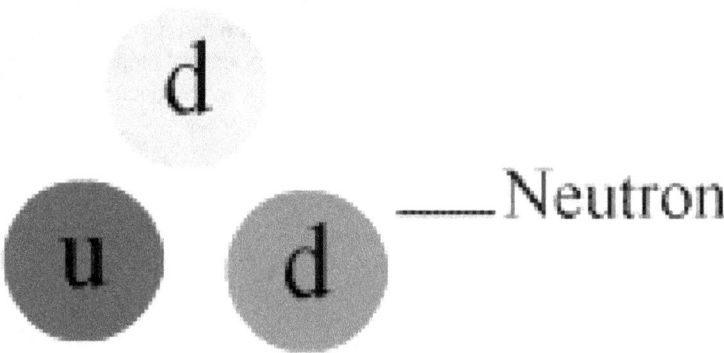

An animation of the nuclear force (or residual strong force) interaction. The small colored double disks are gluons. Anticolors are shown as per this diagram (larger version).

1948, Lattes, Eugene Gardner, and their team first artificially produced pions at the University of California's cyclotron in Berkeley, California, by bombarding carbon atoms with high-speed alpha particles. Further advanced theoretical work was carried out by Riazuddin, who in 1959, used the dispersion relation for Compton scattering of virtual photons on pions to analyze their charge radius.[2]

Nobel Prizes in Physics were awarded to Yukawa in 1949 for his theoretical prediction of the existence of mesons, and to Cecil Powell in 1950 for developing and applying the technique of particle detection using photographic emulsions.

Since the neutral pion is not electrically charged, it is more difficult to detect and observe than the charged pions are.

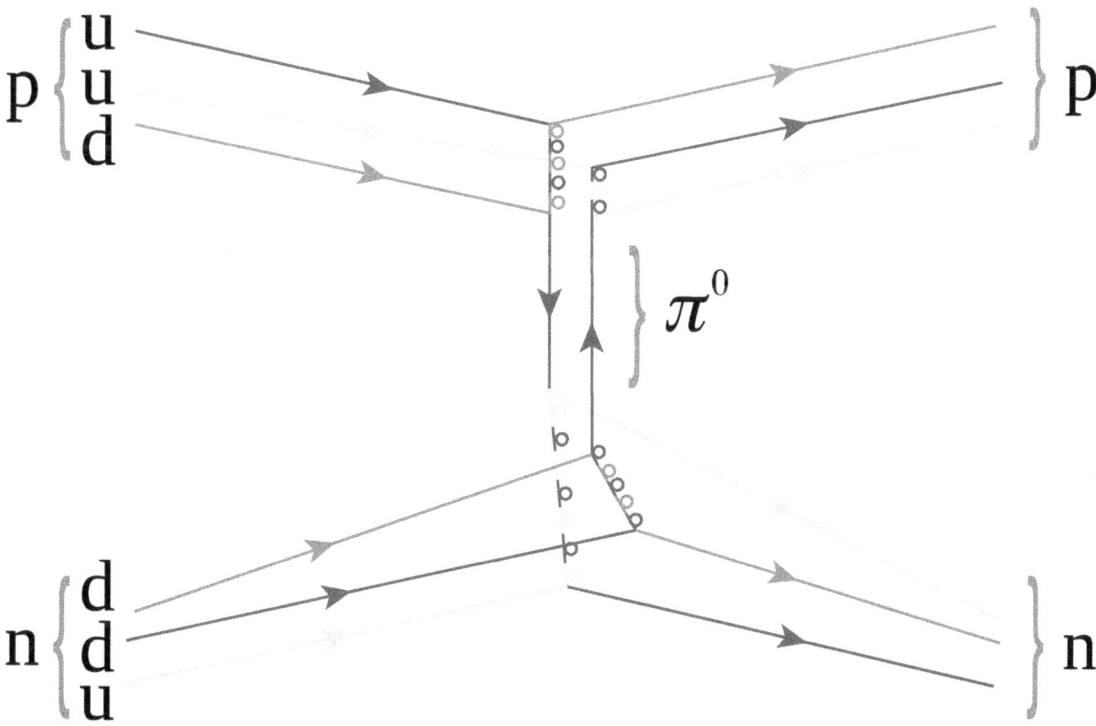

The same process as in the animation with the individual quark constituents shown, to illustrate how the fundamental *strong interaction gives rise to the **nuclear force**. Straight lines are quarks, while multi-colored loops are gluons (the carriers of the fundamental force). Other gluons, which bind together the proton, neutron, and pion "in-flight," are not shown.*

Neutral pions do not leave tracks in photographic emulsions, and neither do they in Wilson cloud chambers. The existence of the neutral pion was inferred from observing its decay products from cosmic rays, a so-called "soft component" of slow electrons with photons. The π0 was identified definitively at the University of California's cyclotron in 1950 by observing its decay into two photons.[3] Later in the same year, they were also observed in cosmic-ray balloon experiments at Bristol University.

The pion also plays a crucial role in cosmology, by imposing an upper limit on the energies of cosmic rays surviving collisions with the cosmic microwave background, through the Greisen–Zatsepin–Kuzmin limit.

In the standard understanding of the strong force interaction (called QCD, "quantum chromodynamics"), pions are understood to be the pseudo-Nambu-Goldstone bosons of spontaneously broken chiral symmetry. This explains why the three kinds of pions' masses are considerably less than the masses of the other mesons, such as the scalar or vector mesons. If their current quarks were massless particles, hypothetically, making the chiral symmetry exact, then the Goldstone theorem would dictate that all pions have zero masses. In reality, since the light quarks actually have minuscule nonzero masses, the pions also have nonzero rest masses, albeit *almost an order of magnitude smaller* than that of the nucleons, roughly[4] $m\pi \approx \sqrt{v}\, m_q / f\pi \approx \sqrt{m_q}\, 45$ MeV, where m are the relevant current quark masses in MeV, 5–10 MeVs.

The use of pions in medical radiation therapy, such as for cancer, was explored at a number of research institutions, including the Los Alamos National Laboratory's Meson Physics Facility, which treated 228 patients between 1974 and 1981 in New Mexico,[5] and the TRIUMF laboratory in Vancouver, British Columbia.

32.2   Theoretical overview

The pion can be thought of as one of the particles that mediate the interaction between a pair of nucleons. This interaction is attractive: it pulls the nucleons together. Written in a non-relativistic form, it is called the Yukawa potential. The pion, being spinless, has kinematics described by the Klein–Gordon equation. In the terms of quantum field theory, the effective field theory Lagrangian describing the pion-nucleon interaction is called the Yukawa interaction.

The nearly identical masses of π± and π0 imply that there must be a symmetry at play; this symmetry is called the SU(2) flavour symmetry or isospin. The reason that there are three pions, π+, π− and π0, is that these are understood to belong to the triplet representation or the adjoint representation **3** of SU(2). By contrast, the up and down quarks transform according to the fundamental representation **2** of SU(2), whereas the anti-quarks transform according to the conjugate representation **2***.

With the addition of the strange quark, one can say that the pions participate in an SU(3) flavour symmetry, belonging to the adjoint representation **8** of SU(3). The other members of this octet are the four kaons and the eta meson.

Pions are pseudoscalars under a parity transformation. Pion currents thus couple to the axial vector current and pions participate in the chiral anomaly.

32.3 Basic properties

Pions are mesons with zero spin, and they are composed of first-generation quarks. In the quark model, an up quark and an anti-down quark make up a π+, whereas a down quark and an anti-up quark make up the π−, and these are the antiparticles of one another. The neutral pion π0 is a combination of an up quark with an anti-up quark or a down quark with an anti-down quark. The two combinations have identical quantum numbers, and hence they are only found in superpositions. The lowest-energy superposition of these is the π0, which is its own antiparticle. Together, the pions form a triplet of isospin. Each pion has isospin ($I = 1$) and third-component isospin equal to its charge ($I_z = +1$, 0 or -1).

32.3.1 Charged pion decays

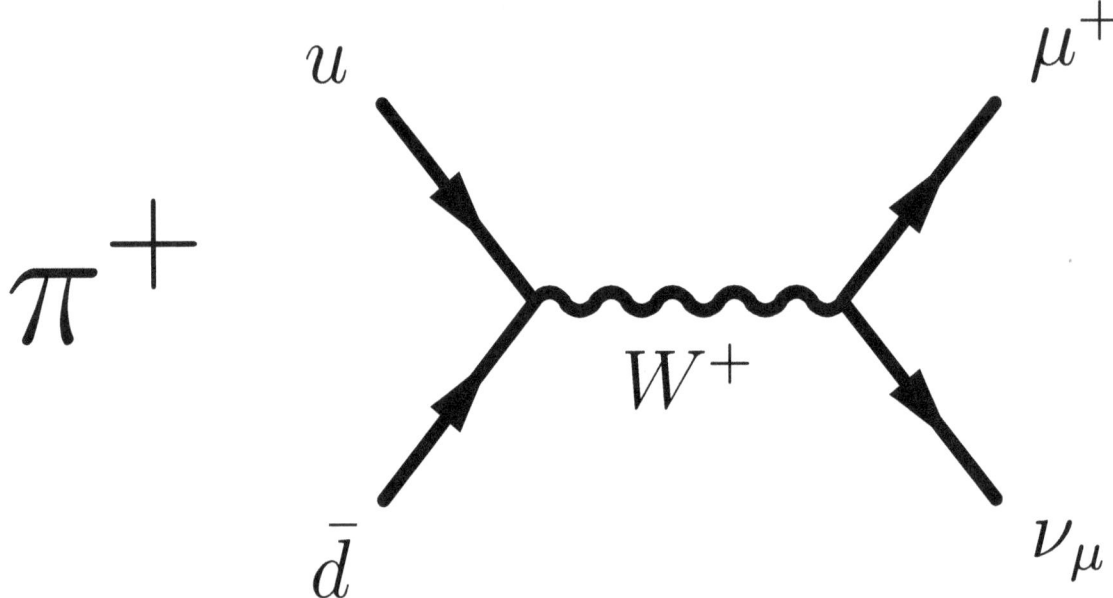

Feynman diagram of the dominating leptonic pion decay.

The π± mesons have a mass of 139.6 MeV/c^2 and a mean lifetime of 2.6×10^{-8} s. They decay due to the weak interaction. The primary decay mode of a pion, with probability 0.999877, is a purely leptonic decay into an anti-muon and a muon neutrino:

The second most common decay mode of a pion, with probability 0.000123, is also a leptonic decay into an electron and the corresponding electron antineutrino. This "electronic mode" was discovered at CERN in 1958:[6]

The suppression of the electronic mode, with respect to the muonic one, is given approximately (to within radiative corrections) by the ratio of the half-widths of the pion–electron and the pion–muon decay reactions:

$$R_\pi = (m_e/m_\mu)^2 \left(\frac{m_\pi^2 - m_e^2}{m_\pi^2 - m_\mu^2} \right)^2 = 1.283 \times 10^{-4}$$

and is a spin effect known as the helicity suppression. Its mechanism is as follows: The negative pion has spin zero, therefore the lepton and antineutrino must be emitted with opposite spins (and opposite linear momenta) to preserve net zero spin (and conserve linear momentum). However, the antineutrino, due to very high speed, is always right-handed, so this implies that the lepton must be emitted with spin in the direction of its linear momentum (i.e., also right-handed). If, however, leptons were massless, they would only exist in the left-handed form, just as the neutrino does (due to parity violation), and this decay mode would be prohibited. Therefore, suppression of the electron decay channel comes from the fact that the electron's mass is much smaller than the muon's. The electron is thus relatively massless compared with the muon, and thus the electronic mode is *almost* prohibited.[7]

Hence, electronic mode decay favors the left-handed symmetry and inhibits this decay channel. Measurements of the above ratio have been considered for decades to be tests of the *V − A structure* (vector minus axial vector or left-handed lagrangian) of the charged weak current and of lepton universality. Experimentally this ratio is $1.230(4) \times 10^{-4}$.[8]

Besides the purely leptonic decays of pions, some structure-dependent radiative leptonic decays (that is, decay to the usual leptons plus a gamma ray) have also been observed.

Also observed, for charged pions only, is the very rare "pion beta decay" (with probability of about 10^{-8}) into a neutral pion plus an electron and electron antineutrino (or for positive pions, a neutral pion, positron, and electron neutrino).

The rate at which pions decay is a prominent quantity in many sub-fields of particle physics, such as chiral perturbation theory. This rate is parametrized by the pion decay constant ($f\pi$), related to the wave function overlap of the quark and antiquark, which is about 130 MeV.[9]

32.3.2   Neutral pion decays

The $\pi 0$ meson has a mass of 135.0 MeV/c^2 and a mean lifetime of 8.4×10^{-17} s. It decays via the electromagnetic force, which explains why its mean lifetime is much smaller than that of the charged pion (which can only decay via the weak force). The main π^0 decay mode, with a branching ratio of BR=0.98823, is into two photons:

The decay $\pi^0 \rightarrow 3\gamma$ (as well as decays into any odd number of photons) is forbidden by the C-symmetry of the electromagnetic interaction. The intrinsic C-parity of the π^0 is **+1**, while the C-parity of a system of **n** photons is $(-1)^n$.

The second largest π^0 decay mode (BR=0.01174) is the Dalitz decay (named after Richard Dalitz), which is a two-photon decay with an internal photon conversion resulting a photon and an electron-positron pair in the final state:

The third largest established decay mode (BR=3.34×10^{-5}) is the double Dalitz decay, with both photons undergoing internal conversion which leads to further suppression of the rate:

The fourth largest established decay mode is the loop-induced and therefore suppressed (and additionally helicity-suppressed) leptonic decay mode (BR=6.46×10^{-8}):

The neutral pion has also been observed to decay into positronium with a branching fraction of the order of 10^{-9}. No other decay modes have been established experimentally. The branching fractions above are the PDG central values, and their uncertainties are not quoted.

[a] ^ Make-up inexact due to non-zero quark masses.[12]

32.4 See also

- Pionium

- List of particles

- Quark model

- Static forces and virtual-particle exchange

- César Lattes

32.5 References

[1] M. Ackermann; et al. (2013). "Detection of the Characteristic Pion-Decay Signature in Supernova Remnants". *Science* **339** (6424): 807–811. arXiv:1302.3307. Bibcode:2013Sci...339..807A. doi:10.1126/science.1231160.

[2] Riazuddin(1959). "Charge Radius of Pion".*Physical Review***114**(4): 1184–1186. Bibcode:1959PhRv..114.1184R.doi:10.184.

[3] R. Bjorklund; W. E. Crandall; B. J. Moyer; H. F. York (1950). "High Energy Photons from Proton-Nucleon Collisions". *Physical Review* **77** (2): 213–218. Bibcode:1950PhRv...77..213B. doi:10.1103/PhysRev.77.213.

[4] Gell-Mann, M.; Renner, B. (1968). "Behavior of Current Divergences under SU_{3}×SU_{3}". *Physical Review* **175** (5): 2195. Bibcode:1968PhRv..175.2195G. doi:10.1103/PhysRev.175.2195.

[5] von Essen, C. F.; Bagshaw, M. A.; Bush, S. E.; Smith, A. R.; Kligerman, M. M. (1987). "Long-term results of pion therapy at Los Alamos". *International Journal of Radiation Oncology*Biology*Physics* **13** (9): 1389–98. doi:10.1016/0360-3016(87)90235-5. PMID 3114189.

[6] Fazzini, T.; Fidecaro, G.; Merrison, A.; Paul, H.; Tollestrup, A. (1958). "Electron Decay of the Pion". *Physical Review Letters* **1** (7): 247. doi:10.1103/PhysRevLett.1.247.

[7] Mesons at Hyperphysics

[8] C. Amsler *et al.*. (2008): Particle listings – π±

[9] LEPTONIC DECAYS OF CHARGED PSEUDO- SCALAR MESONS J. L. Rosner and S. Stone. Particle Data Group. December 18, 2013

[10] C. Amsler *et al.*. (2008): Quark Model

[11] C. Amsler *et al.*. (2008): Particle listings – π0

[12] D. J. Griffiths (1987). *Introduction to Elementary Particles*. John Wiley & Sons. ISBN 0-471-60386-4.

32.6 Further reading

- Gerald Edward Brown and A. D. Jackson, *The Nucleon-Nucleon Interaction*, (1976) North-Holland Publishing, Amsterdam ISBN 0-7204-0335-9

32.7 External links

- Mesons at the Particle Data Group

Chapter 33

Kaon

For other uses, see Kaon (disambiguation).

In particle physics, a **kaon** /ˈkeɪ.ɒn/, also called a **K meson** and denoted K,[nb 1] is any of a group of four mesons distinguished by a quantum number called strangeness. In the quark model they are understood to be bound states of a strange quark (or antiquark) and an up or down antiquark (or quark).

Kaons have proved to be a copious source of information on the nature of fundamental interactions since their discovery in cosmic rays in 1947. They were essential in establishing the foundations of the Standard Model of particle physics, such as the quark model of hadrons and the theory of quark mixing (the latter was acknowledged by a Nobel Prize in Physics in 2008). Kaons have played a distinguished role in our understanding of fundamental conservation laws: CP violation, a phenomenon generating the observed matter–antimatter asymmetry of the universe, was discovered in the kaon system in 1964 (which was acknowledged by a Nobel Prize in 1980). Moreover, direct CP violation was also discovered in the kaon decays in the early 2000s.

33.1 Basic properties

The four kaons are :

1. K−, negatively charged (containing a strange quark and an up antiquark) has mass 493.667±0.013 MeV and mean lifetime (1.2384±0.0024)×10^{-8} s.

2. K+ (antiparticle of above) positively charged (containing an up quark and a strange antiquark) must (by CPT invariance) have mass and lifetime equal to that of K−. The mass difference is 0.032±0.090 MeV, consistent with zero. The difference in lifetime is (0.11±0.09)×10^{-8} s.

3. K0, neutrally charged (containing a down quark and a strange antiquark) has mass 497.648±0.022 MeV. It has mean squared charge radius of −0.076±0.01 fm^2.

4. K0, neutrally charged (antiparticle of above) (containing a strange quark and a down antiquark) has the same mass.

It is clear from the quark model assignments that the kaons form two doublets of isospin; that is, they belong to the fundamental representation of SU(2) called the **2**. One doublet of strangeness +1 contains the K+ and the K0. The antiparticles form the other doublet (of strangeness −1).

[a] ∧ Strong eigenstate. No definite lifetime (see kaon notes below)
[b] ∧ Weak eigenstate. Makeup is missing small CP–violating term (see notes on neutral kaons below).
[c] ∧ The mass of the K0
L and K0

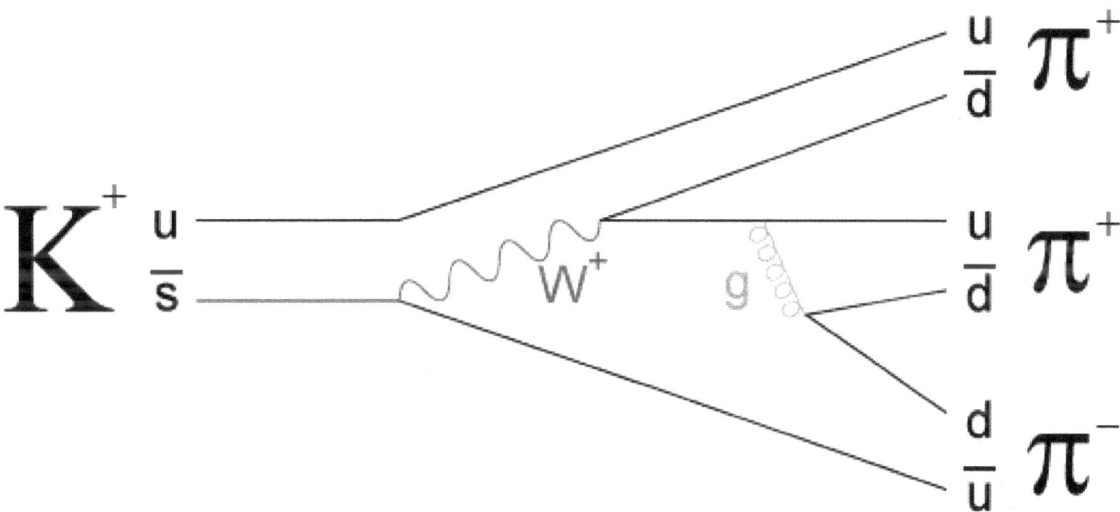

The decay of a kaon (K+) into three pions (2 π+, 1 π−) is a process that involves both weak and strong interactions.
Weak interactions : The strange antiquark (s) of the kaon transmutes into an up antiquark (u) by the emission of a W+ boson; the W+ boson subsequently decays into a down antiquark (d) and an up quark (u).
Strong interactions : An up quark (u) emits a gluon (g) which decays into a down quark (d) and a down antiquark (d).

S are given as that of the K0. However, it is known that a difference between the masses of the K0
L and K0
S on the order of 3.5×10^{-12} MeV/c^2 exists.[4]

Although the K0 and its antiparticle K0 are usually produced via the strong force, they decay weakly. Thus, once created the two are better thought of as superpositions of two weak eigenstates which have vastly different lifetimes:

1. The long-lived neutral kaon is called the K
   L ("K-long"), decays primarily into three pions, and has a mean lifetime of 5.18×10^{-8} s.

2. The short-lived neutral kaon is called the K
   S ("K-short"), decays primarily into two pions, and has a mean lifetime 8.958×10^{-11} s.

(See discussion of neutral kaon mixing below.)

An experimental observation made in 1964 that K-longs rarely decay into two pions was the discovery of CP violation (see below).

Main decay modes for K+:

Decay modes for the K− are charge conjugates of the ones above.

33.2 Strangeness

Main article: Strangeness

The discovery of hadrons with the internal quantum number "strangeness" marks the beginning of a most exciting epoch in particle physics that even now, fifty years later, has not yet found its conclusion ... by and large experiments have driven the development, and that major discoveries came unexpectedly or even against expectations expressed by theorists. — I.I. Bigi and A.I. Sanda, *CP violation*, (ISBN 0-521-44349-0)

In 1947, G. D. Rochester and Clifford Charles Butler of the University of Manchester published two cloud chamber photographs of cosmic ray-induced events, one showing what appeared to be a neutral particle decaying into two charged pions, and one which appeared to be a charged particle decaying into a charged pion and something neutral. The estimated mass of the new particles was very rough, about half a proton's mass. More examples of these "V-particles" were slow in coming.

The first breakthrough was obtained at Caltech, where a cloud chamber was taken up Mount Wilson, for greater cosmic ray exposure. In 1950, 30 charged and 4 neutral V-particles were reported. Inspired by this, numerous mountaintop observations were made over the next several years, and by 1953, the following terminology was adopted: "L-meson" meant muon or pion. "K meson" meant a particle intermediate in mass between the pion and nucleon. "Hyperon" meant any particle heavier than a nucleon.

The decays were extremely slow; typical lifetimes are of the order of 10^{-10} s. However, production in pion-proton reactions proceeds much faster, with a time scale of 10^{-23} s. The problem of this mismatch was solved by Abraham Pais who postulated the new quantum number called "strangeness" which is conserved in strong interactions but violated by the weak interactions. Strange particles appear copiously due to "associated production" of a strange and an antistrange particle together. It was soon shown that this could not be a multiplicative quantum number, because that would allow reactions which were never seen in the new synchrotrons which were commissioned in Brookhaven National Laboratory in 1953 and in the Lawrence Berkeley Laboratory in 1955.

33.3 Parity violation

Two different decays were found for charged strange mesons:

The intrinsic parity of a pion is $P = -1$, and parity is a multiplicative quantum number. Therefore, the two final states have different parity ($P = +1$ and $P = -1$, respectively). It was thought that the initial states should also have different parities, and hence be two distinct particles. However, with increasingly precise measurements, no difference was found between the masses and lifetimes of each, respectively, indicating that they are the same particle. This was known as the **τ–θ puzzle**. It was resolved only by the discovery of parity violation in weak interactions. Since the mesons decay through weak interactions, parity is not conserved, and the two decays are actually decays of the same particle,[5] now called the K+.

33.4 CP violation in neutral meson oscillations

Initially it was thought that although parity was violated, CP (charge parity) symmetry was conserved. In order to understand the discovery of CP violation, it is necessary to understand the mixing of neutral kaons; this phenomenon does not require CP violation, but it is the context in which CP violation was first observed.

33.4.1 Neutral kaon mixing

Since neutral kaons carry strangeness, they cannot be their own antiparticles. There must be then two different neutral kaons, differing by two units of strangeness. The question was then how to establish the presence of these two mesons. The solution used a phenomenon called **neutral particle oscillations**, by which these two kinds of mesons can turn from one into another through the weak interactions, which cause them to decay into pions (see the adjacent figure).

These oscillations were first investigated by Murray Gell-Mann and Abraham Pais together. They considered the CP-invariant time evolution of states with opposite strangeness. In matrix notation one can write

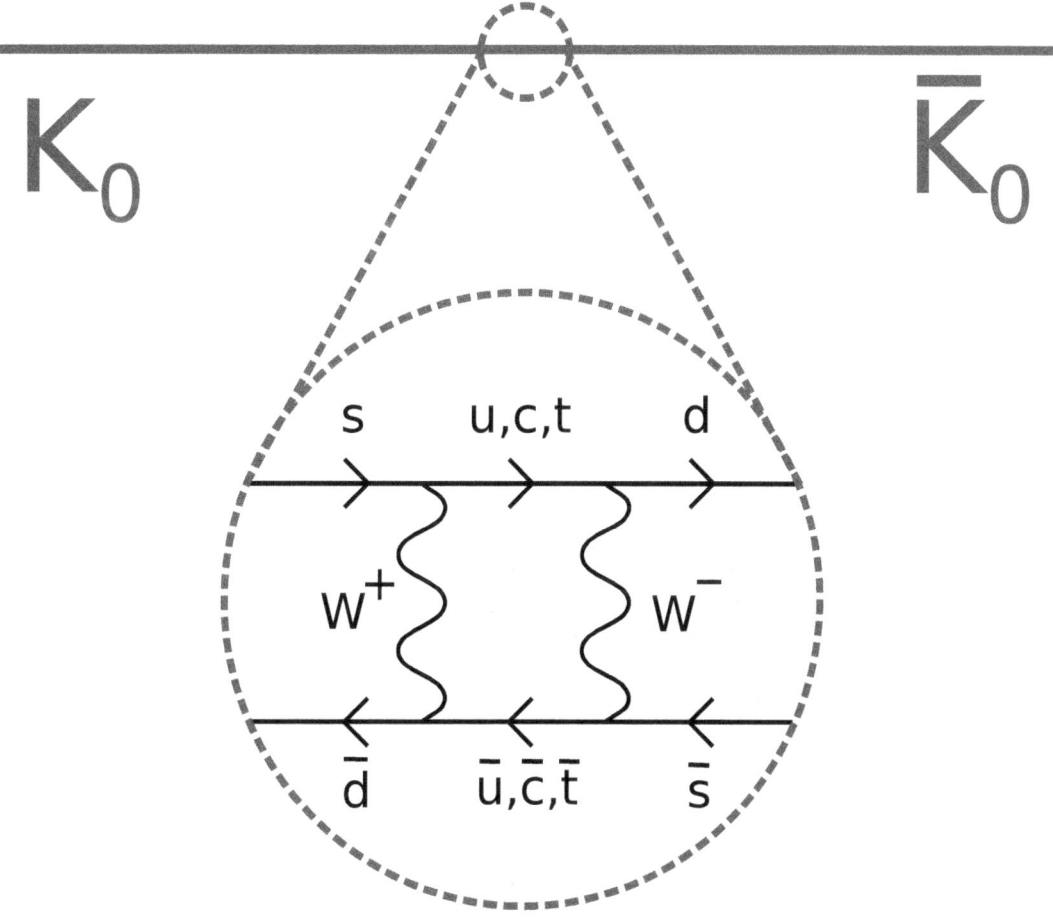

Two different neutral K mesons, carrying different strangeness, can turn from one into another through the weak interactions, since these interactions do not conserve strangeness. The strange quark in the K0 turns into a down quark by successively emitting two W-bosons of opposite charge. The down antiquark in the K0 turns into a strange antiquark by absorbing them.

$$\psi(t) = U(t)\psi(0) = \mathrm{e}^{iHt}\begin{pmatrix} a \\ b \end{pmatrix}, \qquad H = \begin{pmatrix} M & \Delta \\ \Delta & M \end{pmatrix}$$

where ψ is a quantum state of the system specified by the amplitudes of being in each of the two basis states (which are a and b at time $t = 0$). The diagonal elements (M) of the Hamiltonian are due to strong interaction physics which conserves strangeness. The two diagonal elements must be equal, since the particle and antiparticle have equal masses in the absence of the weak interactions. The off-diagonal elements, which mix opposite strangeness particles, are due to weak interactions; CP symmetry requires them to be real.

The consequence of the matrix H being real is that the probabilities of the two states will forever oscillate back and forth. However, if any part of the matrix were imaginary, as is forbidden by CP symmetry, then part of the combination will diminish over time. The diminishing part can be either one component (a) or the other (b), or a mixture of the two.

Mixing

The eigenstates are obtained by diagonalizing this matrix. This gives new eigenvectors, which we can call $\mathbf{K}_1$ which is the difference of the two states of opposite strangeness, and $\mathbf{K}_2$, which is the sum. The two are eigenstates of **CP** with opposite eigenvalues; $\mathbf{K}_1$ has $\mathbf{CP} = +1$, and $\mathbf{K}_2$ has $\mathbf{CP} = -1$ Since the two-pion final state also has $\mathbf{CP} = +1$, only the $\mathbf{K}_1$ can decay this way. The $\mathbf{K}_2$ must decay into three pions. Since the mass of $\mathbf{K}_2$ is just a little larger than the sum of the masses of three pions, this decay proceeds very slowly, about 600 times slower than the decay of $\mathbf{K}_1$ into two pions. These two different modes of decay were observed by Leon Lederman and his coworkers in 1956, establishing the existence of the two weak eigenstates (states with definite lifetimes under decays via the weak force) of the neutral kaons.

These two weak eigenstates are called the K
L (K-long) and K
S (K-short). CP symmetry, which was assumed at the time, implies that K
S = $\mathbf{K}_1$ and K
L = $\mathbf{K}_2$.

Oscillation

Main article: Neutral particle oscillation

An initially pure beam of K0 will turn into its antiparticle while propagating, which will turn back into the original particle, and so on. This is called particle oscillation. On observing the weak decay *into leptons*, it was found that a K0 always decayed into an electron, whereas the antiparticle K0 decayed into the positron. The earlier analysis yielded a relation between the rate of electron and positron production from sources of pure K0 and its antiparticle K0. Analysis of the time dependence of this semileptonic decay showed the phenomenon of oscillation, and allowed the extraction of the mass splitting between the K
S and K
L. Since this is due to weak interactions it is very small, 10^{-15} times the mass of each state.

Regeneration

A beam of neutral kaons decays in flight so that the short-lived K
S disappears, leaving a beam of pure long-lived K
L. If this beam is shot into matter, then the K0 and its antiparticle K0 interact differently with the nuclei. The K0 undergoes quasi-elastic scattering with nucleons, whereas its antiparticle can create hyperons. Due to the different interactions of the two components, quantum coherence between the two particles is lost. The emerging beam then contains different linear superpositions of the K0 and K0. Such a superposition is a mixture of K
L and K
S; the K
S is regenerated by passing a neutral kaon beam through matter. Regeneration was observed by Oreste Piccioni and his collaborators at Lawrence Berkeley National Laboratory. Soon thereafter, Robert Adair and his coworkers reported excess K
S regeneration, thus opening a new chapter in this history.

33.4.2 CP violation

While trying to verify Adair's results, J. Christenson, James Cronin, Val Fitch and Rene Turlay of Princeton University found decays of K
L into two pions ($\mathbf{CP} = +1$) in an experiment performed in 1964 at the Alternating Gradient Synchrotron at the Brookhaven laboratory.[6] As explained in an earlier section, this required the assumed initial and final states to have different values of $\mathbf{CP}$, and hence immediately suggested CP violation. Alternative explanations such as non-linear quantum mechanics and a new unobserved particle were soon ruled out, leaving CP violation as the only possibility. Cronin and Fitch received the Nobel Prize in Physics for this discovery in 1980.

It turns out that although the K
L and K

S are weak eigenstates (because they have definite lifetimes for decay by way of the weak force), they are *not quite* **CP** eigenstates. Instead, for small ε (and up to normalization),

K
L = **K**$_2$ + ε**K**$_1$

and similarly for K
S. Thus occasionally the K
L decays as a **K**$_1$ with **CP** = +1, and likewise the K
S can decay with **CP** = −1. This is known as **indirect CP violation**, CP violation due to mixing of K0 and its antiparticle. There is also a **direct CP violation** effect, in which the CP violation occurs during the decay itself. Both are present, because both mixing and decay arise from the same interaction with the W boson and thus have CP violation predicted by the CKM matrix.

33.5   See also

- Hadrons, mesons, hyperons and flavour
- Strange quark and the quark model
- Parity (physics), charge conjugation, time reversal symmetry, CPT invariance and CP violation
- Neutrino oscillation
- Neutral particle oscillation

33.6   Notes and references

Notes

[1] The positively charged kaon used to be called τ$^+$ and θ$^+$, as it was supposed to be two different particles until the 1960s. See the parity violation section.

References

[1] J. Beringer *et al.* (2012): Particle listings – K±

[2] J. Beringer *et al.* (2012): Particle listings – K0

[3] J. Beringer *et al.* (2012): Particle listings – K0
    S

[4] J. Beringer *et al.* (2012): Particle listings – K0
    L

[5] Lee, T. D.; Yang, C. N. (1 October 1956). "Question of Parity Conservation in Weak Interactions". *Physical Review* **104** (1): 254. Bibcode:1956PhRv..104..254L. doi:10.1103/PhysRev.104.254. One way out of the difficulty is to assume that parity is not strictly conserved, so that Θ+ and τ+ are two different decay modes of the same particle, which necessarily has a single mass value and a single lifetime.

[6] http://journals.aps.org/prl/pdf/10.1103/PhysRevLett.13.138

33.6.1 Bibliography

- C.Amsler; Doser, M; Antonelli, M; Asner, D; Babu, K; Baer, H; Band, H; Barnett, R; Bergren, E; Bergren, E.; Beringer, J.; Bernardi, G.; Bertl, W.; Bichsel, H.; Biebel, O.; Bloch, P.; Blucher, E.; Blusk, S.; Cahn, R. N.; Carena, M.; Caso, C.; Ceccucci, A.; Chakraborty, D.; Chen, M.-C.; Chivukula, R. S.; Cowan, G.; Dahl, O.; d'Ambrosio, G.; Damour, T.; et al. (2008). "Review of Particle Physics". *Physics Letters B* (Particle Data Group) **667** (1): 1–1340. Bibcode:2008PhLB..667....1P. doi:10.1016/j.physletb.2008.07.018.

- S. Eidelman; et al. (2004). "Review of Particle Physics 2004 – Strange Mesons". Particle Data Group.

  Particle Data Group; Eidelman, S.; Hayes, K. G.; Olive, K. A.; Aguilar-Benitez, M.; Amsler, C.; Asner, D.; Babu, K. S.; Barnett, R. M.; Beringer, J.; Burchat, P. R.; Carone, C. D.; Caso, S.; Conforto, G.; Dahl, O.; d'Ambrosio, G.; Doser, M.; Feng, J. L.; Gherghetta, T.; Gibbons, L.; Goodman, M.; Grab, C.; Groom, D. E.; Gurtu, A.; Hagiwara, K.; Hernández-Rey, J. J.; Hikasa, K.; Honscheid, K.; Jawahery, H.; et al. (2004). "Review of Particle Physics*1". *Physics Letters B* **592** (1): 1. arXiv:astro-ph/0406663. Bibcode:2004PhLB..592....1P. doi:10.1016/j.physletb.2004.06.001.

- *The quark model*, by J.J.J. Kokkedee

- M.S. Sozzi (2008). *Discrete symmetries and CP violation*. Oxford University Press. ISBN 978-0-19-929666-8.

- I.I. Bigi, A.I. Sanda (2000). *CP violation*. Cambridge University Press. ISBN 0-521-44349-0.

- D.J. Griffiths (1987). *Introduction to Elementary Particle*. John Wiley & Sons. ISBN 0-471-60386-4.

Chapter 34

Nuclear force

This article is about the force that holds nucleons together in a nucleus. For the force that holds quarks together in a nucleon, see Strong interaction.

The **nuclear force** (or **nucleon–nucleon interaction** or **residual strong force**) is the force between protons and

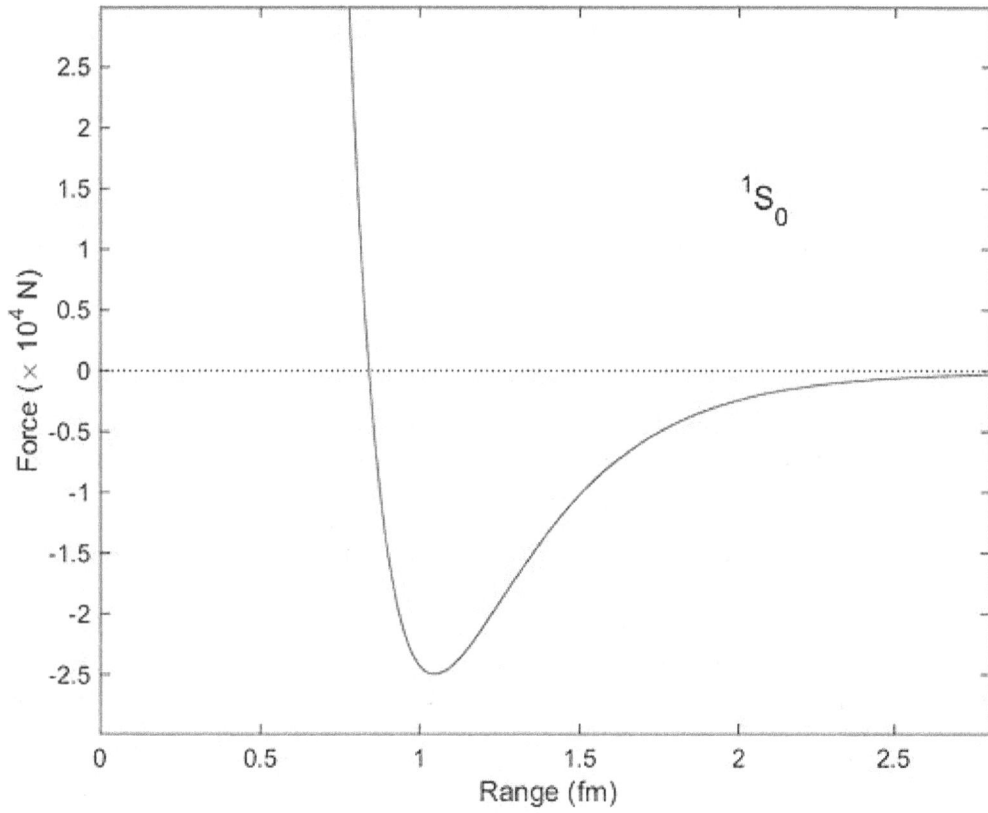

Force (in units of 10,000 N) between two nucleons as a function of distance as computed from the Reid potential (1968).[1] The spins of the neutron and proton are aligned, and they are in the S angular momentum state. The attractive (negative) force has a maximum at a distance of about 1 fm with a force of about 25,000 N. Particles much closer than a distance of 0.8 fm experience a large repulsive (positive) force. Particles separated by a distance greater than 1 fm are still attracted (Yukawa potential), but the force falls as an exponential function of distance.

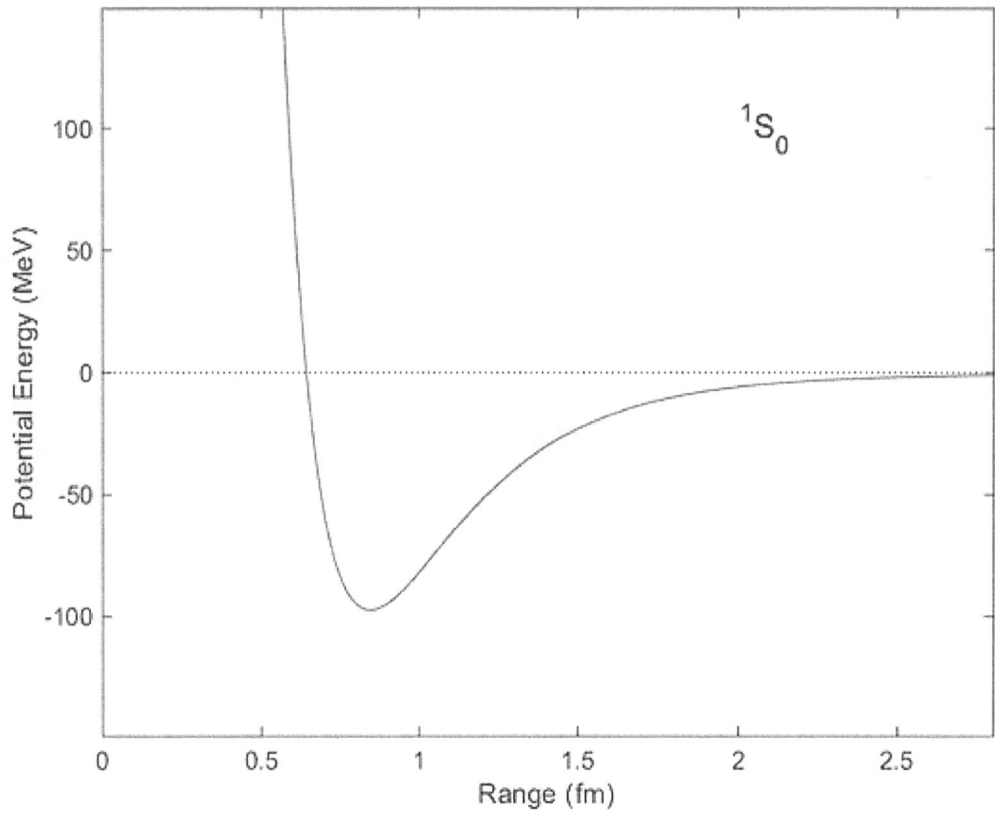

Corresponding potential energy (in units of MeV) of two nucleons as a function of distance as computed from the Reid potential. The potential well is a minimum at a distance of about 0.8 fm. With this potential nucleons can become bound with a negative "binding energy."

neutrons, subatomic particles that are collectively called nucleons. The nuclear force is responsible for binding protons and neutrons into atomic nuclei. Neutrons and protons are affected by the nuclear force almost identically. Since protons have charge +1 e, they experience a Coulomb repulsion that tends to push them apart, but at short range the nuclear force is sufficiently attractive as to overcome the electromagnetic repulsive force. The mass of a nucleus is less than the sum total of the individual masses of the protons and neutrons which form it. The difference in mass between bound and unbound nucleons is known as the mass defect. Energy is released when some large nuclei break apart, and it is this energy that used in nuclear power and nuclear weapons.[2][3]

The nuclear force is powerfully attractive between nucleons at distances of about 1 femtometer (fm, or 1.0×10^{-15} metres) between their centers, but rapidly decreases to insignificance at distances beyond about 2.5 fm. At distances less than 0.7 fm, the nuclear force becomes repulsive. This repulsive component is responsible for the physical size of nuclei, since the nucleons can come no closer than the force allows. By comparison, the size of an atom, measured in angstroms (Å, or 1.0×10^{-10} m), is five orders of magnitude larger. The nuclear force is not simple, however, since it depends on the nucleon spins, has a tensor component, and may depend on the relative momentum of the nucleons.[4]

A quantitative description of the nuclear force relies on partially empirical equations that model the internucleon potential energies, or potentials. (Generally, forces within a system of particles can be more simply modeled by describing the system's potential energy; the negative gradient of a potential is equal to the vector force.) The constants for the equations are phenomenological, that is, determined by fitting the equations to experimental data. The internucleon potentials attempt to describe the properties of nucleon–nucleon interaction. Once determined, any given potential can be used in, e.g., the Schrödinger equation to determine the quantum mechanical properties of the nucleon system.

The discovery of the neutron in 1932 revealed that atomic nuclei were made of protons and neutrons, held together by an attractive force. By 1935 the nuclear force was conceived to be transmitted by particles called mesons. This theoretical development included a description of the Yukawa potential, an early example of a nuclear potential. Mesons, predicted by theory, were discovered experimentally in 1947. By the 1970s, the quark model had been developed, which showed that the mesons and nucleons were composed of quarks and gluons. By this new model, the nuclear force, resulting from the exchange of mesons between neighboring nucleons, is a residual effect of the strong force.

34.1   Description

The nuclear force is only felt between particles composed of quarks, or hadrons. At small separations between nucleons (less than ~ 0.7 fm between their centers, depending upon spin alignment) the force becomes repulsive, which keeps the nucleons at a certain average separation, even if they are of different types. This repulsion arises from the Pauli exclusion force for identical nucleons (such as two neutrons or two protons). A Pauli exclusion force also occurs between quarks of the same type within nucleons, when the nucleons are different (a proton and a neutron, for example). The nuclear force also has a "tensor" component which depends on whether or not the spins (angular momentum vectors) of the nucleons are aligned (point in the same direction) or anti-aligned (i.e., point in opposite directions in space).

At distances larger than 0.7 fm the force becomes attractive between spin-aligned nucleons, becoming maximal at a center–center distance of about 0.9 fm. Beyond this distance the force drops exponentially, until beyond about 2.0 fm separation, the force is negligible. Nucleons have a radius of about 0.8 fm.[5]

At short distances (less than 1.7 fm or so), the nuclear force is stronger than the Coulomb force between protons; it thus overcomes the repulsion of protons inside the nucleus. However, the Coulomb force between protons has a much larger range due to its decay as the inverse square of charge separation, and Coulomb repulsion thus becomes the only significant force between protons when their separation exceeds about 2 to 2.5 fm.

For two particles that are the same (such as two neutrons or two protons) the force is not enough to bind the particles, since the spin vectors of two particles of the same type must point in opposite directions when the particles are near each other and are (save for spin) in the same quantum state. This requirement for fermions stems from the Pauli exclusion principle. For fermion particles of different types (such as a proton and neutron), particles may be close to each other and have aligned spins without violating the Pauli exclusion principle, and the nuclear force may bind them (in this case, into a deuteron), since the nuclear force is much stronger for spin-aligned particles. But if the particles' spins are anti-aligned the nuclear force is too weak to bind them, even if they are of different types.

To disassemble a nucleus into unbound protons and neutrons requires work against the nuclear force. Conversely, energy is released when a nucleus is created from free nucleons or other nuclei: the nuclear binding energy. Because of mass–energy equivalence (i.e. Einstein's famous formula $E = mc^2$), releasing this energy causes the mass of the nucleus to be lower than the total mass of the individual nucleons, leading to the so-called "mass defect".[6]

The nuclear force is nearly independent of whether the nucleons are neutrons or protons. This property is called *charge independence*. The force depends on whether the spins of the nucleons are parallel or antiparallel, and it has a noncentral or *tensor* component. This part of the force does not conserve orbital angular momentum, which is a constant of motion under central forces.

The symmetry resulting in the strong force, proposed by Werner Heisenberg, is that protons and neutrons are identical in every respect, other than their charge. This is not completely true, because neutrons are a tiny bit heavier, but it is an approximate symmetry. Protons and neutrons are therefore viewed as the same particle, but with different isospin quantum number. The strong force is invariant under SU(2) transformations, just as particles with "regular spin" are. Isospin and "regular" spin are related under this SU(2) symmetry group. There are only strong attractions when the total isospin is 0, as is confirmed by experiment.[7]

The information on nuclear force are obtained by scattering experiments and the study of light nuclei binding energy.

The nuclear force occurs by the exchange of virtual light mesons, such as the virtual pions, as well as two types of virtual mesons with spin (vector mesons), the rho mesons and the omega mesons. The vector mesons account for the spin-dependence of the nuclear force in this "virtual meson" picture.

The nuclear force is separate from what historically was known as the weak nuclear force. The weak interaction is one

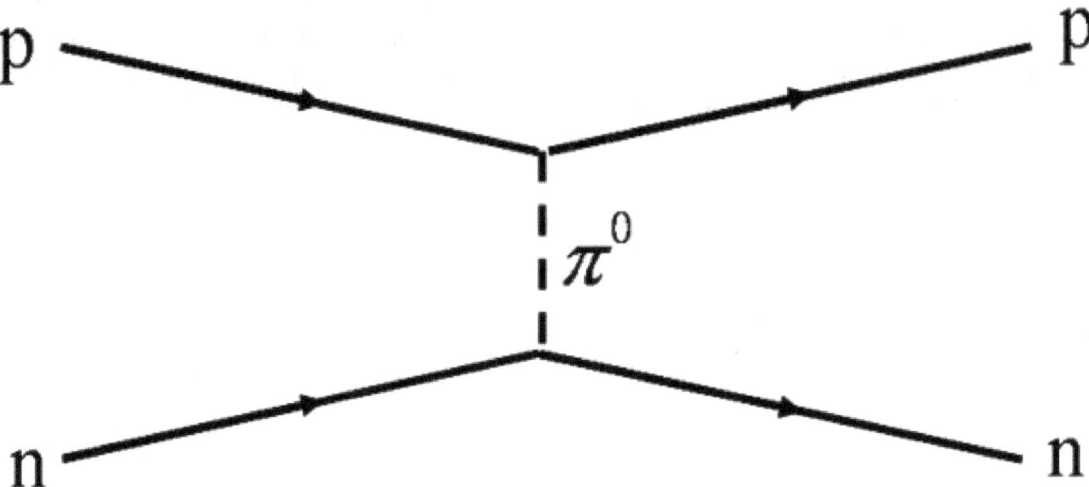

A Feynman diagram of a strong proton–neutron interaction mediated by a neutral pion. Time proceeds from left to right.

of the four fundamental interactions, and it refers to such processes as beta decay. The weak force plays no role in the interaction of nucleons, though it is responsible for the decay of neutrons to protons and vice versa.

34.2 History

The nuclear force has been at the heart of nuclear physics ever since the field was born in 1932 with the discovery of the neutron by James Chadwick. The traditional goal of nuclear physics is to understand the properties of atomic nuclei in terms of the 'bare' interaction between pairs of nucleons, or nucleon–nucleon forces (NN forces).

Within months after the discovery of the neutron, Werner Heisenberg[8][9][10] and Dmitri Ivanenko[11] had proposed proton–neutron models for the nucleus.[12] Heisenberg approached the description of protons and neutrons in the nucleus through quantum mechanics, an approach that was not at all obvious at the time. Heisenberg's theory for protons and neutrons in the nucleus was a "major step toward understanding the nucleus as a quantum mechanical system."[13] Heisenberg introduced the first theory of nuclear exchange forces that bind the nucleons. He considered protons and neutrons to be different quantum states of the same particle, i.e., nucleons distinguished by the value of their nuclear isospin quantum numbers.

One of the earliest models for the nucleus was the liquid drop model developed in the 1930s. One property of nuclei is that the average binding energy per nucleon is approximately the same for all stable nuclei, which is similar to a liquid drop. The liquid drop model treated the nucleus as a drop of incompressible nuclear fluid, with nucleons behaving like molecules in a liquid. The model was first proposed by George Gamow and then developed by Niels Bohr, Werner Heisenberg and Carl Friedrich von Weizsäcker. This crude model did not explain all the properties of the nucleus, but it did explain the spherical shape of most nuclei. The model also gave good predictions for the nuclear binding energy of nuclei.

In 1934, Hideki Yukawa made the earliest attempt to explain the nature of the nuclear force. According to his theory, massive bosons (mesons) mediate the interaction between two nucleons. Although, in light of quantum chromodynamics (QCD), meson theory is no longer perceived as fundamental, the meson-exchange concept (where hadrons are treated as elementary particles) continues to represent the best working model for a quantitative NN potential. The Yukawa potential (also called a screened Coulomb potential) is a potential of the form

$$V_{\text{Yukawa}}(r) = -g^2 \frac{e^{-\mu r}}{r},$$

where g is a magnitude scaling constant, i.e., the amplitude of potential, μ is the Yukawa particle mass, r is the radial distance to the particle. The potential is monotone increasing, implying that the force is always attractive. The constants are determined empirically. The Yukawa potential depends only on the distance between particles, r, hence it models a central force.

Throughout the 1930s a group at Columbia University led by I. I. Rabi developed magnetic resonance techniques to determine the magnetic moments of nuclei. These measurements led to the discovery in 1939 that the deuteron also possessed an electric quadrupole moment.[14][15] This electrical property of the deuteron had been interfering with the measurements by the Rabi group. The deuteron, composed of a proton and a neutron, is one of the simplest nuclear systems. The discovery meant that the physical shape of the deuteron was not symmetric, which provided valuable insight into the nature of the nuclear force binding nucleons. In particular, the result showed that the nuclear force was not a central force, but had a tensor character.[1] Hans Bethe identified the discovery of the deuteron's quadrupole moment as one of the important events during the formative years of nuclear physics.[14]

Historically, the task of describing the nuclear force phenomenologically was formidable. The first semi-empirical quantitative models came in the mid-1950s,[1] such as the Woods–Saxon potential (1954). There was substantial progress in experiment and theory related to the nuclear force in the 1960s and 1970s. One influential model was the Reid potential (1968).[1] In recent years, experimenters have concentrated on the subtleties of the nuclear force, such as its charge dependence, the precise value of the πNN coupling constant, improved phase shift analysis, high-precision NN data, high-precision NN potentials, NN scattering at intermediate and high energies, and attempts to derive the nuclear force from QCD.

34.3 The nuclear force as a residual of the strong force

The nuclear force is a residual effect of the more fundamental strong force, or strong interaction. The strong interaction is the attractive force that binds the elementary particles called quarks together to form the nucleons themselves. This more powerful force is mediated by particles called gluons. Gluons hold quarks together with a force like that of electric charge, but of far greater strength. Quarks, gluons and their dynamics are mostly confined within nucleons, but residual influences extend slightly beyond nucleon boundaries to give rise to the nuclear force.

The nuclear forces arising between nucleons are analogous to the forces in chemistry between neutral atoms or molecules called London forces. Such forces between atoms are much weaker than the attractive electrical forces that hold the atoms themselves together (i.e., that bind electrons to the nucleus), and their range between atoms is shorter, because they arise from small separation of charges inside the neutral atom. Similarly, even though nucleons are made of quarks in combinations which cancel most gluon forces (they are "color neutral"), some combinations of quarks and gluons nevertheless leak away from nucleons, in the form of short-range nuclear force fields that extend from one nucleon to another nearby nucleon. These nuclear forces are very weak compared to direct gluon forces ("color forces" or strong forces) inside nucleons, and the nuclear forces extend only over a few nuclear diameters, falling exponentially with distance. Nevertheless, they are strong enough to bind neutrons and protons over short distances, and overcome the electrical repulsion between protons in the nucleus.

Sometimes, the nuclear force is called the **residual strong force**, in contrast to the strong interactions which arise from QCD. This phrasing arose during the 1970s when QCD was being established. Before that time, the *strong nuclear force* referred to the inter-nucleon potential. After the verification of the quark model, *strong interaction* has come to mean QCD.

34.4 Nucleon–nucleon potentials

Two-nucleon systems such as the deuteron, the nucleus of a deuterium atom, as well as proton–proton or neutron–proton scattering are ideal for studying the NN force. Such systems can be described by attributing a *potential* (such as the Yukawa potential) to the nucleons and using the potentials in a Schrödinger equation. The form of the potential is derived phenomenologically, although for the long-range interaction, meson-exchange theories help to construct the potential. The parameters of the potential are determined by fitting to experimental data such as the deuteron binding energy or NN

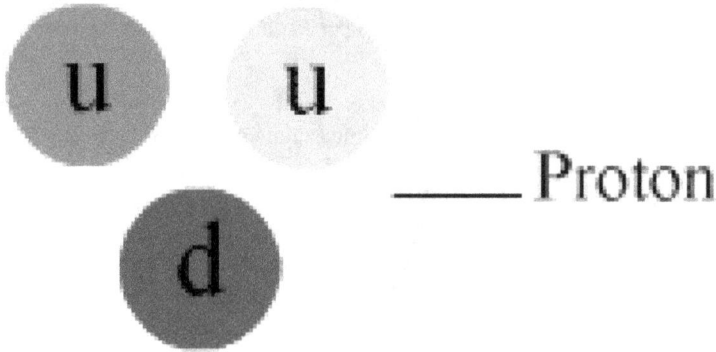

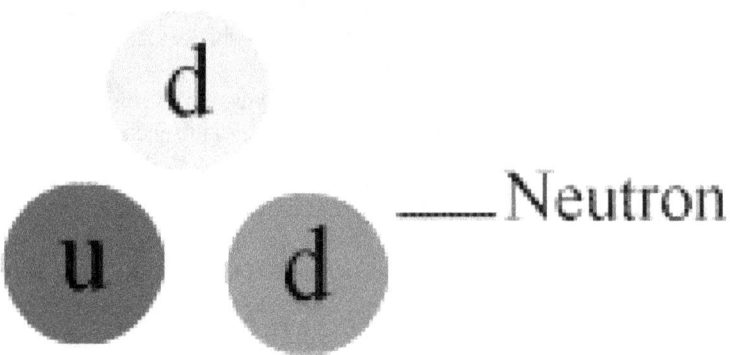

An animation of the interaction. The colored double circles are gluons. Anticolors are shown as per this diagram (larger version).

elastic scattering cross sections (or, equivalently in this context, so-called *NN* phase shifts).

The most widely used *NN* potentials are the Paris potential, the Argonne AV18 potential ,[16] the CD-Bonn potential and the Nijmegen potentials.

A more recent approach is to develop effective field theories for a consistent description of nucleon–nucleon and three-nucleon forces. In particular, chiral symmetry breaking can be analyzed in terms of an effective field theory (called chiral perturbation theory) which allows perturbative calculations of the interactions between nucleons with pions as exchange particles.

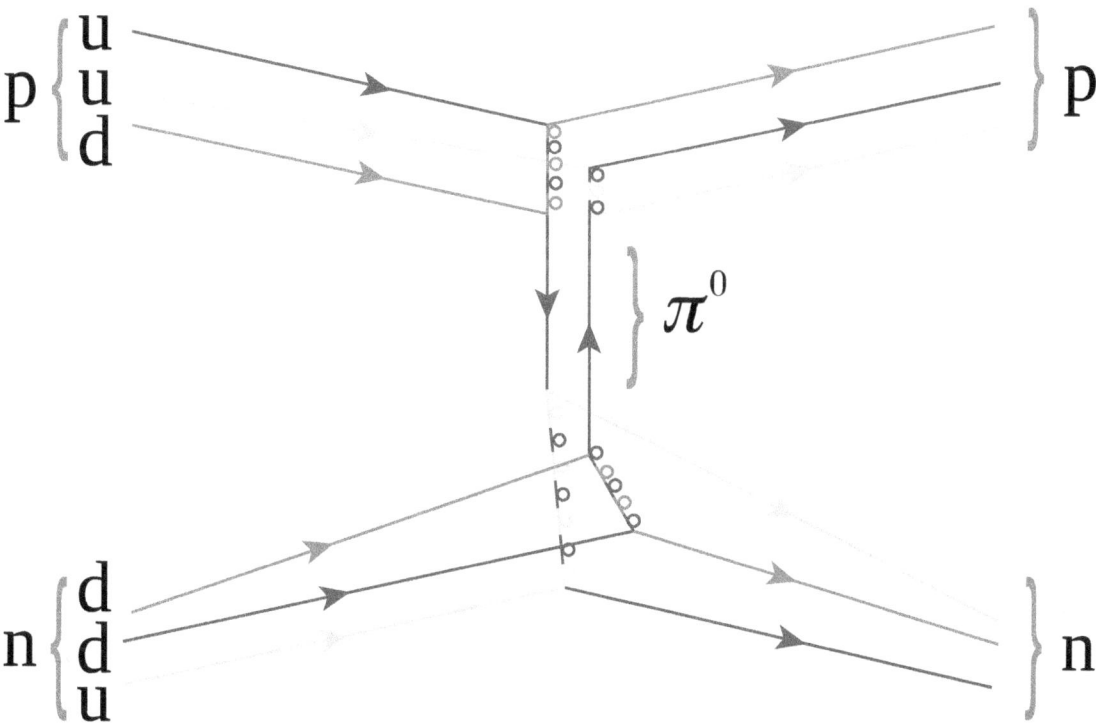

The same diagram as that above with the individual quark constituents shown, to illustrate how the fundamental *strong interaction gives rise to the **nuclear force**. Straight lines are quarks, while multi-colored loops are gluons (the carriers of the fundamental force). Other gluons, which bind together the proton, neutron, and pion "in-flight," are not shown.*

34.4.1   From nucleons to nuclei

The ultimate goal of nuclear physics would be to describe all nuclear interactions from the basic interactions between nucleons. This is called the *microscopic* or *ab initio* approach of nuclear physics. There are two major obstacles to overcome before this dream can become reality:

- Calculations in many-body systems are difficult and require advanced computation techniques.

- There is evidence that three-nucleon forces (and possibly higher multi-particle interactions) play a significant role. This means that three-nucleon potentials must be included into the model.

This is an active area of research with ongoing advances in computational techniques leading to better first-principles calculations of the nuclear shell structure. Two- and three-nucleon potentials have been implemented for nuclides up to $A = 12$.

34.4.2   Nuclear potentials

A successful way of describing nuclear interactions is to construct one potential for the whole nucleus instead of considering all its nucleon components. This is called the *macroscopic* approach. For example, scattering of neutrons from nuclei can be described by considering a plane wave in the potential of the nucleus, which comprises a real part and an imaginary part. This model is often called the **optical model** since it resembles the case of light scattered by an opaque glass sphere.

Nuclear potentials can be *local* or *global*: local potentials are limited to a narrow energy range and/or a narrow nuclear mass range, while global potentials, which have more parameters and are usually less accurate, are functions of the energy and the nuclear mass and can therefore be used in a wider range of applications.

34.5 See also

- Strong interaction

- Standard Model

34.6 References

[1] Reid,R.V. (1968). "Local phenomenological nucleon–nucleon potentials".*Annals of Physics***50**: 411–448. Bibcode:1968. doi:10.1016/0003-4916(68)90126-7.

[2] Binding Energy, Mass Defect, Furry Elephant physics educational site, retr 2012 7 1

[3] Chapter 4 NUCLEAR PROCESSES, THE STRONG FORCE, M. Ragheb 1/30/2013, University of Illinois

[4] Kenneth S. Krane (1988). *Introductory Nuclear Physics*. Wiley & Sons. ISBN 0-471-80553-X.

[5] Povh, B.; Rith, K.; Scholz, C.; Zetsche, F. (2002). *Particles and Nuclei: An Introduction to the Physical Concepts*. Berlin: Springer-Verlag. p. 73. ISBN 978-3-540-43823-6.

[6] Stern, Dr. Swapnil Nikam (February 11, 2009). "Nuclear Binding Energy". *"From Stargazers to Starships"*. NASA website. Retrieved 2010-12-30.

[7] Griffiths, David, Introduction to Elementary Particles

[8] Heisenberg, W. (1932). "Über den Bau der Atomkerne. I". *Z. Phys.* **77**: 1–11. doi:10.1007/BF01342433.

[9] Heisenberg, W. (1932). "Über den Bau der Atomkerne. II". *Z. Phys.* **78** (3–4): 156–164. doi:10.1007/BF01337585.

[10] Heisenberg, W. (1933). "Über den Bau der Atomkerne. III". *Z. Phys.* **80** (9–10): 587–596. doi:10.1007/BF01335696.

[11] Iwanenko, D.D., The neutron hypothesis, Nature **129** (1932) 798.

[12] Miller A. I. *Early Quantum Electrodynamics: A Sourcebook*, Cambridge University Press, Cambridge, 1995, ISBN 0521568919, pp. 84–88.

[13] Brown, L.M.; Rechenberg, H. (1996). *The Origin of the Concept of Nuclear Forces*. Bristol and Philadelphia: Institute of Physics Publishing. ISBN 0750303735.

[14] John S. Rigden (1987). *Rabi, Scientist and Citizen*. New York: Basic Books, Inc. pp. 99–114. ISBN 9780674004351. Retrieved May 9, 2015.

[15] Kellogg, J.M.; Rabi, I.I.; Ramsey, N.F.; Zacharias, J.R. (1939). "An electrical quadrupole moment of the deuteron". *Physical Review* **55**: 318–319. Bibcode:1939PhRv...55..318K. doi:10.1103/physrev.55.318. Retrieved May 9, 2015.

[16] Wiringa, R. B.; Stoks, V. G. J.; Schiavilla, R. (1995). "Accurate nucleon–nucleon potential with charge-independence breaking". *Physical Review C* **51**: 38. arXiv:nucl-th/9408016. Bibcode:1995PhRvC..51...38W. doi:10.1103/PhysRevC.51.38.

34.7 Bibliography

- Gerald Edward Brown and A. D. Jackson, *The Nucleon–Nucleon Interaction*, (1976) North-Holland Publishing, Amsterdam ISBN 0-7204-0335-9

- R. Machleidt and I. Slaus, "The nucleon–nucleon interaction", *J. Phys.* G **27** (2001) R69 *(topical review)*.

- E.A. Nersesov, *Fundamentals of atomic and nuclear physics*, (1990), Mir Publishers, Moscow, ISBN 5-06-001249-2

- P. Navrátil and W.E. Ormand, "Ab initio shell model with a genuine three-nucleon force for the p-shell nuclei", Phys. Rev. C **68**, 034305 (2003).

Chapter 35

Glueball

In particle physics, a **glueball** (or **gluonium**) is a hypothetical composite particle.[1] It consists solely of gluon particles, without valence quarks. Such a state is possible because gluons carry color charge and experience the strong interaction. Glueballs are extremely difficult to identify in particle accelerators, because they mix with ordinary meson states.[2]

Theoretical calculations show that glueballs should exist at energy ranges accessible with current collider technology. However, due to the aforementioned difficulty (among others), they have (as of 2015) so far not been observed and identified with certainty,[3] although phenomenological calculations have suggested that an experimentally identified glueball candidate, denoted $f_0(1710)$, has properties consistent with those expected of a Standard Model glueball.[4] The prediction that glueballs exist is one of the most important predictions of the Standard Model of particle physics that has not yet been confirmed experimentally.[5]

35.1 Properties of glueballs

In principle, it is theoretically possible for all properties of glueballs to be calculated exactly and derived directly from the equations and fundamental physical constants of quantum chromodynamics (QCD) without further experimental input. So, the predicted properties of these hypothetical particles can be described in exquisite detail using only Standard Model physics which have wide acceptance in the theoretical physics literature. But, the fact that QCD calculations are so difficult that solutions to these equations are almost always numerical approximations (reached by several very different methodologies) and the considerable uncertainty in the measurement of some of the relevant key physical constants can lead to variation in theoretical predictions of glueball properties like mass and branching ratios in glueball decays.

35.1.1 Constituent particles and color charge

Theoretical studies of glueballs have focused on glueballs consisting of either two gluons or three gluons, by analogy to mesons and baryons that have two and three quarks respectively. As in the case of mesons and baryons, glueballs would be QCD color charge neutral (aka isospin = 0). The baryon number of a glueball is zero.

35.1.2 Total angular momentum

Two gluon glueballs can have total angular momentum (J) of 0 (which are scalar or pseudo-scalar) or 2 (tensor). Three gluon glueballs can have total angular momentum (J) of 1 (vector boson) or 3. All glueballs have integer total angular momentum which implies that they are bosons rather than fermions.

Glueballs are the only particles predicted by the Standard Model with total angular momentum (J) (sometimes called "intrinsic spin") that could be either 2 or 3 in their ground states, although mesons made of two quarks with J=0 and

J=1 with similar masses have been observed and excited states of other mesons can have these values of total angular momentum.

Fundamental particles with ground states having J=0 or J=2 are easily distinguished from glueballs. The hypothetical graviton, while having a total angular momentum J=2 would be massless and lack color charge, and so would be easily distinguished from glueballs. The Standard Model Higgs boson for which an experimentally measured mass of about 125-126 GeV/c^2 has been determined (although the status of the measured particle as a true Standard Model Higgs boson has not been definitively established), is the only fundamental particle with J=0 in the Standard Model, also lacks color charge and hence does not engage in strong force interactions. The Higgs boson is about 25-80 times as heavy as the mass of the various glueball states predicted by the Standard Model.

35.1.3 Electric charge

All glueballs would have electric charge, Q(e), of zero as gluons themselves do not have an electric charge.

35.1.4 Mass and parity

Glueballs are predicted by quantum chromodynamics to be massive, notwithstanding the fact that gluons themselves have zero rest mass in the Standard Model. Glueballs with all four possible combinations of quantum numbers P (parity) and C (c-parity) for every possible total angular momentum have been considered, producing at least fifteen possible glueball states including excited glueball states that share the same quantum numbers but have differing masses with the lightest states having masses as low as 1.4 GeV/c^2 (for a glueball with quantum numbers J=0, P=+, C=+), and the heaviest states having masses as great as almost 5 GeV/c^2 (for a glueball with quantum numbers J=0, P=+, C=-).[6]

These masses are on the same order of magnitude as the masses of many experimentally observed mesons and baryons, as well as to the masses of the tau lepton, charm quark, bottom quark, some hydrogen isotopes, and some helium isotopes.

35.1.5 Stability and decay channels

Just as all Standard Model mesons and baryons, except the proton, are unstable in isolation, all glueballs are predicted by the Standard Model to be unstable in isolation, with various QCD calculations predicting the total decay width (which is functionally related to half-life) for various glueball states. QCD calculations also make predictions regarding the expected decay patterns of glueballs.[7][8] For example, glueballs would not have radiative or two photon decays, but would have decays into pairs of pions, pairs of kaons, or pairs of eta mesons.[7]

35.2   Practical impact on macroscopic low energy physics

Because Standard Model glueballs are so ephemeral (decaying almost immediately into more stable decay products) and are only generated in high energy physics, glueballs only arise synthetically in the natural conditions found on Earth that humans can easily observe. They are scientifically notable mostly because they are a testable prediction of the Standard Model, and not because of phenomenological impact on macroscopic processes, or their engineering applications.

35.3   Lattice QCD simulations

Lattice field theory provides a way to study the glueball spectrum theoretically and from first principles. Some of the first quantities calculated using lattice QCD methods (in 1980) were glueball mass estimates.[10] Morningstar and Peardon[11] computed in 1999 the masses of the lightest glueballs in QCD without dynamical quarks. The three lowest states are tabulated below. The presence of dynamical quarks would slightly alter these data, but also makes the computations more difficult. Since that time calculations within QCD (lattice and sum rules) find the lightest glueball to be a scalar with mass in the range of about 1000–1700 MeV.[12]

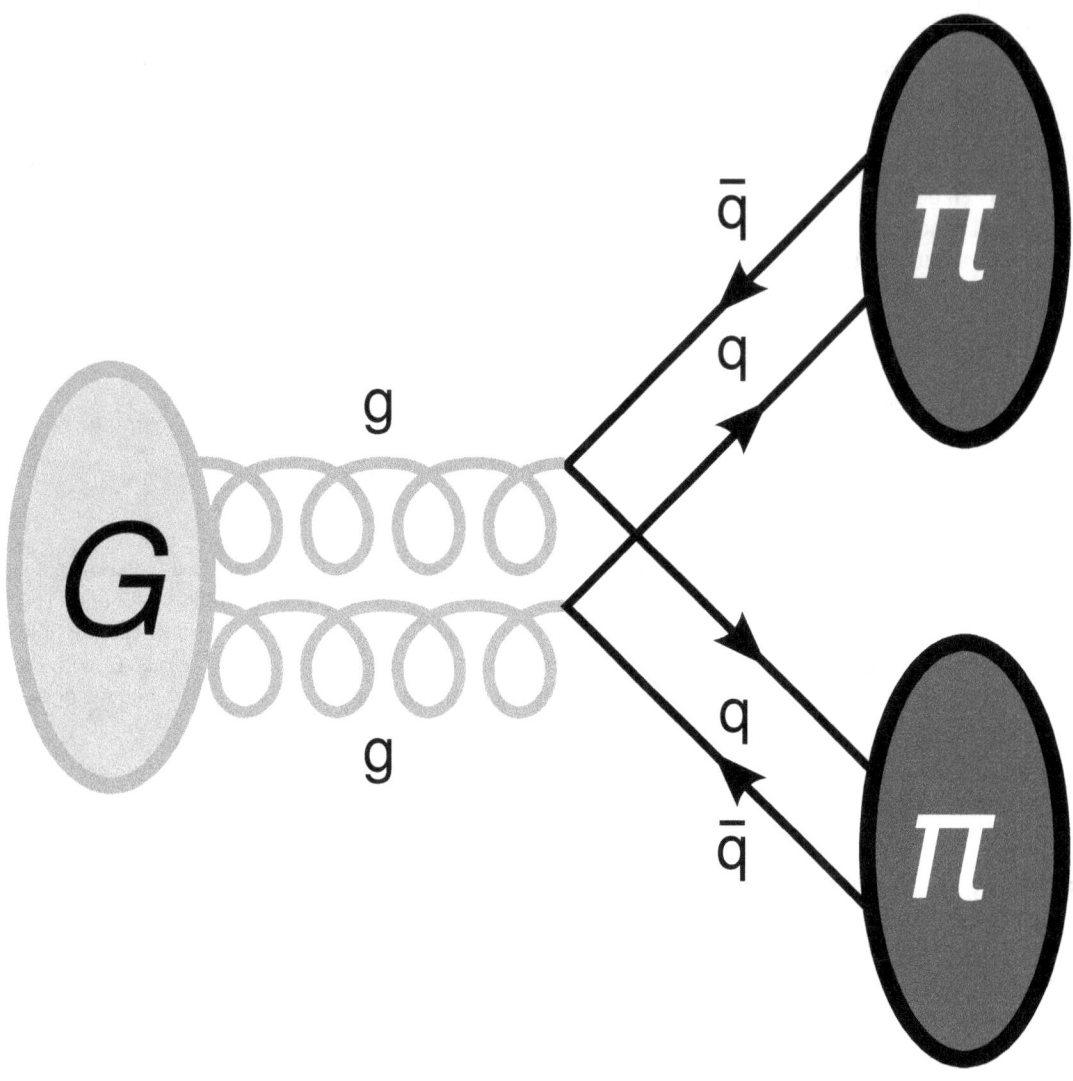

Feynman diagram of a glueball (G) decaying to two pions (π). Such decays help the study of and search for glueballs.[9]

35.4 Experimental candidates

Particle accelerator experiments are often able to identify unstable composite particles and assign masses to those particles to a precision of approximately 10 MeV/c^2, without being able to immediately assign to the particle resonance that is observed all of the properties of that particle. Scores of such particles have been detected, although particles detected in some experiments but not others can be viewed as doubtful. Some of the candidate particle resonances that could be glueballs, although the evidence is not definitive, include the following:

Vector, Pseudo-Vector, or Tensor Glueball Candidates:

- X(3020) observed by the BaBar collaboration is a candidate for an excited state of the 2-+, 1+- or 1-- glueball states with a mass of about 3.02 GeV/c^2.[5]

Scalar Glueball Candidates:

- $f_0(500)$ also known as σ -- the properties of this particle are possibly consistent with a 1000 MeV or 1500 MeV mass glueball.[13]

- $f_0(980)$ -- the structure of this composite particle is consistent with the existence of a light glueball.[13]

- $f_0(1370)$ -- existence of this resonance is disputed but is a candidate for a glueball-meson mixing state[13]

- $f_0(1500)$ -- existence of this resonance is undisputed but its status as a glueball-meson mixing state or pure glueball is not well established.[13]

- $f_0(1710)$ -- existence of this resonance is undisputed but its status as a glueball-meson mixing state or pure glueball is not well established.[13]

Other Glueball Candidates:

- Gluon jets at the LEP experiment show a 40% excess over theoretical expectations of electromagnetically neutral clusters which suggests that electromagnetically neutral particles expected in gluon rich environments such as glueballs are likely to be present.[13]

Many of these candidates have been the subject of active investigation for at least eighteen years.[7] The GlueX experiment, scheduled to begin in 2014, has been specifically designed to produce more definitive experimental evidence glueballs.[14]

35.5   See also

- Exotic meson

- GlueX

- Gluon

- Yang–Mills theory

35.6   References

[1]   • Frank Close and Phillip R. Page, "Glueballs", *Scientific American*, vol. 279 no. 5 (November 1998) pp. 80–85

[2]   Vincent Mathieu; Nikolai Kochelev; Vicente Vento (2009). "The Physics of Glueballs". *International Journal of Modern Physics E* **18**: 1–49. arXiv:0810.4453. Bibcode:2009IJMPE..18....1M. doi:10.1142/S0218301309012124. Glueball on arxiv.org

[3] Wolfgang Ochs, "The Status of Glueballs"J.Phys.G:Nuclear and Particle Physics40,67(2013)DOI:10.1088/0954-3899/ http://arxiv.org/pdf/1301.5183v3.pdf

[4] Frederic Brünner; Anton Rebhan (2015-09-21). "Nonchiral Enhancement of Scalar Glueball Decay in the Witten-Sakai-Sugimoto Model". *Phys. Rev. Lett.* **115** (13). arXiv:1504.05815. doi:10.1103/PhysRevLett.115.131601.

[5] Y.K. Hsiao, C.Q. Geng, "Identifying Glueball at 3.02 GeV in Baryonic B Decays" (Version 2: October 9, 2013) http://arxiv.org/abs/1302.3331

[6] Wolfgang Ochs, "The Status of Glueballs"J.Phys.G:Nuclear and Particle Physics40,6(2013)DOI:10.1088/0954-3899/40/ http://arxiv.org/pdf/1301.5183v3.pdf

[7] Walter Taki, "Search for Glueballs" (1996) http://www.slac.stanford.edu/cgi-wrap/getdoc/ssi96-006.pdf

[8] See, e.g., Walaa I. Eshraim, Stanislaus Janowski, "Branching ratios of the pseudoscalar glueball with a mass of 2.6 GeV", prepared for Proceedings of Confinement X - Conference on Quark Confinement and the Hadron Spectrum (Munich/Germany, 8–12 October 2012) (pre-print published January 15, 2013) http://arxiv.org/abs/1301.3345

[9] T. Cohen, F. J. Llanes-Estrada, J. R. Pelaez, J. Ruiz de Elvira (2014). "Non-ordinary light meson couplings and the 1/Nc expansion". arXiv:1405.4831 [hep-ph].

[10] B. Berg. Plaquette-plaquette correlations in the su(2) lattice gauge theory. Phys. Lett., B97:401, 1980.

[11] Colin J. Morningstar; Mike Peardon (1999). "Glueball spectrum from an anisotropic lattice study". *Physical Review D* **60** (3): 034509. arXiv:hep-lat/9901004. Bibcode:1999PhRvD..60c4509M. doi:10.1103/PhysRevD.60.034509.

[12] Wolfgang Ochs, "The status of glueballs" Source: JOURNAL OF PHYSICS G-NUCLEAR AND PARTICLE PHYSICS Volume: 40 Issue: 4 Article Number: 043001 DOI: 10.1088/0954-3899/40/4/043001 Published: APR 2013

[13] Wolfgang Ochs(2013). "The status of glueballs".*Journal of Physics G***40**(4): 043001. arXiv:1301.5183. Bibcode:2013JPhG. doi:10.1088/0954-3899/40/4/043001.

[14] "The Physics of GlueX".

Chapter 36

Quantum number

"Q-number" redirects here. For the Q-theory concept, see Q-analog.

Quantum numbers describe values of conserved quantities in the dynamics of a quantum system. In the case of quantum numbers of electrons, they can be defined as "the sets of numerical values which give acceptable solutions to the Schrödinger wave equation for the hydrogen atom". Perhaps the most important aspect of quantum mechanics is the quantization of observable quantities, since quantum numbers are discrete sets of integers or half-integers, although they could approach infinity in some cases. This is distinguished from classical mechanics where the values can range continuously. Quantum numbers often describe specifically the energy levels of electrons in atoms, but other possibilities include angular momentum, spin, etc. Any quantum system can have one or more quantum numbers; it is thus difficult to list all possible quantum numbers.[1]

36.1   How many quantum numbers?

The question of *how many quantum numbers are needed to describe any given system* has no universal answer. Hence for each system one must find the answer for a full analysis of the system. A quantized system requires at least one quantum number. The dynamics of any quantum system are described by a quantum Hamiltonian, H. There is one quantum number of the system corresponding to the energy, i.e., the eigenvalue of the Hamiltonian. There is also one quantum number for each operator O that commutes with the Hamiltonian. These are all the quantum numbers that the system can have. Note that the operators O defining the quantum numbers should be independent of each other. Often, there is more than one way to choose a set of independent operators. Consequently, in different situations different sets of quantum numbers may be used for the description of the same system.

36.2   Spatial and angular momentum numbers

There are four quantum numbers which can describe an electron in an atom completely.

- Principal quantum number (n)

- Azimuthal quantum number (ℓ)

- Magnetic quantum number (m)

- Spin quantum number (s)

36.2.1 Traditional nomenclatures

Many different models have been proposed throughout the history of quantum mechanics, but the most prominent system of nomenclature spawned from the Hund-Mulliken molecular orbital theory of Friedrich Hund, Robert S. Mulliken, and contributions from Schrödinger, Slater and John Lennard-Jones. This system of nomenclature incorporated Bohr energy levels, Hund-Mulliken orbital theory, and observations on electron spin based on spectroscopy and Hund's rules.[2]

This model describes electrons using four quantum numbers, n, ℓ, $m\ell$, ms, given below. It is also the common nomenclature in the classical description of nuclear particle states (e.g. protons and neutrons). Molecular orbitals require different quantum numbers, because the Hamiltonian and its symmetries are quite different.

1. **The principal quantum number (n)** describes the electron shell, or energy level, of an atom. The value of n ranges from 1 to the shell containing the outermost electron of that atom, i.e.[3]

   $n = 1, 2, \ldots$.

   For example, in caesium (Cs), the outermost valence electron is in the shell with energy level 6, so an electron in caesium can have an n value from 1 to 6.

   For particles in a time-independent potential (see Schrödinger equation), it also labels the nth eigenvalue of Hamiltonian (H), i.e. the energy, E with the contribution due to angular momentum (the term involving $\mathbf{J}^2$) left out. This number therefore has a dependence only on the distance between the electron and the nucleus (i.e., the radial coordinate, $\mathbf{r}$). The average distance increases with $\mathbf{n}$, and hence quantum states with different principal quantum numbers are said to belong to different shells.

2. **The azimuthal quantum number (ℓ)** (also known as the **angular quantum number** or **orbital quantum number**) describes the subshell, and gives the magnitude of the orbital angular momentum through the relation

   $L^2 = \hbar^2 \, \ell \, (\ell + 1)$.

   In chemistry and spectroscopy, "$\ell = 0$" is called an s orbital, "$\ell = 1$" a p orbital, "$\ell = 2$" a d orbital, and "$\ell = 3$" an f orbital.

   The value of ℓ ranges from 0 to $n - 1$, because the first p orbital ($\ell = 1$) appears in the second electron shell ($n = 2$), the first d orbital ($\ell = 2$) appears in the third shell ($n = 3$), and so on:[4]

   $\ell = 0, 1, 2, \ldots, n - 1$.

   A quantum number beginning in 3, 0, ... describes an electron in the s orbital of the third electron shell of an atom. In chemistry, this quantum number is very important, since it specifies the shape of an atomic orbital and strongly influences chemical bonds and bond angles.

3. **The magnetic quantum number ($m\ell$)** describes the specific orbital (or "cloud") within that subshell, and yields the *projection* of the orbital angular momentum *along a specified axis*:

   $Lz = m\ell \, \hbar$.

   The values of $m\ell$ range from $-\ell$ to ℓ, with integer steps between them:[5]

   The s subshell ($\ell = 0$) contains only one orbital, and therefore the $m\ell$ of an electron in an s orbital will always be 0. The p subshell ($\ell = 1$) contains three orbitals (in some systems, depicted as three "dumbbell-shaped" clouds), so the $m\ell$ of an electron in a p orbital will be -1, 0, or 1. The d subshell ($\ell = 2$) contains five orbitals, with $m\ell$ values of -2, -1, 0, 1, and 2.

4. **The spin projection quantum number (ms)** describes the spin (intrinsic angular momentum) of the electron within that orbital, and gives the projection of the spin angular momentum S along the specified axis:

   $Sz = ms \, \hbar$.

   In general, the values of ms range from $-s$ to s, where s is the spin quantum number, an intrinsic property of particles:[6]

$ms = -s, -s + 1, -s + 2,...,s - 2, s - 1, s.$

An electron has spin number $s = \frac{1}{2}$, consequently ms will be $\pm\frac{1}{2}$, referring to "spin up" and "spin down" states. Each electron in any individual orbital must have different quantum numbers because of the Pauli exclusion principle, therefore an orbital never contains more than two electrons.

Note that there is no universal fixed value for $m\ell$ and ms values. Rather, the $m\ell$ and ms values are random. The only requirement is that the naming schematic used within a particular set of calculations or descriptions must be consistent (e.g. the orbital occupied by the first electron in a p orbital could be described as $m\ell = -1$ or $m\ell = 0$, or $m\ell = 1$, but the $m\ell$ value of the other electron in that orbital must be different; yet, the $m\ell$ assigned to electrons in other orbitals again can be $m\ell = -1$ or $m\ell = 0$, or $m\ell = 1$).

These rules are summarized as follows:

Example: The quantum numbers used to refer to the outermost valence electrons of the Carbon (C) atom, which are located in the 2p atomic orbital, are; $n = 2$ (2nd electron shell), $\ell = 1$ (p orbital subshell), $m\ell = 1, 0$ or -1, $ms = \frac{1}{2}$ (parallel spins).

Results from spectroscopy indicated that up to two electrons can occupy a single orbital. However two electrons can never have the same exact quantum state nor the same set of quantum numbers according to Hund's rules, which addresses the Pauli exclusion principle. A fourth quantum number with two possible values was added as an *ad hoc* assumption to resolve the conflict; this supposition could later be explained in detail by relativistic quantum mechanics and from the results of the renowned Stern–Gerlach experiment.

36.3   Total angular momenta numbers

36.3.1   Total momentum of a particle

For more details on this topic, see Clebsch–Gordan coefficients.
See also: Azimuthal quantum number § Total angular momentum of an electron in the atom

When one takes the spin-orbit interaction into consideration, the L and S operators no longer commute with the Hamiltonian, and their eigenvalues therefore change over time. Thus another set of quantum numbers should be used. This set includes[7][8]

1. **The total angular momentum quantum number:**

   $j = |\ell \pm s|$

   which gives the total angular momentum through the relation

   $J^2 = \hbar^2 \, j \, (j + 1).$

2. **The projection of the total angular momentum along a specified axis:**

   $mj = -j, -j + 1, -j + 2,...,j - 2, j - 1, j$

   analogous to the above, and satisfies

   $mj = m\ell + ms$ and $|m\ell + ms| \leq j.$

3. **Parity**

This is the eigenvalue under reflection, and is positive (+1) for states which came from even ℓ and negative (−1) for states which came from odd ℓ. The former is also known as **even parity** and the latter as **odd parity**, and is given by

$$P = (-1)^\ell.$$

For example, consider the following eight states, defined by their quantum numbers:

The quantum states in the system can be described as linear combination of these eight states. However, in the presence of spin-orbit interaction, if one wants to describe the same system by eight states which are eigenvectors of the Hamiltonian (i.e. each represents a state which does not mix with others over time), we should consider the following eight states:

36.3.2 Nuclear angular momentum quantum numbers

In nuclei, the entire assembly of protons and neutrons (nucleons) has a resultant angular momentum due to the angular momenta of each nucleon, usually denoted **I**. If the total angular momentum of a neutron is $jn = \ell + s$ and for a proton is $jp = \ell + s$ (where s for protons and neutrons happens to be ½ again) then the **nuclear angular momentum quantum numbers** I are given by:

$$I = |jn - jp|, \; |jn - jp| + 1, \; |jn - jp| + 2,..., \; (jn + jp) - 2, \; (jn + jp) - 1, \; (jn + jp)$$

Parity with the number I is used to label nuclear angular momentum states, examples for some isotopes of Hydrogen (H), Carbon (C), and Sodium (Na) are;[9]

The reason for the unusual fluctuations in I, even by differences of just one nucleon, are due to the odd/even numbers of protons and neutrons - pairs of nucleons have a total angular momentum of zero (just like electrons in orbitals), leaving an odd/even numbers of unpaired nucleons. The property of nuclear spin is an important factor for the operation of NMR spectroscopy in organic chemistry,[8] and MRI in nuclear medicine,[9] due to the nuclear magnetic moment interacting with an external magnetic field.

36.4 Elementary particles

For a more complete description of the quantum states of elementary particles, see Standard model and Flavour (particle physics).

Elementary particles contain many quantum numbers which are usually said to be intrinsic to them. However, it should be understood that the elementary particles are quantum states of the standard model of particle physics, and hence the quantum numbers of these particles bear the same relation to the Hamiltonian of this model as the quantum numbers of the Bohr atom does to its Hamiltonian. In other words, each quantum number denotes a symmetry of the problem. It is more useful in quantum field theory to distinguish between spacetime and internal symmetries.

Typical quantum numbers related to spacetime symmetries are spin (related to rotational symmetry), the parity, C-parity and T-parity (related to the Poincaré symmetry of spacetime). Typical **internal symmetries** are lepton number and baryon number or the electric charge. (For a full list of quantum numbers of this kind see the article on flavour.)

A minor but often confusing point is as follows: most conserved quantum numbers are additive, so in an elementary particle reaction, the *sum* of the quantum numbers should be the same before and after the reaction. However, some, usually called a *parity*, are multiplicative; i.e., their *product* is conserved. All multiplicative quantum numbers belong to a symmetry (like parity) in which applying the symmetry transformation twice is equivalent to doing nothing (involution). These are all examples of an abstract group called $\mathbf{Z}_2$.

36.5   See also

- Electron configuration

36.6   References and external links

[1] McGraw Hill Encyclopaedia of Physics (2nd Edition), C.B. Parker, 1994, ISBN 0-07-051400-3

[2] Chemistry, Matter, and the Universe, R.E. Dickerson, I. Geis, W.A. Benjamin Inc. (USA), 1976, ISBN 0-19-855148-7

[3] Concepts of Modern Physics (4th Edition), A. Beiser, Physics, McGraw-Hill (International), 1987, ISBN 0-07-100144-1

[4] Molecular Quantum Mechanics Parts I and II: An Introduction to QUANTUM CHEMISRTY (Volume 1), P.W. Atkins, Oxford University Press, 1977, ISBN 0-19-855129-0

[5] Quantum Physics of Atoms, Molecules, Solids, Nuclei, and Particles (2nd Edition), R. Eisberg, R. Resnick, John Wiley & Sons, 1985, ISBN 978-0-471-87373-0

[6] Quantum Mechanics (2nd edition), Y. Peleg, R. Pnini, E. Zaarur, E. Hecht, Schuam's Outlines, McGraw Hill (USA), 2010, ISBN 978-0-07-162358-2

[7] Molecular Quantum Mechanics Parts I and II: An Introduction to QUANTUM CHEMISTRY (Volume 1), P.W. Atkins, Oxford University Press, 1977, ISBN 0-19-855129-0

[8] Molecular Quantum Mechanics Part III: An Introduction to QUANTUM CHEMISTRY (Volume 2), P.W. Atkins, Oxford University Press, 1977

[9] Introductory Nuclear Physics, K.S. Krane, 1988, John Wiley & Sons Inc, ISBN 978-0-471-80553-3

36.6.1   General principles

- Dirac, Paul A.M. (1982). *Principles of quantum mechanics*. Oxford University Press. ISBN 0-19-852011-5.

36.6.2   Atomic physics

- Quantum numbers for the hydrogen atom

- Lecture notes on quantum numbers

36.6.3   Particle physics

- Griffiths, David J. (2004). *Introduction to Quantum Mechanics (2nd ed.)*. Prentice Hall. ISBN 0-13-805326-X.

- Halzen, Francis and Martin, Alan D. (1984). *QUARKS AND LEPTONS: An Introductory Course in Modern Particle Physics*. John Wiley & Sons. ISBN 0-471-88741-2.

- The particle data group

Chapter 37

Poincaré group

For the Poincaré group (fundamental group) of a topological space, see Fundamental group.

The **Poincaré group**, named after Henri Poincaré,[1] is the group of Minkowski spacetime isometries.[2][3] It is a ten-generator non-abelian Lie group of fundamental importance in physics.

37.1   Overview

A Minkowski spacetime isometry has the property that the interval between events is left invariant. For example, if everything was postponed by two hours including two events and the path you took to go from one to the other, then the time interval between the events recorded by a stop-watch you carried with you would be the same. Or if everything was shifted five miles to the west, or turned 60 degrees to the right, you would also see no change in the interval. It turns out that the proper length of an object is also unaffected by such a shift. A time or space reversal (a reflection) is also an isometry of this group.

In Minkowski space (i.e. ignoring the effects of gravity), there are ten degrees of freedom of the isometries, which may be thought of as translation through time or space (four degrees, one per dimension); reflection through a plane (three degrees, the freedom in orientation of this plane); or a "boost" in any of the three spatial directions (three degrees). Composition of transformations is the operator of the Poincaré group, with proper rotations being produced as the composition of an even number of reflections.

In classical physics, the Galilean group is a comparable ten-parameter group that acts on absolute time and space. Instead of boosts, it features shear mappings to relate co-moving frames of reference.

37.2   Details

The Poincaré group is the group of Minkowski spacetime isometries. It is a ten-dimensional noncompact Lie group. The abelian group of translations is a normal subgroup, while the Lorentz group is also a subgroup, the stabilizer of the origin. The Poincaré group itself is the minimal subgroup of the affine group which includes all translations and Lorentz transformations. More precisely, it is a semidirect product of the translations and the Lorentz group,

$$\mathbf{R}^{1,3} \rtimes \mathrm{SO}(1,3) \,.$$

Another way of putting this is that the Poincaré group is a group extension of the Lorentz group by a vector representation of it; it is sometimes dubbed, informally, as the "*inhomogeneous Lorentz group*". In turn, it can also be obtained as a group contraction of the de Sitter group SO(4,1) ~ Sp(2,2), as the de Sitter radius goes to infinity.

Its positive energy unitary irreducible representations are indexed by mass (nonnegative number) and spin (integer or half integer) and are associated with particles in quantum mechanics (see Wigner's classification).

In accordance with the Erlangen program, the geometry of Minkowski space is defined by the Poincaré group: Minkowski space is considered as a homogeneous space for the group.

The **Poincaré algebra** is the Lie algebra of the Poincaré group. It is a Lie algebra extension of the Lie algebra of the Lorentz group. More specifically, the proper ($\det\Lambda=1$), orthochronous ($\Lambda^0{}_0 \geq 1$) part of the Lorentz subgroup (its identity component), $SO^+(1,3)$, is connected to the identity and is thus provided by the exponentiation $\exp(ia_\mu P^\mu)\exp(i\omega_{\mu\nu}M^{\mu\nu}/2)$ of this Lie algebra. In component form, the Poincaré algebra is given by the commutation relations:[4][5]

where P is the generator of translations, M is the generator of Lorentz transformations, and η is the $(+,-,-,-)$ Minkowski metric (see Sign convention).

The bottom commutation relation is the ("homogeneous") Lorentz group, consisting of rotations, $J_i = -\epsilon_{imn}M^{mn}/2$, and boosts, $K_i = M_{i0}$. In this notation, the entire Poincaré algebra is expressible in noncovariant (but more practical) language as

$$[J_m, P_n] = i\epsilon_{mnk}P_k \ ,$$

$$[J_i, P_0] = 0 \ ,$$

$$[K_i, P_k] = i\eta_{ik}P_0 \ ,$$

$$[K_i, P_0] = -iP_i \ ,$$

$$[J_m, J_n] = i\epsilon_{mnk}J_k \ ,$$

$$[J_m, K_n] = i\epsilon_{mnk}K_k \ ,$$

$$[K_m, K_n] = -i\epsilon_{mnk}J_k \ ,$$

where the bottom line commutator of two boosts is often referred to as a "Wigner rotation". Note the important simplification $[J_m + i\,K_m\ ,\ J_n - i\,K_n] = 0$, which permits reduction of the Lorentz subalgebra to **su(2)**⊕**su(2)** and efficient treatment of its associated representations.

The Casimir invariants of this algebra are $P_\mu P^\mu$ and $W_\mu W^\mu$ where W_μ is the Pauli–Lubanski pseudovector; they serve as labels for the representations of the group.

The Poincaré group is the full symmetry group of any relativistic field theory. As a result, all elementary particles fall in representations of this group. These are usually specified by the *four-momentum* squared of each particle (i.e. its mass squared) and the intrinsic quantum numbers J^{PC}, where J is the spin quantum number, P is the parity and C is the charge-conjugation quantum number. In practice, charge conjugation and parity are violated by many quantum field theories; where this occurs, P and C are forfeited. Since CPT symmetry is invariant in quantum field theory, a time-reversal quantum number may be constructed from those given.

As a topological space, the group has four connected components: the component of the identity; the time reversed component; the spatial inversion component; and the component which is both time-reversed and spatially inverted.

37.3   Poincaré symmetry

Poincaré symmetry is the full symmetry of special relativity. It includes:

- *translations* (displacements) in time and space (***P***), forming the abelian Lie group of translations on space-time;

- *rotations* in space, forming the non-Abelian Lie group of three-dimensional rotations (***J***);

- *boosts*, transformations connecting two uniformly moving bodies (K).

The last two symmetries, J and K, together make the Lorentz group (see also Lorentz invariance); the semi-direct product of the translations group and the Lorentz group then produce the Poincaré group. Objects which are invariant under this group are then said to possess **Poincaré invariance** or **relativistic invariance**.

37.4 See also

- Euclidean group
- Representation theory of the Poincaré group
- Wigner's classification
- Symmetry in quantum mechanics
- Center of mass (relativistic)
- Pauli–Lubanski pseudovector
- Particle physics and representation theory

37.5 Notes

[1] Poincaré,Henri, "Sur la dynamique de l'électron",*Rendiconti del Circolo matematico di Palermo***21**: 129–176,doi:10.1007/ (Wikisource translation: On the Dynamics of the Electron).

[2] Minkowski, Hermann, "Die Grundgleichungen für die elektromagnetischen Vorgänge in bewegten Körpern", *Nachrichten von der Gesellschaft der Wissenschaften zu Göttingen, Mathematisch-Physikalische Klasse*: 53–111 (Wikisource translation: The Fundamental Equations for Electromagnetic Processes in Moving Bodies).

[3] Minkowski, Hermann, "Raum und Zeit", *Physikalische Zeitschrift* **10**: 75–88

[4] N.N. Bogolubov (1989). *General Principles of Quantum Field Theory* (2nd ed.). Springer. p. 272. ISBN 0-7923-0540-X.

[5] T. Ohlsson (2011). *Relativistic Quantum Physics: From Advanced Quantum Mechanics to Introductory Quantum Field Theory*. Cambridge University Press. p. 10. ISBN 1-13950-4320.

37.6 References

- Wu-Ki Tung (1985). *Group Theory in Physics*. World Scientific Publishing. ISBN 9971-966-57-3.

- Weinberg, Steven (1995). *The Quantum Theory of Fields* **1**. Cambridge: Cambridge University press. ISBN 978-0-521-55001-7.

- L.H. Ryder (1996). *Quantum Field Theory* (2nd ed.). Cambridge University Press. p. 62. ISBN 0-52147-8146.

Chapter 38

Parity (physics)

In quantum mechanics, a **parity transformation** (also called **parity inversion**) is the flip in the sign of *one* spatial coordinate. In three dimensions, it is also often described by the simultaneous flip in the sign of all three spatial coordinates (a point reflection):

$$\mathbf{P} : \begin{pmatrix} x \\ y \\ z \end{pmatrix} \mapsto \begin{pmatrix} -x \\ -y \\ -z \end{pmatrix}.$$

It can also be thought of as a test for chirality of a physical phenomenon, in that a parity inversion transforms a phenomenon into its mirror image. A parity transformation on something achiral, on the other hand, can be viewed as an identity transformation. All fundamental interactions of elementary particles, with the exception of the weak interaction, are symmetric under parity. The weak interaction is chiral and thus provides a means for probing chirality in physics. In interactions that are symmetric under parity, such as electromagnetism in atomic and molecular physics, parity serves as a powerful controlling principle underlying quantum transitions.

A matrix representation of **P** (in any number of dimensions) has determinant equal to −1, and hence is distinct from a rotation, which has a determinant equal to 1. In a two-dimensional plane, a simultaneous flip of all coordinates in sign is *not* a parity transformation; it is the same as a 180°-rotation.

38.1 Simple symmetry relations

Under rotations, classical geometrical objects can be classified into scalars, vectors, and tensors of higher rank. In classical physics, physical configurations need to transform under representations of every symmetry group.

Quantum theory predicts that states in a Hilbert space do not need to transform under representations of the group of rotations, but only under projective representations. The word *projective* refers to the fact that if one projects out the phase of each state, where we recall that the overall phase of a quantum state is not an observable, then a projective representation reduces to an ordinary representation. All representations are also projective representations, but the converse is not true, therefore the projective representation condition on quantum states is weaker than the representation condition on classical states.

The projective representations of any group are isomorphic to the ordinary representations of a central extension of the group. For example, projective representations of the 3-dimensional rotation group, which is the special orthogonal group SO(3), are ordinary representations of the special unitary group SU(2) (see Representation theory of SU(2)). Projective representations of the rotation group that are not representations are called spinors, and so quantum states may transform not only as tensors but also as spinors.

If one adds to this a classification by parity, these can be extended, for example, into notions of

- *scalars* ($P = 1$) and *pseudoscalars* ($P = -1$) which are rotationally invariant.

- *vectors* ($P = -1$) and *axial vectors* (also called *pseudovectors*) ($P = 1$) which both transform as vectors under rotation.

One can define **reflections** such as

$$V_x : \begin{pmatrix} x \\ y \\ z \end{pmatrix} \mapsto \begin{pmatrix} -x \\ y \\ z \end{pmatrix},$$

which also have negative determinant and form a valid parity transformation. Then, combining them with rotations (or successively performing x-, y-, and z-reflections) one can recover the particular parity transformation defined earlier. The first parity transformation given does not work in an even number of dimensions, though, because it results in a positive determinant. In odd number of dimensions only the latter example of a parity transformation (or any reflection of an odd number of coordinates) can be used.

Parity forms the abelian group Z_2 due to the relation $\mathbf{P}^2 = 1$. All Abelian groups have only one-dimensional irreducible representations. For Z_2, there are two irreducible representations: one is even under parity ($\mathbf{P}\varphi = \varphi$), the other is odd ($\mathbf{P}\varphi = -\varphi$). These are useful in quantum mechanics. However, as is elaborated below, in quantum mechanics states need not transform under actual representations of parity but only under projective representations and so in principle a parity transformation may rotate a state by any phase.

38.2 Classical mechanics

Newton's equation of motion $\mathbf{F} = m\mathbf{a}$ (if the mass is constant) equates two vectors, and hence is invariant under parity. The law of gravity also involves only vectors and is also, therefore, invariant under parity.

However, angular momentum $\mathbf{L}$ is an axial vector,

$$\mathbf{L} = \mathbf{r} \times \mathbf{p},$$
$$\mathbf{P}(\mathbf{L}) = (-\mathbf{r}) \times (-\mathbf{p}) = \mathbf{L}.$$

In classical electrodynamics, the charge density ϱ is a scalar, the electric field, $\mathbf{E}$, and current $\mathbf{j}$ are vectors, but the magnetic field, $\mathbf{H}$ is an axial vector. However, Maxwell's equations are invariant under parity because the curl of an axial vector is a vector.

38.3 Effect of spatial inversion on some variables of classical physics

38.3.1 Even

Classical variables, predominantly scalar quantities, which do not change upon spatial inversion include:

t , the time when an event occurs

m , the mass of a particle

E , the energy of the particle

P , power (rate of work done)

ρ , the electric charge density

V , the electric potential (voltage)

ρ , energy density of the electromagnetic field

L , the angular momentum of a particle (both orbital and spin) (axial vector)

B , the magnetic field (axial vector)

H , the auxiliary magnetic field

M , the magnetization

T_{ij} Maxwell stress tensor.

All masses, charges, coupling constants, and other physical constants, except those associated with the weak force

38.3.2   Odd

Classical variables, predominantly vector quantities, which have their sign flipped by spatial inversion include:

h , the helicity

Φ , the magnetic flux

x , the position of a particle in three-space

v , the velocity of a particle

a , the acceleration of the particle

p , the linear momentum of a particle

F , the force exerted on a particle

J , the electric current density

E , the electric field

D , the electric displacement field

P , the electric polarization

A , the electromagnetic vector potential

S , Poynting vector.

38.4   Quantum mechanics

38.4.1   Possible eigenvalues

In quantum mechanics, spacetime transformations act on quantum states. The parity transformation, **P**, is a unitary operator, in general acting on a state ψ as follows: $\mathbf{P}\psi(r) = e^{i\varphi/2}\psi(-r)$.

One must then have $\mathbf{P}^2\psi(r) = e^{i\varphi}\psi(r)$, since an overall phase is unobservable. The operator $\mathbf{P}^2$, which reverses the parity of a state twice, leaves the spacetime invariant, and so is an internal symmetry which rotates its eigenstates by phases $e^{i\varphi}$. If $\mathbf{P}^2$ is an element e^{iQ} of a continuous U(1) symmetry group of phase rotations, then $e^{-iQ/2}$ is part of this U(1) and so is also a symmetry. In particular, we can define $\mathbf{P}' = \mathbf{P}e^{-iQ/2}$, which is also a symmetry, and so we can choose to call $\mathbf{P}'$ our parity operator, instead of **P**. Note that $\mathbf{P}'^2 = 1$ and so $\mathbf{P}'$ has eigenvalues ± 1. However, when no such symmetry group exists, it may be that all parity transformations have some eigenvalues which are phases other than ± 1.

For electronic wavefunctions, even states are usually indicated by a subscript g for *gerade* (German: even) and odd states by a subscript u for *ungerade* (German: odd). For example, the lowest energy level of the hydrogen molecule ion (H_2^+) is labelled $1\sigma_g$ and the next-lowest $1\sigma_u$.[1]

38.4.2 Consequences of parity symmetry

When parity generates the Abelian group $\mathbb{Z}_2$, one can always take linear combinations of quantum states such that they are either even or odd under parity (see the figure). Thus the parity of such states is ± 1. The parity of a multiparticle state is the product of the parities of each state; in other words parity is a multiplicative quantum number

In quantum mechanics, Hamiltonians are invariant (symmetric) under a parity transformation if $\mathbf{P}$ commutes with the Hamiltonian. In non-relativistic quantum mechanics, this happens for any potential which is scalar, i.e., $V = V(r)$, hence the potential is spherically symmetric. The following facts can be easily proven:

- If $|A\rangle$ and $|B\rangle$ have the same parity, then $\langle A| \mathbf{X} |B\rangle = 0$ where $\mathbf{X}$ is the position operator.
- For a state $|L, L_z\rangle$ of orbital angular momentum $\mathbf{L}$ with z-axis projection L_z, $\mathbf{P}|L, L_z\rangle = (-1)^L|L, L_z\rangle$.
- If $[\mathbf{H}, \mathbf{P}] = 0$, then atomic dipole transitions only occur between states of opposite parity.[2]
- If $[\mathbf{H}, \mathbf{P}] = 0$, then a non-degenerate eigenstate of $\mathbf{H}$ is also an eigenstate of the parity operator; i.e., a non-degenerate eigenfunction of $\mathbf{H}$ is either invariant to $\mathbf{P}$ or is changed in sign by $\mathbf{P}$.

Some of the non-degenerate eigenfunctions of $\mathbf{H}$ are unaffected (invariant) by parity $\mathbf{P}$ and the others will be merely reversed in sign when the Hamiltonian operator and the parity operator commute:

$$\mathbf{P}\, \Psi = c\, \Psi,$$

where c is a constant, the eigenvalue of $\mathbf{P}$,

$$\mathbf{P}^2\Psi = c\mathbf{P}\, \Psi.$$

38.5 Quantum field theory

The intrinsic parity assignments in this section are true for relativistic quantum mechanics as well as quantum field theory.

If we can show that the vacuum state is invariant under parity ($\mathbf{P}|0\rangle = |0\rangle$), the Hamiltonian is parity invariant ($[\mathbf{H}, \mathbf{P}] = 0$) and the quantization conditions remain unchanged under parity, then it follows that every state has good parity, and this parity is conserved in any reaction.

To show that quantum electrodynamics is invariant under parity, we have to prove that the action is invariant and the quantization is also invariant. For simplicity we will assume that canonical quantization is used; the vacuum state is then invariant under parity by construction. The invariance of the action follows from the classical invariance of Maxwell's equations. The invariance of the canonical quantization procedure can be worked out, and turns out to depend on the transformation of the annihilation operator:

$$\mathbf{P}a(\mathbf{p}, \pm)\mathbf{P}^+ = -a(-\mathbf{p}, \pm)$$

where $\mathbf{p}$ denotes the momentum of a photon and $\pm$ refers to its polarization state. This is equivalent to the statement that the photon has odd intrinsic parity. Similarly all vector bosons can be shown to have odd intrinsic parity, and all axial-vectors to have even intrinsic parity.

There is a straightforward extension of these arguments to scalar field theories which shows that scalars have even parity, since

$$\mathbf{P}a(\mathbf{p})\mathbf{P}^+ = a(-\mathbf{p}).$$

This is true even for a complex scalar field. (*Details of spinors are dealt with in the article on the* Dirac equation, *where it is shown that fermions and antifermions have opposite intrinsic parity.*)

With fermions, there is a slight complication because there is more than one spin group.

38.6 Parity in the standard model

38.6.1 Fixing the global symmetries

See also: $(-1)^F$

In the Standard Model of fundamental interactions there are precisely three global internal U(1) symmetry groups available, with charges equal to the baryon number B, the lepton number L and the electric charge Q. The product of the parity operator with any combination of these rotations is another parity operator. It is conventional to choose one specific combination of these rotations to define a standard parity operator, and other parity operators are related to the standard one by internal rotations. One way to fix a standard parity operator is to assign the parities of three particles with linearly independent charges B, L and Q. In general one assigns the parity of the most common massive particles, the proton, the neutron and the electron, to be +1.

Steven Weinberg has shown that if $\mathbf{P}^2 = (-1)^F$, where F is the fermion number operator, then, since the fermion number is the sum of the lepton number plus the baryon number, $F = B + L$, for all particles in the Standard Model and since lepton number and baryon number are charges Q of continuous symmetries e^{iQ}, it is possible to redefine the parity operator so that $\mathbf{P}^2 = 1$. However, if there exist Majorana neutrinos, which experimentalists today believe is quite possible, their fermion number is equal to one because they are neutrinos while their baryon and lepton numbers are zero because they are Majorana, and so $(-1)^F$ would not be embedded in a continuous symmetry group. Thus Majorana neutrinos would have parity $\pm i$.

38.6.2 Parity of the pion

In 1954, a paper by William Chinowsky and Jack Steinberger demonstrated that the pion has negative parity.[3] They studied the decay of an "atom" made from a deuteron (2
1H+) and a negatively charged pion ($\pi-$) in a state with zero orbital angular momentum $L = 0$ into two neutrons (n).

Neutrons are fermions and so obey Fermi–Dirac statistics, which implies that the final state is antisymmetric. Using the fact that the deuteron has spin one and the pion spin zero together with the antisymmetry of the final state they concluded that the two neutrons must have orbital angular momentum $L = 1$. The total parity is the product of the intrinsic parities of the particles and the extrinsic parity of the spherical harmonic function $(-1)^L$. Since the orbital momentum changes from zero to one in this process, if the process is to conserve the total parity then the products of the intrinsic parities of the initial and final particles must have opposite sign. A deuteron nucleus is made from a proton and a neutron, and so using the aforementioned convention that protons and neutrons have intrinsic parities equal to +1 they argued that the parity of the pion is equal to minus the product of the parities of the two neutrons divided by that of the proton and neutron in the deuteron, $(-1)(1)^2/(1)^2$, which is equal to minus one. Thus they concluded that the pion is a pseudoscalar particle.

38.6.3 Parity violation

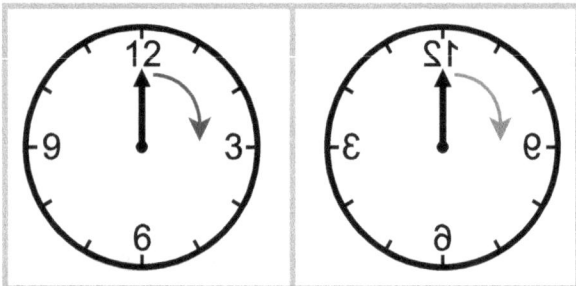

Top: P-symmetry: A clock built like its mirrored image will behave like the mirrored image of the original clock.
Bottom: P-asymmetry: A clock built like its mirrored image will *not* behave like the mirrored image of the original clock.

Although parity is conserved in electromagnetism, strong interactions and gravity, it turns out to be violated in weak interactions. The Standard Model incorporates **parity violation** by expressing the weak interaction as a chiral gauge interaction. Only the left-handed components of particles and right-handed components of antiparticles participate in weak interactions in the Standard Model. This implies that parity is not a symmetry of our universe, unless a hidden mirror sector exists in which parity is violated in the opposite way.

By the mid-20th Century, it had been suggested by several scientists that parity might not be conserved (in different contexts), but without solid evidence these suggestions were not considered important. Then, in 1956, a careful review and analysis by theoretical physicists Tsung Dao Lee and Chen Ning Yang[4] went further, showing that while parity conservation had been verified in decays by the strong or electromagnetic interactions, it was untested in the weak interaction. They proposed several possible direct experimental tests. They were mostly ignored, but Lee was able to convince his Columbia colleague Chien-Shiung Wu to try it. She needed special cryogenic facilities and expertise, so the experiment was done at the National Bureau of Standards.

In 1957 C. S. Wu, E. Ambler, R. W. Hayward, D. D. Hoppes, and R. P. Hudson found a clear violation of parity conservation in the beta decay of cobalt-60.[5] As the experiment was winding down, with double-checking in progress, Wu informed Lee and Yang of their positive results, and saying the results need further examination, she asked them not to publicize the results first. However, Lee revealed the results to his Columbia colleagues on 4 January 1957 at a "Friday Lunch" gathering of the Physics Department of Columbia. Three of them, R. L. Garwin, Leon Lederman, and R. Weinrich modified an existing cyclotron experiment, and they immediately verified the parity violation.[6] They delayed publication of their results until after Wu's group was ready, and the two papers appeared back to back in the same physics journal.

After the fact, it was noted that an obscure 1928 experiment had in effect reported parity violation in weak decays, but since the appropriate concepts had not yet been developed, those results had no impact.[7] The discovery of parity violation immediately explained the outstanding τ–θ puzzle in the physics of kaons.

In 2010, it was reported that physicists working with the Relativistic Heavy Ion Collider (RHIC) had created a short-lived parity symmetry-breaking bubble in quark-gluon plasmas. An experiment conducted by several physicists including Yale's Jack Sandweiss as part of the STAR collaboration, suggested that parity may also be violated in the strong interaction.[8]

38.6.4 Intrinsic parity of hadrons

To every particle one can assign an **intrinsic parity** as long as nature preserves parity. Although weak interactions do not, one can still assign a parity to any hadron by examining the strong interaction reaction that produces it, or through decays not involving the weak interaction, such as rho meson decay to pions.

38.7   See also

- Electroweak theory

- Standard Model

- Mirror matter

38.8   References

General

- Perkins, Donald H. (2000). *Introduction to High Energy Physics*. ISBN 9780521621960.

- Sozzi, M. S. (2008). *Discrete symmetries and CP violation*. Oxford University Press. ISBN 978-0-19-929666-8.

- Bigi, I. I.; Sanda, A. I. (2000). *CP Violation*. Cambridge Monographs on Particle Physics, Nuclear Physics and Cosmology. Cambridge University Press. ISBN 0-521-44349-0.

- Weinberg, S. (1995). *The Quantum Theory of Fields*. Cambridge University Press. ISBN 0-521-67053-5.

Specific

[1] Levine, I.N. *Quantum Chemistry* (Prentice-Hall, 4th edn. 1991), p.355

[2] Bransden, B. H.; Joachain, C. J. (2003). *Physics of Atoms and Molecules* (2nd ed.). Prentice Hall. p. 204. ISBN 978-0-582-35692-4.

[3] Chinowsky, W.; Steinberger, J. (1954). "Absorption of Negative Pions in Deuterium: Parity of the Pion". *Physical Review* **95** (6): 1561–1564. Bibcode:1954PhRv...95.1561C. doi:10.1103/PhysRev.95.1561.

[4] Lee, T. D.; Yang, C. N. (1956). "Question of Parity Conservation in Weak Interactions". *Physical Review* **104** (1): 254–258. Bibcode:1956PhRv..104..254L. doi:10.1103/PhysRev.104.254.

[5] Wu, C. S.; Ambler, E; Hayward, R. W.; Hoppes, D. D.; Hudson, R. P. (1957). "Experimental Test of Parity Conservation in Beta Decay". *Physical Review* **105** (4): 1413–1415. Bibcode:1957PhRv..105.1413W. doi:10.1103/PhysRev.105.1413.

[6] Garwin, R. L.; Lederman, L. M.; Weinrich, M. (1957). "Observations of the Failure of Conservation of Parity and Charge Conjugation in Meson Decays: The Magnetic Moment of the Free Muon". *Physical Review* **105** (4): 1415–1417. Bibcode:1957Ph .doi:10.1103/PhysRev.105.1415.

[7] Roy, A. (2005). "Discovery of parity violation". *Resonance* **10** (12): 164–175. doi:10.1007/BF02835140.

[8] Muzzin, S. T. (19 March 2010). "For One Tiny Instant, Physicists May Have Broken a Law of Nature". *PhysOrg*. Retrieved 2011-08-05.

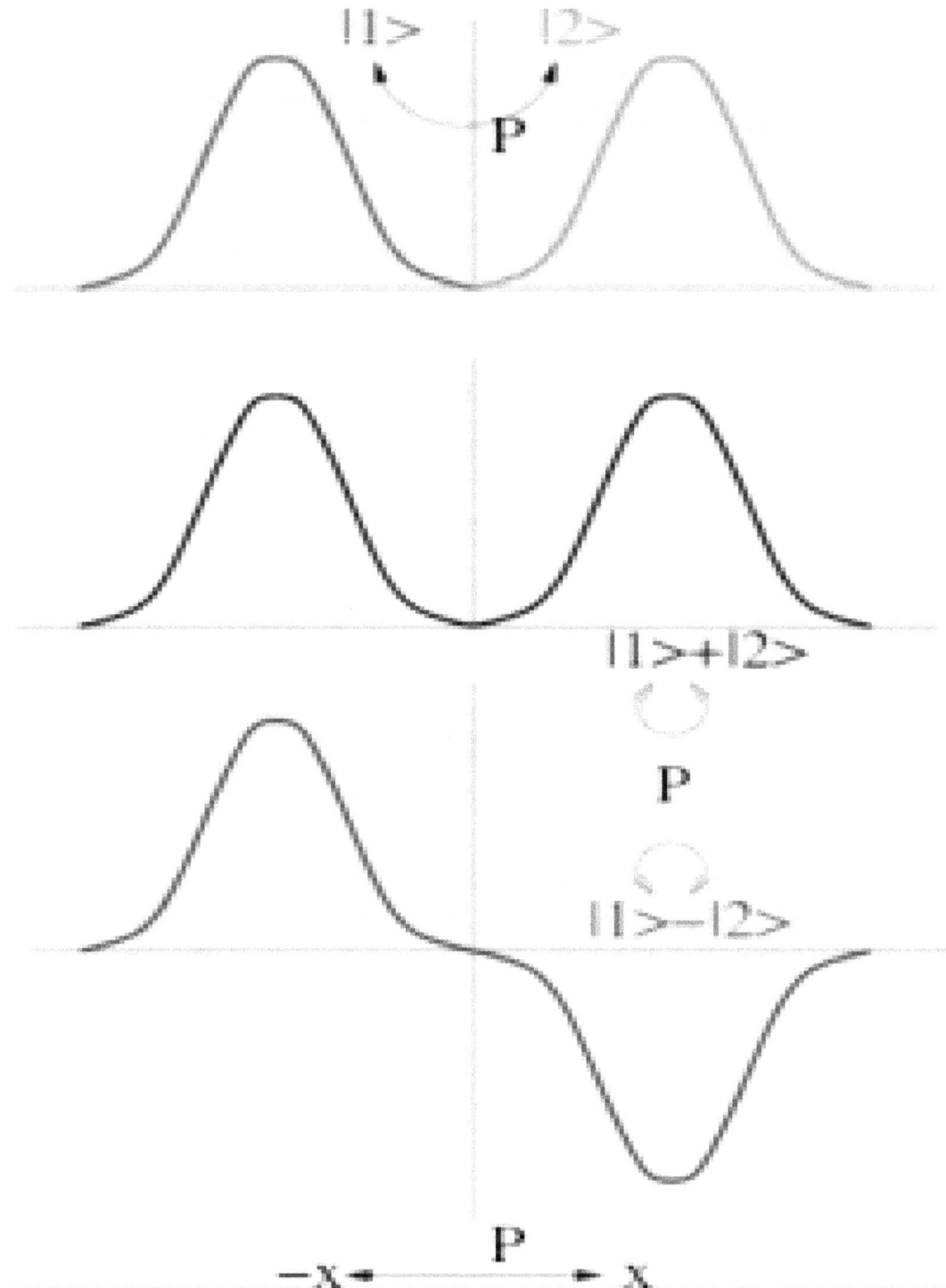

Two dimensional representations of parity are given by a pair of quantum states which go into each other under parity. However, this representation can always be reduced to linear combinations of states, each of which is either even or odd under parity. One says that all irreducible representationsof parity are one-dimensional.

Chapter 39

C parity

In physics, the **C parity** or **charge parity** is a multiplicative quantum number of some particles that describes their behavior under the symmetry operation of charge conjugation.

Charge conjugation changes the sign of all quantum charges (that is, additive quantum numbers), including the electrical charge, baryon number and lepton number, and the flavor charges strangeness, charm, bottomness, topness and Isospin (I_3). In contrast, it doesn't affect the mass, linear momentum or spin of a particle.

39.1   Formalism

Consider an operation $\mathcal{C}$ that transforms a particle into its antiparticle,

$$\mathcal{C}\,|\psi\rangle = |\bar{\psi}\rangle.$$

Both states must be normalizable, so that

$$1 = \langle\psi|\psi\rangle = \langle\bar{\psi}|\bar{\psi}\rangle = \langle\psi|\mathcal{C}^\dagger\mathcal{C}|\psi\rangle,$$

which implies that $\mathcal{C}$ is unitary,

$$\mathcal{C}\mathcal{C}^\dagger = \mathbf{1}.$$

By acting on the particle twice with the $\mathcal{C}$ operator,

$$\mathcal{C}^2|\psi\rangle = \mathcal{C}|\bar{\psi}\rangle = |\psi\rangle,$$

we see that $\mathcal{C}^2 = \mathbf{1}$ and $\mathcal{C} = \mathcal{C}^{-1}$. Putting this all together, we see that

$$\mathcal{C} = \mathcal{C}^\dagger,$$

meaning that the charge conjugation operator is Hermitian and therefore a physically observable quantity.

39.1.1 Eigenvalues

For the eigenstates of charge conjugation,

$$\mathcal{C}\left|\psi\right\rangle = \eta_C\left|\psi\right\rangle$$

As with parity transformations, applying $\mathcal{C}$ twice must leave the particle's state unchanged,

$$\mathcal{C}^2\left|\psi\right\rangle = \eta_C\mathcal{C}\left|\psi\right\rangle = \eta_C^2\left|\psi\right\rangle = \left|\psi\right\rangle$$

allowing only eigenvalues of $\eta_C = \pm 1$ the so-called *C-parity* or *charge parity* of the particle.

39.1.2 Eigenstates

The above implies that $\mathcal{C}\left|\psi\right\rangle$ and $\left|\psi\right\rangle$ have exactly the same quantum charges, so only truly neutral systems – those where all quantum charges and the magnetic moment are zero – are eigenstates of charge parity, that is, the photon and particle-antiparticle bound states like the neutral pion, η or the positronium.

39.2 Multiparticle systems

For a system of free particles, the C parity is the product of C parities for each particle.

In a pair of bound bosons there is an additional component due to the orbital angular momentum. For example, in a bound state of two pions, $\pi^+\,\pi^-$ with an orbital angular momentum **L**, exchanging π^+ and π^- inverts the relative position vector, which is identical to a parity operation. Under this operation, the angular part of the spatial wave function contributes a phase factor of $(-1)^L$, where L is the angular momentum quantum number associated with **L**.

$$\mathcal{C}\left|\pi^+\,\pi^-\right\rangle = (-1)^L\left|\pi^+\,\pi^-\right\rangle$$

With a two-fermion system, two extra factors appear: one comes from the spin part of the wave function, and the second from the exchange of a fermion by its antifermion.

$$\mathcal{C}\left|f\,\bar{f}\right\rangle = (-1)^L(-1)^{S+1}(-1)\left|f\,\bar{f}\right\rangle = (-1)^{L+S}\left|f\,\bar{f}\right\rangle$$

Bound states can be described with the spectroscopic notation $^{2S+1}L_J$ (see term symbol), where S is the total spin quantum number, L the total orbital momentum quantum number and J the total angular momentum quantum number. Example: the *positronium* is a bound state electron-positron similar to an hydrogen atom. The *parapositronium* and *ortopositronium* correspond to the states $^1S_0$ and $^3S_1$.

- With $S = 0$ spins are anti-parallel, and with $S = 1$ they are parallel. This gives a multiplicity $(2S+1)$ of 1 or 3, respectively

- The total orbital angular momentum quantum number is $L = 0$ (S, in spectroscopic notation)

- Total angular momentum quantum number is $J = 0, 1$

- C parity $\eta C = (-1)^{L+S} = +1, -1$, respectively. Since charge parity is preserved, annihilation of these states in photons ($\eta C(\gamma) = -1$) must be:

39.3 Experimental tests of C-parity conservation

- $\pi^0 \to 3\gamma$: The neutral pion, π^0 , is observed to decay to two photons, $\gamma+\gamma$. We can infer that the pion therefore has $\eta_C = (-1)^2 = 1$, but each additional γ introduces a factor of -1 to the overall C parity of the pion. The decay to 3γ would violate C parity conservation. A search for this decay was conducted[1] using pions created in the reaction $\pi^- + p \to \pi^0 + n$.

- $\eta \to \pi^+\pi^-\pi^0$ [2] Decay of the Eta meson.

- $p\bar{p}$ annihilations[3]

39.4 References

[1] MacDonough, J.; et al. (1988). *Phys. Review* **D38**: 2121. Missing or empty |title= (help)

[2] Gormley, M.; et al. (1968). *Phys. Rev. Lett.* **21**: 402. Bibcode:1968PhRvL..21..402G. doi:10.1103/PhysRevLett.21.402. Missing or empty |title= (help)

[3] Baltay, C; et al. (1965). *Phys. Rev. Lett.* **14**: 591. Bibcode:1965PhRvL..14..591R. doi:10.1103/PhysRevLett.14.591. Missing or empty |title= (help)

Chapter 40

Flavour (particle physics)

In particle physics, **flavour** or **flavor** refers to a species of an elementary particle. The Standard Model counts six flavours of quarks and six flavours of leptons. They are conventionally parameterized with *flavour quantum numbers* that are assigned to all subatomic particles, including composite ones. For hadrons, these quantum numbers depend on the numbers of constituent quarks of each particular flavour.

40.1 Intuitive description

Elementary particles are not eternal and indestructible. Unlike in classical mechanics, where forces only change a particle's momentum, the weak force can alter the essence of a particle, even an elementary particle. This means that it can convert one quark to another quark with different mass and electric charge, and the same for leptons. From the point of view of quantum mechanics, changing the flavour of a particle by the weak force is no different in principle from changing its spin by electromagnetic interaction, and should be described with quantum numbers as well. In particular, flavour states may undergo quantum superposition.

In atomic physics the principal quantum number of an electron specifies the electron shell in which it resides, which determines the energy level of the whole atom. In an analogous way, the five flavour quantum numbers of a quark specify which of six flavours (u, d, s, c, b, t) it has, and when these quarks are combined this results in different types of baryons and mesons with different masses, electric charges, and decay modes.

40.2 Flavour symmetry

If there are two or more particles which have identical interactions, then they may be interchanged without affecting the physics. Any (complex) linear combination of these two particles give the same physics, as long as they are orthogonal or perpendicular to each other. In other words, the theory possesses symmetry transformations such as $M \begin{pmatrix} u \\ d \end{pmatrix}$, where u and d are the two fields, and M is any 2×2 unitary matrix with a unit determinant. Such matrices form a Lie group called SU(2) (see special unitary group). This is an example of flavour symmetry.

In quantum chromodynamics, flavour is a global symmetry. In the electroweak theory, on the other hand, this symmetry is broken, and flavour changing processes exist, such as quark decay or neutrino oscillations.

40.3 Flavour quantum numbers

40.3.1 Leptons

All leptons carry a lepton number $L = 1$. In addition, leptons carry weak isospin, T_3, which is $-1/2$ for the three charged leptons (i.e. electron, muon and tau) and $+1/2$ for the three associated neutrinos. Each doublet of a charged lepton and a neutrino consisting of opposite T_3 are said to constitute one generation of leptons. In addition, one defines a quantum number called weak hypercharge, Y_W, which is -1 for all left-handed leptons.[1] Weak isospin and weak hypercharge are gauged in the Standard Model.

Leptons may be assigned the six flavour quantum numbers: electron number, muon number, tau number, and corresponding numbers for the neutrinos. These are conserved in strong and electromagnetic interactions, but violated by weak interactions. Therefore, such flavour quantum numbers are not of great use. A separate quantum number for each generation is more useful: electronic lepton number (+1 for electrons and electron neutrinos), muonic lepton number (+1 for muons and muon neutrinos), and tauonic lepton number (+1 for tau leptons and tau neutrinos). However, even these numbers are not absolutely conserved, as neutrinos of different generations can mix; that is, a neutrino of one flavour can transform into another flavour. The strength of such mixings is specified by a matrix called the Pontecorvo–Maki–Nakagawa–Sakata matrix (PMNS matrix).

40.3.2 Quarks

All quarks carry a baryon number $B = 1/3$. They also all carry weak isospin, $T_3 = \pm 1/2$. The positive-T_3 quarks (up, charm, and top quarks) are called *up-type quarks* and negative-T_3 quarks (down, strange, and bottom quarks) are called *down-type quarks*. Each doublet of up and down type quarks constitutes one generation of quarks.

For all the quark flavour quantum numbers (strangeness, charm, topness and bottomness) the convention is that the flavour charge and the electric charge of a quark have the same sign. Thus any flavour carried by a charged meson has the same sign as its charge. Quarks have the following flavour quantum numbers:

- Isospin, less ambiguously known as "isobaric spin", which has value $I_3 = 1/2$ for the up quark and $I_3 = -1/2$ for the down quark.

- Strangeness (S): Defined as $S = -(n_s - n_{\bar{s}})$, where n_s represents the number of strange quarks (s) and $n_{\bar{s}}$ represents the number of strange antiquarks (s). This quantum number was introduced by Murray Gell-Mann. This definition gives the strange quark a strangeness of -1 for the above-mentioned reason.

- Charm (C): Defined as $C = (n_c - n_{\bar{c}})$, where n_c represents the number of charm quarks (c) and $n_{\bar{c}}$ represents the number of charm antiquarks. Is $+1$ for the charm quark.

- Bottomness (B′): Also called 'beauty'. Defined as $B' = -(n_b - n_{\bar{b}})$, where n_b represents the number of bottom quarks (b) and $n_{\bar{b}}$ represents the number of bottom antiquarks.

- Topness (T): Also called 'truth'. Defined as $T = (n_t - n_{\bar{t}})$, where n_t represents the number of top quarks (t) and $n_{\bar{t}}$ represents the number of top antiquarks. However, because of the extremely short half-life of the top quark, by the time it can interact strongly it has already decayed to another flavour of quark (usually to a bottom quark). For that reason the top quark doesn't hadronize, that is it never forms any meson or baryon.

These five quantum numbers, together with baryon number (which is not a flavour quantum number) completely specify numbers of all 6 quark flavours separately (as $n_q - n_{\bar{q}}$, i.e. an antiquark is counted with the minus sign). They are conserved by both the electromagnetic and strong interactions (but not the weak interaction). From them can be built the derived quantum numbers:

- Hypercharge (Y): $Y = B + S + C + B' + T$

- Electric charge: $Q = I_3 + 1/2\,Y$ (see Gell-Mann–Nishijima formula)

The terms "strange" and "strangeness" predate the discovery of the quark, but continued to be used after its discovery for the sake of continuity (i.e. the strangeness of each type of hadron remained the same); strangeness of anti-particles being referred to as +1, and particles as −1 as per the original definition. Strangeness was introduced to explain the rate of decay of newly discovered particles, such as the kaon, and was used in the Eightfold Way classification of hadrons and in subsequent quark models. These quantum numbers are preserved under strong and electromagnetic interactions, but not under weak interactions.

For first-order weak decays, that is processes involving only one quark decay, these quantum numbers (e.g. charm) can only vary by 1 ($|C| = \pm 1$); $\Delta B' = \pm 1$. Since first-order processes are more common than second-order processes (involving two quark decays), this can be used as an approximate "selection rule" for weak decays.

A quark of a given flavour is an eigenstate of the weak interaction part of the Hamiltonian: it will interact in a definite way with the W and Z bosons. On the other hand, a fermion of a fixed mass (an eigenstate of the kinetic and strong interaction parts of the Hamiltonian) is normally a superposition of various flavours. As a result, the flavour content of a quantum state may change as it propagates freely. The transformation from flavour to mass basis for quarks is given by the Cabibbo–Kobayashi–Maskawa matrix (CKM matrix). This matrix is analogous to the PMNS matrix for neutrinos, and defines the strength of flavour changes under weak interactions of quarks.

The CKM matrix allows for CP violation if there are at least three generations.

40.3.3 Antiparticles and hadrons

Flavour quantum numbers are additive. Hence antiparticles have flavour equal in magnitude to the particle but opposite in sign. Hadrons inherit their flavour quantum number from their valence quarks: this is the basis of the classification in the quark model. The relations between the hypercharge, electric charge and other flavour quantum numbers hold for hadrons as well as quarks.

40.4 Quantum chromodynamics

Flavour symmetry is closely related to chiral symmetry. This part of the article is best read along with the one on chirality.

Quantum chromodynamics (QCD) contains six flavours of quarks. However, their masses differ and as a result they are not strictly interchangeable with each other. The up and down flavours are close to having equal masses, and the theory of these two quarks possesses an approximate SU(2) symmetry (isospin symmetry).

Under some circumstances, the masses of the quarks can be neglected entirely. One can then make flavour transformations independently on the left- and right-handed parts of each quark field. The flavour group is then a chiral group $SU_L(N_f)$ × $SU_R(N_f)$.

If all quarks had non-zero but equal masses, then this chiral symmetry is broken to the *vector symmetry* of the "diagonal flavour group" $SU(N_f)$, which applies the same transformation to both helicities of the quarks. Such a reduction of the symmetry is called *explicit symmetry breaking*. The amount of explicit symmetry breaking is controlled by the current quark masses in QCD.

Even if quarks are massless, chiral flavour symmetry can be spontaneously broken if the vacuum of the theory contains a chiral condensate (as it does in low-energy QCD). This gives rise to an effective mass for the quarks, often identified with the valence quark mass in QCD.

40.4.1 Symmetries of QCD

Analysis of experiments indicate that the current quark masses of the lighter flavours of quarks are much smaller than the QCD scale, ΛQCD, hence chiral flavour symmetry is a good approximation to QCD for the up, down and strange quarks. The success of chiral perturbation theory and the even more naive chiral models spring from this fact. The valence quark masses extracted from the quark model are much larger than the current quark mass. This indicates that QCD has spontaneous chiral symmetry breaking with the formation of a chiral condensate. Other phases of QCD may break the chiral flavour symmetries in other ways.

40.5 Conservation laws

All of the various charges discussed above are conserved by the fact that the charge operator is best understood as the generator of a symmetry that commutes with the Hamiltonian. Thus, the eigenvalues of the various charge operators are conserved.

Absolutely conserved flavour quantum numbers are: (including the baryon number for completeness)

- electric charge (Q)

- weak isospin (I_3)

- baryon number (B)

- lepton number (L)

In some theories, the individual baryon and lepton number conservation can be violated, if the difference between them ($B - L$) is conserved (see chiral anomaly). All other flavour quantum numbers are violated by the electroweak interactions. Strong interactions conserve all flavours.

40.6 History

Some of the historical events that lead to the development of flavour symmetry are discussed in the article on isospin.

40.7 See also

- Standard Model (mathematical formulation)

- Cabibbo–Kobayashi–Maskawa matrix

- Strong CP problem and chirality (physics)

- Chiral symmetry breaking and quark matter

- Quark flavour tagging, such as B-tagging, is an example of particle identification in experimental particle physics.

40.8 References

[1] See table in S. Raby, R. Slanky (1997). "Neutrino Masses: How to add them to the Standard Model" (PDF). *Los Alamos Science* (25): 64.

40.9 Further reading

- Lessons in Particle Physics Luis Anchordoqui and Francis Halzen, University of Wisconsin, 18th Dec. 2009

40.10 External links

- The particle data group.

Chapter 41

Isospin

In nuclear physics and particle physics, **isospin** (*isotopic spin*, *isobaric spin*) is a quantum number related to the strong interaction. Particles that are affected equally by the strong force but have different charges (e.g. protons and neutrons) can be treated as being different states of the same particle with isospin values related to the number of charge states.[1]

Although it does not have the units of angular momentum and is not a type of spin, the formalism that describes it is mathematically similar to that of angular momentum in quantum mechanics, which means it can be coupled in the same manner. For example, a proton-neutron pair can be coupled in a state of total isospin 1 or 0.[2] It is a dimensionless quantity and the name derives from the fact that the mathematical structures used to describe it are very similar to those used to describe the intrinsic angular momentum (spin).

This term was derived from *isotopic spin*, a confusing term to which nuclear physicists prefer *isobaric spin*, which is more precise in meaning. Isospin symmetry is a subset of the flavour symmetry seen more broadly in the interactions of baryons and mesons. Isospin symmetry remains an important concept in particle physics, and a close examination of this symmetry historically led directly to the discovery and understanding of quarks and of the development of Yang–Mills theory.

41.1   Motivation for isospin

Isospin was introduced by Werner Heisenberg in 1932[3] to explain symmetries of the then newly discovered neutron:

- The mass of the neutron and the proton are almost identical: they are nearly degenerate, and both are thus often called nucleons. Although the proton has a positive charge, and the neutron is neutral, they are almost identical in all other respects.

- The strength of the strong interaction between any pair of nucleons is the same, independent of whether they are interacting as protons or as neutrons.

Thus, isospin was introduced as a concept well before the development in the 1960s of the quark model which provides our modern understanding. The name *isospin* however, was introduced by Eugene Wigner in 1937.[4]

Protons and neutrons, baryons of spin $\frac{1}{2}$, were grouped together as nucleons because they both have nearly the same mass and interact in nearly the same way. Thus, it was convenient to treat them as being different states of the same particle. Since a spin $\frac{1}{2}$ particle has two states, the two were said to be of isospin $\frac{1}{2}$. The proton and neutron were then associated with different isospin projections $I_3 = +\frac{1}{2}$ and $-\frac{1}{2}$ respectively. When constructing a physical theory of nuclear forces, one could then simply assume that it does not depend on isospin.

These considerations would also prove useful in the analysis of meson-nucleon interactions after the discovery of the pions in 1947. The three pions ($\pi+$, $\pi0$, $\pi-$) could be assigned to an isospin triplet with $I = 1$ and $I_3 = +1$, 0 or -1. By assuming that isospin was conserved by nuclear interactions, the new mesons were more easily accommodated by nuclear theory.

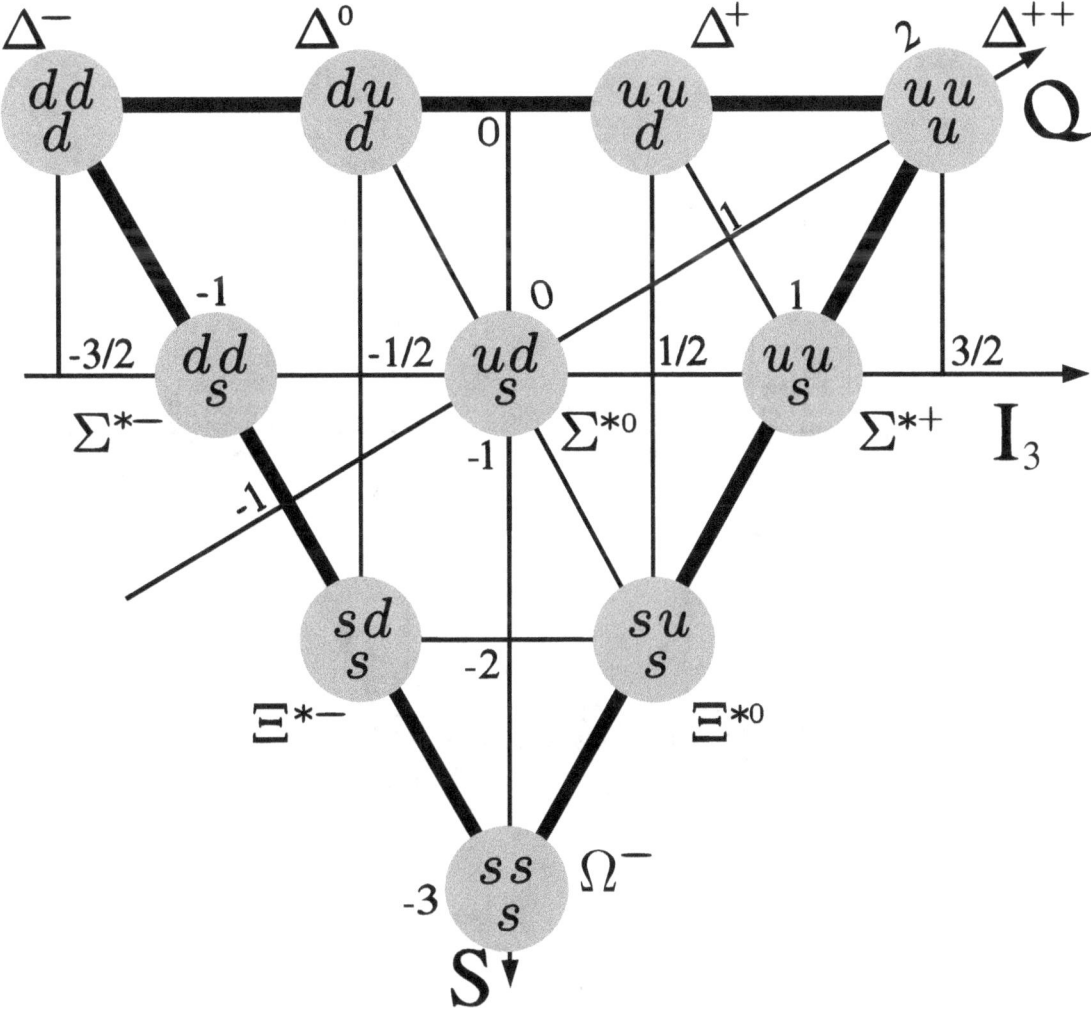

Combinations of three u, d or s-quarks forming baryons with spin-$^3/_2$ form the baryon decuplet.

As further particles were discovered, they were assigned into isospin multiplets according to the number of different charge states seen: 2 doublets, $I = -^1/_2$ and $I = ^1/_2$ of K mesons (K−, K0),(K+, K0), a triplet $I = 1$ of Sigma baryons (Σ+, Σ0, Σ−) a singlet $I = 0$ Lambda baryon (Λ0), a quartet $I = ^3/_2$ Delta baryons (Δ++, Δ+, Δ0, Δ−), and so on. This multiplet structure was combined with strangeness in Murray Gell-Mann's eightfold way, ultimately leading to the quark model and quantum chromodynamics.

41.2   Modern understanding of isospin

Observation of the light baryons (those made of up, down and strange quarks) lead us to believe that some of these particles are so similar in terms of their strong interactions that they can be treated as different states of the same particle. In the modern understanding of quantum chromodynamics, this is because up and down quarks are very similar in mass, and have the same strong interactions. Particles made of the same numbers of up and down quarks have similar masses and are grouped together. For examples, the particles known as the Delta baryons—baryons of spin $^3/_2$ made of a mix of three up and down quarks—are grouped together because they all have nearly the same mass (approximately 1232 MeV/c^2), and interact in nearly the same way.

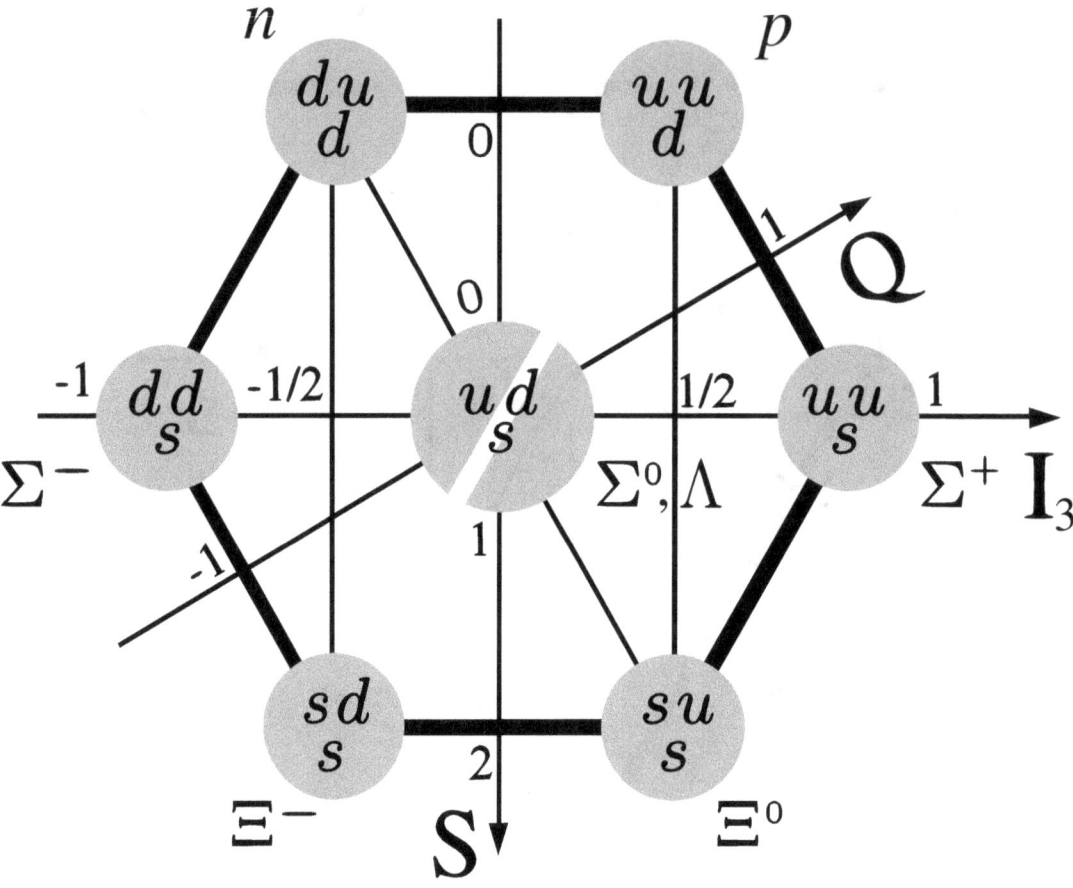

Combinations of three u, d or s-quarks forming baryons with spin-$\frac{1}{2}$ form the baryon octet

However, because the up and down quarks have different charges ($\frac{2}{3}$ e and $-\frac{1}{3}$ e respectively), the four Deltas also have different charges (Δ++ (uuu), Δ+ (uud), Δ0 (udd), Δ− (ddd)). These Deltas could be treated as the same particle and the difference in charge being due to the particle being in different states. Isospin was devised as a parallel to spin to associate an isospin projection (denoted I_3) to each charged state. Since there were four Deltas, four projections were needed. Because isospin was modeled on spin, the isospin projections were made to vary in increments of 1 and to have four increments of 1, you needed an isospin value of $\frac{3}{2}$ (giving the projections $I_3 = \frac{3}{2}, \frac{1}{2}, -\frac{1}{2}, -\frac{3}{2}$). Thus, all the Deltas were said to have isospin $I = \frac{3}{2}$ and each individual charge had different I_3 (e.g. the Δ++ was associated with $I_3 = +\frac{3}{2}$). In the isospin picture, the four Deltas and the two nucleons were thought to be the different states of two particles. In the quark model, the Deltas can be thought of as the excited states of the nucleons.

After the quark model was elaborated, it was noted that the isospin projection was related to the up and down quark content of particles. The relation is

$$I_3 = \frac{1}{2}\left[(n_u - n_{\bar{u}}) - (n_d - n_{\bar{d}})\right]$$

where n_u and n_d are the numbers of up and down quarks respectively, and n_u and n_d are the numbers of up and down antiquarks respectively.

By this, the value of I_3 of the nucleons proton (symbol p) and neutron (symbol n) is determined by their quark composition, *uud* for the proton and *udd* for the neutron.

41.3   Isospin symmetry

In quantum mechanics, when a Hamiltonian has a symmetry, that symmetry manifests itself through a set of states that have the same energy; that is, the states are degenerate. In particle physics, the near mass-degeneracy of the neutron and proton points to an approximate symmetry of the Hamiltonian describing the strong interactions. The neutron does have a slightly higher mass due to isospin breaking; this is due to the difference in the masses of the up and down quarks and the effects of the electromagnetic interaction. However, the appearance of an approximate symmetry is still useful, since the small breakings can be described by a perturbation theory, which gives rise to slight differences between the near-degenerate states.

41.3.1   SU(2)

See also: Representation theory of SU(2)

Heisenberg's contribution was to note that the mathematical formulation of this symmetry was in certain respects similar to the mathematical formulation of spin, whence the name "isospin" derives. To be precise, the isospin symmetry is given by the invariance of the Hamiltonian of the strong interactions under the action of the Lie group SU(2). The neutron and the proton are assigned to the doublet (the spin-$\frac{1}{2}$, **2**, or fundamental representation) of SU(2). The pions are assigned to the triplet (the spin-1, **3**, or adjoint representation) of SU(2). Though, there is a difference from the theory of spin: the group action does not preserve flavor.

Like the case for regular spin, the isospin operator **I** is vector-valued: it has three components I_x, I_y, I_z which are coordinates in the same 3-dimensional vector space where the **3** representation acts. Note that it has nothing to do with the physical space, except similar mathematical formalism. Isospin is described by two quantum numbers: I, the total isospin, and I_3, an eigenvalue of the I_z projection for which flavor states are eigenstates, not an *arbitrary projection* as in the case of spin. In other words, each I_3 state specifies certain flavor state of a multiplet. The third coordinate (z), to which the "3" subscript refers, is chosen due to notational conventions which relate bases in **2** and **3** representation spaces. Namely, for the spin-$\frac{1}{2}$ case, components of **I** are equal to Pauli matrices divided by 2 and $I_z = \frac{1}{2}\,\tau_3$, where

$$\tau_3 = \begin{pmatrix} 1 & 0 \\ 0 & -1 \end{pmatrix}$$

While the forms of these matrices are the isomorphic to those of spin, *these* Pauli matrices only acts within the Hilbert space of isospin, not that of spin, and therefore is common to denote them with **τ** rather than **σ** to avoid confusion.

The power of isospin symmetry and related methods such as the Eightfold Way come from the observation that families of particles with similar masses tend to correspond to the invariant subspaces associated with the irreducible representations of the Lie algebra $\mathfrak{su}(2)$. In this context, an invariant subspace is spanned by basis vectors which correspond to particles in a family. Under the action of the Lie algebra $\mathfrak{su}(2)$, which generates rotations in isospin space, elements corresponding to definite particle states or superpositions of states can be rotated into each other, but can never leave the space (since the subspace is in fact invariant). This is reflective of the symmetry present. The fact that unitary matrices will commute with the Hamiltonian means that the physical quantities calculated do not change even under unitary transformation. In the case of isospin, this machinery is used to reflect the fact that the strong force behaves the same under the exchange of the up and down quark (and by extension the exchange of the proton and the neutron).

41.4   Relationship to flavor

The discovery and subsequent analysis of additional particles, both mesons and baryons, made it clear that the concept of isospin symmetry could be broadened to an even larger symmetry group, now called flavor symmetry. Once the kaons and their property of strangeness became better understood, it started to become clear that these, too, seemed to be a part of an enlarged symmetry that contained isospin as a subgroup. The larger symmetry was named the Eightfold Way by Murray Gell-Mann, and was promptly recognized to correspond to the adjoint representation of SU(3). To better understand the origin of this symmetry, Gell-Mann proposed the existence of up, down and strange quarks which would belong to the fundamental representation of the SU(3) flavor symmetry.

Although isospin symmetry is very slightly broken, SU(3) symmetry is more badly broken, due to the much higher mass of the strange quark compared to the up and down. The discovery of charm, bottomness and topness could lead to further expansions up to SU(6) flavour symmetry, but the very large masses of these quarks makes such symmetries almost useless. In modern applications, such as lattice QCD, isospin symmetry is often treated as exact while the heavier quarks must be treated separately.

41.5 Quark content and isospin

Up and down quarks each have isospin $I = \frac{1}{2}$, and isospin 3-components (I_3) of $\frac{1}{2}$ and $-\frac{1}{2}$ respectively. All other quarks have $I = 0$. In general

$$I_3 = \frac{1}{2}(n_u - n_d).$$

41.5.1 Hadron nomenclature

Main articles: Baryon and Mesons

Hadron nomenclature is based on isospin.[5]

- Particles of isospin $\frac{3}{2}$ can only be made by a mix of three u and d quarks (Delta baryons).

- Particles of isospin 1 are made of a mix of two u and d quarks (Pi mesons, Rho mesons, Sigma baryons with one heavier quark, etc.).

- Particles of isospin $\frac{1}{2}$ can be made of a mix of three u and d quarks (nucleons) or from one u or d quark with heavier quarks (K mesons, D mesons, Xi baryons, etc.)

- Particles of isospin 0 can be made of one u and one d quark (Eta mesons, Omega mesons, Lambda baryons, etc.), or from no u or d quarks at all (Omega baryons, Phi mesons, etc.), with heavier quarks in all cases.

41.5.2 Isospin symmetry of quarks

In the framework of the Standard Model, the isospin symmetry of the proton and neutron are reinterpreted as the isospin symmetry of the up and down quarks. Technically, the nucleon doublet states are seen to be linear combinations of products of 3-particle isospin doublet states and spin doublet states. That is, the (spin-up) proton wave function, in terms of quark-flavour eigenstates, is described by

$$|p\uparrow\rangle = \tfrac{1}{3\sqrt{2}} \begin{pmatrix} |duu\rangle & |udu\rangle & |uud\rangle \end{pmatrix} \begin{pmatrix} 2 & -1 & -1 \\ -1 & 2 & -1 \\ -1 & -1 & 2 \end{pmatrix} \begin{pmatrix} |\downarrow\uparrow\uparrow\rangle \\ |\uparrow\downarrow\uparrow\rangle \\ |\uparrow\uparrow\downarrow\rangle \end{pmatrix} \;[6]$$

and the (spin-up) neutron by

$$|n\uparrow\rangle = \tfrac{1}{3\sqrt{2}} \begin{pmatrix} |udd\rangle & |dud\rangle & |ddu\rangle \end{pmatrix} \begin{pmatrix} 2 & -1 & -1 \\ -1 & 2 & -1 \\ -1 & -1 & 2 \end{pmatrix} \begin{pmatrix} |\downarrow\uparrow\uparrow\rangle \\ |\uparrow\downarrow\uparrow\rangle \\ |\uparrow\uparrow\downarrow\rangle \end{pmatrix} \;[6]$$

Here, $|u\rangle$ is the up quark flavour eigenstate, and $|d\rangle$ is the down quark flavour eigenstate, while $|\uparrow\rangle$ and $|\downarrow\rangle$ are the eigenstates of S_z. Although these superpositions are the technically correct way of denoting a proton and neutron in terms of quark flavour and spin eigenstates, for brevity, they are often simply referred to as "*uud*" and "*udd*". Note also that the derivation above assumes exact isospin symmetry and is modified by SU(2)-breaking terms.

Similarly, the isospin symmetry of the pions are given by:

$$|\pi^+\rangle = |u\overline{d}\rangle$$
$$|\pi^0\rangle = \frac{1}{\sqrt{2}}\left(|u\overline{u}\rangle - |d\overline{d}\rangle\right)$$
$$|\pi^-\rangle = -|d\overline{u}\rangle$$

41.5.3   Weak isospin

Main article: weak isospin

Isospin is similar to, but should not be confused with weak isospin. Briefly, weak isospin is the gauge symmetry of the weak interaction which connects quark and lepton doublets of left-handed particles in all generations; for example, up and down quarks, top and bottom quarks, electrons and electron neutrinos. By contrast (strong) isospin connects only up and down quarks, acts on both chiralities (left and right) and is a global (not a gauge) symmetry.

41.6   Gauged isospin symmetry

Attempts have been made to promote isospin from a global to a local symmetry. In 1954, Chen Ning Yang and Robert Mills suggested that the notion of protons and neutrons, which are continuously rotated into each other by isospin, should be allowed to vary from point to point. To describe this, the proton and neutron direction in isospin space must be defined at every point, giving local basis for isospin. A gauge connection would then describe how to transform isospin along a path between two points.

This Yang–Mills theory describes interacting vector bosons, like the photon of electromagnetism. Unlike the photon, the SU(2) gauge theory would contain self-interacting gauge bosons. The condition of gauge invariance suggests that they have zero mass, just as in electromagnetism.

Ignoring the massless problem, as Yang and Mills did, the theory makes a firm prediction: the vector particle should couple to all particles of a given isospin *universally*. The coupling to the nucleon would be the same as the coupling to the kaons. The coupling to the pions would be the same as the self-coupling of the vector bosons to themselves.

When Yang and Mills proposed the theory, there was no candidate vector boson. J. J. Sakurai in 1960 predicted that there should be a massive vector boson which is coupled to isospin, and predicted that it would show universal couplings. The rho mesons were discovered a short time later, and were quickly identified as Sakurai's vector bosons. The couplings of the rho to the nucleons and to each other were verified to be universal, as best as experiment could measure. The fact that the diagonal isospin current contains part of the electromagnetic current led to the prediction of rho-photon mixing and the concept of vector meson dominance, ideas which led to successful theoretical pictures of GeV-scale photon-nucleus scattering.

Although the discovery of the quarks led to reinterpretation of the rho meson as a vector bound state of a quark and an antiquark, it is sometimes still useful to think of it as the gauge boson of a hidden local symmetry[7]

41.7   References

[1] http://www.thefreedictionary.com/isospin

[2] Povh, Bogdan; Klaus, Rith; Scholz, Christoph; Zetsche, Frank (2008) [1993]. "2". *Particles and Nuclei*. p. 21. ISBN 978-3-540-79367-0.

[3] Heisenberg,W.(1932). "Über den Bau der Atomkerne".*Zeitschrift für Physik*(in German)**77**: 1–11. Bibcode:1932ZPhy...77H. doi:10.1007/BF01342433.

[4] Wigner, E. (1937). "On the Consequences of the Symmetry of the Nuclear Hamiltonian on the Spectroscopy of Nuclei". *Physical Review* **51** (2): 106–119. Bibcode:1937PhRv...51..106W. doi:10.1103/PhysRev.51.106.

[5] C. Amsler; et al.; (Particle Data Group) (2008). "Review of Particle Physics: Naming scheme for hadrons" (PDF). *Physics Letters B* **667**: 1. Bibcode:2008PhLB..667....1P. doi:10.1016/j.physletb.2008.07.018.

[6] Greiner, W.; Müller, B. (1989). *Quantum Mechanics: Symmetries*. Springer-Verlag. p. 279. ISBN 3-540-58080-8.

[7] Bando, M.; Kugo, T.; Uehara, S.; Yamawaki, K.; Yanagida, T. (1985). "Is the ρ Meson a Dynamical Gauge Boson of Hidden Local Symmetry?". *Physical Review Letters* **54** (12): 1215–1218. Bibcode:1985PhRvL..54.1215B. doi:10.1103/PhysRevLett.54.. PMID10030967.

41.8 Further reading

- Itzykson, C.; Zuber, J.-B. (1980). *Quantum Field Theory*. McGraw-Hill. ISBN 0-07-032071-3.

- Griffiths, D. (1987). *Introduction to Elementary Particles*. John Wiley & Sons. ISBN 0-471-60386-4.

41.9 External links

- **Nuclear Structure and Decay Data - IAEA** Nuclides' Isospin

Chapter 42

G-parity

In theoretical physics, **G-parity** is a multiplicative quantum number that results from the generalization of C-parity to multiplets of particles.

C-parity applies only to neutral systems; in the pion triplet, only π^0 has C-parity. On the other hand, strong interaction does not see electrical charge, so it cannot distinguish amongst π^+, π^0 and π^-. We can generalize the C-parity so it applies to all charge states of a given multiplet:

$$\mathcal{G} \begin{pmatrix} \pi^+ \\ \pi^0 \\ \pi^- \end{pmatrix} = \eta_G \begin{pmatrix} \pi^+ \\ \pi^0 \\ \pi^- \end{pmatrix}$$

where $\eta G = \pm 1$ are the eigenvalues of G-parity. The G-parity operator is defined as

$$\mathcal{G} = \mathcal{C} \, e^{(i\pi I_2)}$$

where $\mathcal{C}$ is the C-parity operator, and I_2 is the operator associated with the 2nd component of the isospin "vector". G-parity is a combination of charge conjugation and a π rad (180°) rotation around the 2nd axis of isospin space. Given that charge conjugation and isospin are preserved by strong interactions, so is G. Weak and electromagnetic interactions, though, are not invariant under G-parity.

Since G-parity is applied on a whole multiplet, charge conjugation has to see the multiplet as a neutral entity. Thus, only multiplets with an average charge of 0 will be eigenstates of G, that is

$$\bar{Q} = \bar{B} = \bar{Y} = 0$$

(see Q, B, Y).

In general

$$\eta_G = \eta_C \, (-1)^I$$

where ηC is a C-parity eigenvalue, and I is the isospin. For fermion-antifermion systems, we have

$$\eta_G = (-1)^{S+L+I}$$

where S is the total spin, L the total orbital angular momentum quantum number. For boson–antiboson systems we have

$$\eta_G = (-1)^{L+I}$$

42.1 See also

- Quark model

42.2 References

- T. D. Lee and C. N. Yang (1956). "Charge conjugation, a new quantum number G, and selection rules concerning a nucleon-antinucleon system". *Il Nuovo Cimento* **3** (4): 749–753. doi:10.1007/BF02744530.

- Charles Goebel(1956). "Selection Rules for NN̄ Annihilation".*Phys.Rev.***103**(1): 258–261. Bibcode:1956PhRv. doi:10.1103/PhysRev.103.258.

Chapter 43

Strangeness

This article is about a concept in particle physics. For the definition of "strangeness", see wikt:strangeness. For other uses, see Strange (disambiguation).

In particle physics, **strangeness** ("S") is a property of particles, expressed as a quantum number, for describing decay of particles in strong and electromagnetic reactions, which occur in a short period of time. The strangeness of a particle is defined as:

$$S = -(n_s - n_{\bar{s}})$$

where n_s represents the number of strange quarks (s) and n_s represents the number of strange antiquarks (s).

The terms *strange* and *strangeness* predate the discovery of the quark, and were adopted after its discovery in order to preserve the continuity of the phrase; strangeness of anti-particles being referred to as +1, and particles as −1 as per the original definition. For all the quark flavor quantum numbers (strangeness, charm, topness and bottomness) the convention is that the flavor charge and the electric charge of a quark have the same sign. With this, any flavor carried by a charged meson has the same sign as its charge.

43.1 Conservation

Strangeness was introduced by Murray Gell-Mann and Kazuhiko Nishijima to explain the fact that certain particles, such as the kaons or certain hyperons, were created easily in particle collisions, yet decayed much more slowly than expected for their large masses and large production cross sections. Noting that collisions seemed to always produce pairs of these particles, it was postulated that a new conserved quantity, dubbed "strangeness", was preserved during their creation, but *not* conserved in their decay.

In our modern understanding, strangeness is conserved during the strong and the electromagnetic interactions, but not during the weak interactions. Consequently, the lightest particles containing a strange quark cannot decay by the strong interaction, and must instead decay via the much slower weak interaction. In most cases these decays change the value of the strangeness by one unit. However, this doesn't necessarily hold in second-order weak reactions, where there are mixes of K0 and K0 mesons. All in all, the amount of strangeness can change in a weak interaction reaction by +1, 0 or −1 (depending on the reaction).

43.2 See also

- Strangeness production

43.3 References

- D.J. Griffiths (1987). *Introduction to Elementary Particles*. John Wiley & Sons. ISBN 0-471-60386-4.

43.4 Further reading

- Lessons in Particle Physics Luis Anchordoqui and Francis Halzen, University of Wisconsin, 18th Dec. 2009

Chapter 44

Baryon number

In particle physics, the **baryon number** is a strictly conserved additive quantum number of a system. It is defined as

$$B = \frac{1}{3}\left(n_{\mathrm{q}} - n_{\bar{\mathrm{q}}}\right),$$

where n_{q} is the number of quarks, and n_{q} is the number of antiquarks. Baryons (three quarks) have a baryon number of +1, mesons (one quark, one antiquark) have a baryon number of 0, and antibaryons (three antiquarks) have a baryon number of −1. Exotic hadrons like pentaquarks (four quarks, one antiquark) and tetraquarks (two quarks, two antiquarks) are also classified as baryons and mesons depending on their baryon number.

44.1 Baryon number vs. quark number

See also: Color charge

Quarks carry not only electric charge, but also charges such as color charge and weak isospin. Because of a phenomenon known as *color confinement*, a hadron cannot have a net color charge; that is, the total color charge of a particle has to be zero ("white"). A quark can have one of three "colors", dubbed "red", "green", and "blue".

For normal hadrons, a white color can thus be achieved in one of three ways:

- A quark of one color with an antiquark of the corresponding anticolor, giving a meson with baryon number 0,

- Three quarks of different colors, giving a baryon with baryon number +1,

- Three antiquarks into an antibaryon with baryon number −1.

The baryon number was defined long before the quark model was established, so rather than changing the definitions, particle physicists simply gave quarks one third the baryon number. Nowadays it might be more accurate to speak of the conservation of **quark number**.

In theory, exotic hadrons can be formed by adding pairs of quark and antiquark, provided that each pair has a matching color/anticolor. For example, a pentaquark (four quarks, one antiquark) could have the individual quark colors: red, green, blue, blue, and antiblue.

44.2 Particles not formed of quarks

Particles without any quarks have a baryon number of zero. Such particles include leptons (electron, muon, tau and their neutrinos) and gauge bosons (photon, W and Z bosons, gluons, and the Higgs boson); or the hypothetical graviton.

44.3 Conservation

See also: Conservation law (physics)

The baryon number is conserved in nearly all the interactions of the Standard Model. 'Conserved' means that the sum of the baryon number of all incoming particles is the same as the sum of the baryon numbers of all particles resulting from the reaction. An exception is the chiral anomaly proposed by some extensions of the standard model. However, sphalerons are not all that common. Electroweak sphalerons can only change the baryon number by 3. No experimental evidence of sphalerons has yet been observed.

The still hypothetical idea of a grand unified theory allows for the changing of a baryon into several leptons (see $B - L$), thus violating the conservation of both baryon and lepton numbers.[1] Proton decay would be an example of such a process taking place, but has never been observed.

44.4 See also

- Lepton number

- Flavour (particle physics)

- Isospin

- Hypercharge

- Proton decay

- $B - L$

44.5 References

[1] Griffiths, David (2008). *Introduction to Elementary Particles* (2nd ed.). New York: John Wiley & Sons. p. 77. ISBN 9783527618477. In the grand unified theories new interactions are contemplated, permitting decays such as p+ → e+ + π0 or
p+ → ν
μ + π+ in which baryon number and lepton number change.

Chapter 45

Ground state

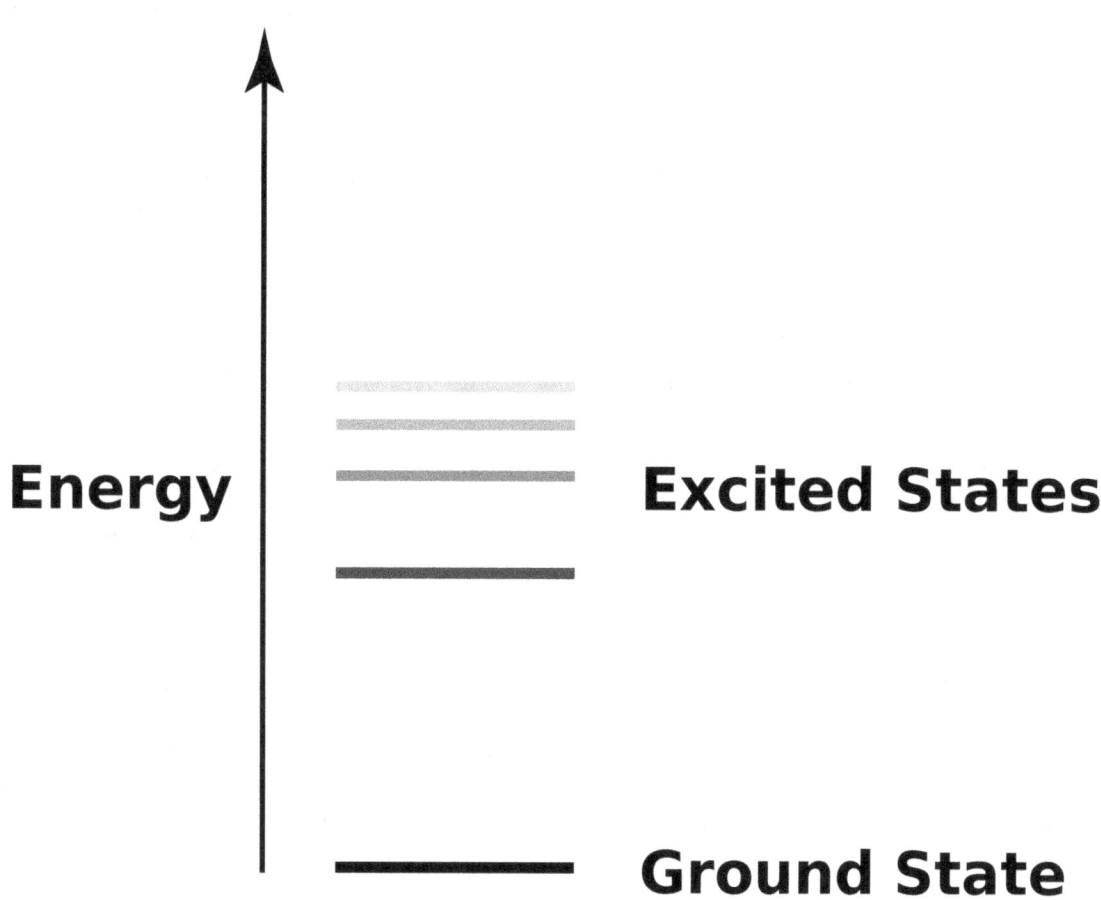

*Energy levels for an electron in an atom: **ground state** and excited states. After absorbing energy, an electron may jump from the ground state to a higher energy excited state.*

The **ground state** of a quantum mechanical system is its lowest-energy state; the energy of the ground state is known as the zero-point energy of the system. An excited state is any state with energy greater than the ground state. The ground state of a quantum field theory is usually called the vacuum state or the vacuum.

If more than one ground state exists, they are said to be degenerate. Many systems have degenerate ground states.

302

Degeneracy occurs whenever there exists a unitary operator which acts non-trivially on a ground state and commutes with the Hamiltonian of the system.

According to the third law of thermodynamics, a system at absolute zero temperature exists in its ground state; thus, its entropy is determined by the degeneracy of the ground state. Many systems, such as a perfect crystal lattice, have a unique ground state and therefore have zero entropy at absolute zero. It is also possible for the highest excited state to have absolute zero temperature for systems that exhibit negative temperature.

45.1   1D ground state has no nodes

In 1D the ground state of the Schrödinger equation has no nodes. This can be proved considering an average energy in the state with a node at $x = 0$, i.e. $\psi(0) = 0$. Consider the average energy in this state

$\langle \psi | H | \psi \rangle = \int dx \left(-\frac{\hbar^2}{2m} \psi^* \frac{d^2 \psi}{dx^2} + V(x) |\psi(x)|^2 \right)$ where $V(x)$ is the potential. Now consider a small interval around $x = 0$, i.e. $x \in [-\epsilon, \epsilon]$. Take a new wavefunction $\psi'(x)$ to be defined as $\psi'(x) = \psi(x), x < -\epsilon$ and $\psi'(x) = -\psi(x), x > \epsilon$ and constant for $x \in [-\epsilon, \epsilon]$. If epsilon is small enough then this is always possible to do so that $\psi'(x)$ is continuous. So assuming $\psi(x) \approx -cx$ around $x = 0$, we can write the new function as

$$\psi'(x) = N \begin{cases} |\psi(x)| & |x| > \epsilon \\ c\epsilon & |x| \leq \epsilon \end{cases}$$

where $N = \frac{1}{\sqrt{1+|c|^2\epsilon^3/3}}$ is the norm. Note that the kinetic energy density $|d\psi'/dx|^2 < |d\psi/dx|^2$ everywhere because of the normalization. Now consider the potential energy. For definiteness let us choose $V(x) \geq 0$. Then it is clear that outside the interval $x \in [-\epsilon, \epsilon]$ the potential energy density is smaller for the ψ' because $|\psi'| < |\psi|$ there. On the other hand, in the interval $x \in [-\epsilon, \epsilon]$ we have

$V_{avg}^\epsilon{}' = \int_{-\epsilon}^\epsilon dx \, V(x)|\psi'|^2 = \frac{\epsilon^3|c|^2}{1+|c|^2\epsilon^3/3} \int_{-\epsilon}^\epsilon V(x) \approx \frac{2\epsilon^4|c|^2}{3} V(0) + \dots$

which is correct to this order of ϵ and $\dots$ indicate higher order corrections. On the other hand, the potential energy in the ψ state is

$V_{avg}^\epsilon = \int_{-\epsilon}^\epsilon dx \, V(x)|\psi|^2 = \int_{-\epsilon}^\epsilon dx \, |c|^2|x|^2 V(x) \approx \frac{2\epsilon^4|c|^2}{3} V(0) + \dots$. which is the same as that of the ψ' state to the order shown. Therefore, the potential energy unchanged to leading order in ϵ by deforming the state with a node ψ into a state without a node ψ'. We can do this by removing all nodes thereby reducing the energy, which implies that the ground state energy must not have a node. This completes the proof.

45.2   Examples

- The wave function of the ground state of a particle in a one-dimensional well is a half-period sine wave which goes to zero at the two edges of the well. The energy of the particle is given by $\frac{h^2 n^2}{8mL^2}$, where h is the Planck constant, m is the mass of the particle, n is the energy state ($n = 1$ corresponds to the ground-state energy), and L is the width of the well.

- The wave function of the ground state of a hydrogen atom is a spherically-symmetric distribution centred on the nucleus, which is largest at the center and reduces exponentially at larger distances. The electron is most likely to be found at a distance from the nucleus equal to the Bohr radius. This function is known as the 1s atomic orbital. For hydrogen (H), an electron in the ground state has energy −13.6 eV, relative to the ionization threshold. In other words, 13.6 eV is the energy input required for the electron to no longer be bound to the atom.

- The exact definition of one second of time since 1997 has been the duration of 9,192,631,770 periods of the radiation corresponding to the transition between the two hyperfine levels of the ground state of the caesium−133 atom at rest at a temperature of 0 K.[1]

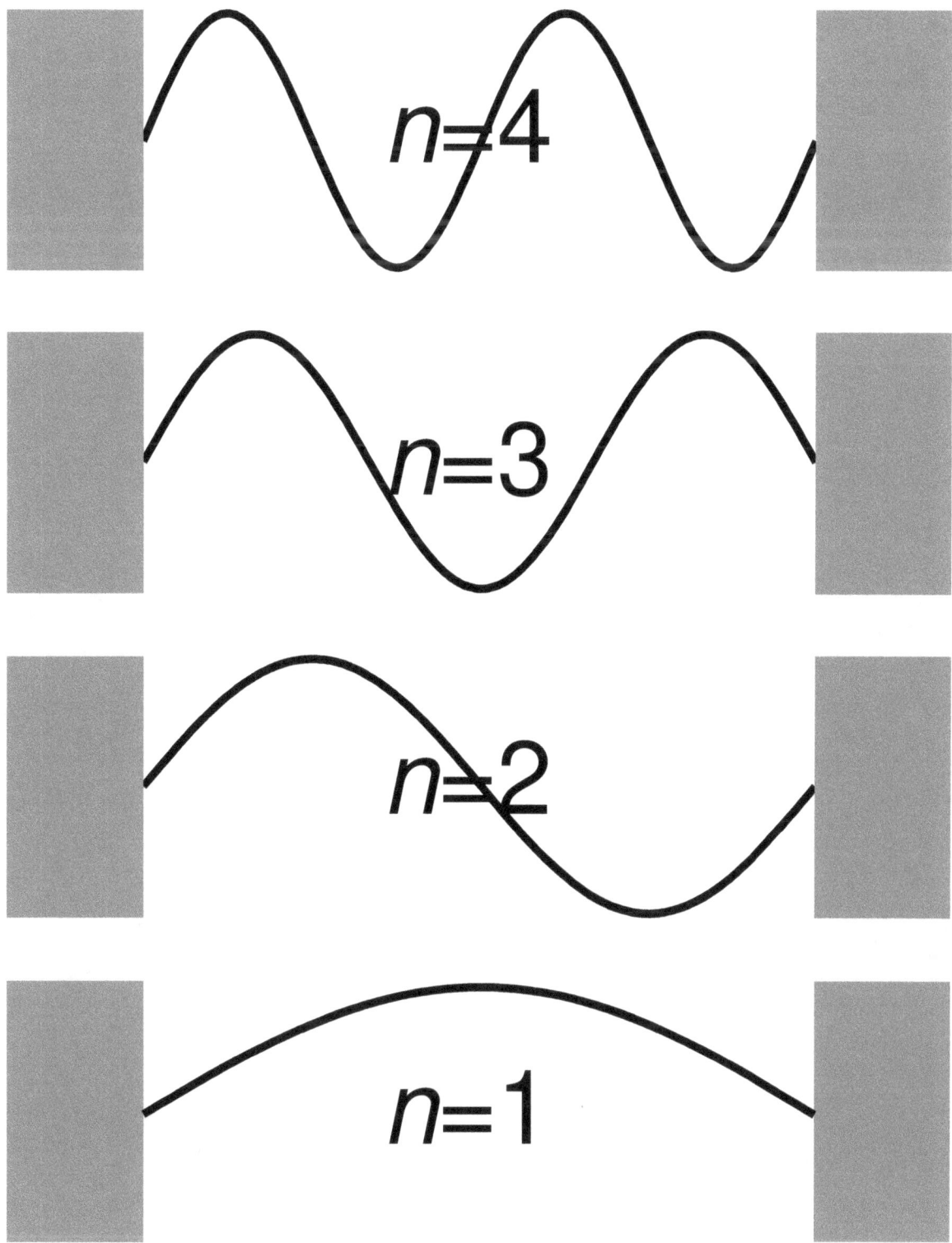

Initial wave functions for the first four states of a one-dimensional particle in a box

45.3   Notes

[1] "Unit of time (second)". *SI Brochure*. BIPM. Retrieved 2013-12-22.

45.4 Bibliography

- Feynman, Richard; Leighton, Robert; Sands, Matthew (1965). "see section 2-5 for energy levels, 19 for the hydrogen atom". *The Feynman Lectures on Physics* **3**.

Chapter 46

Excited state

"Excited" redirects here. For other uses, see Excited (disambiguation).
Excitation is an elevation in energy level above an arbitrary baseline energy state. In physics there is a specific technical

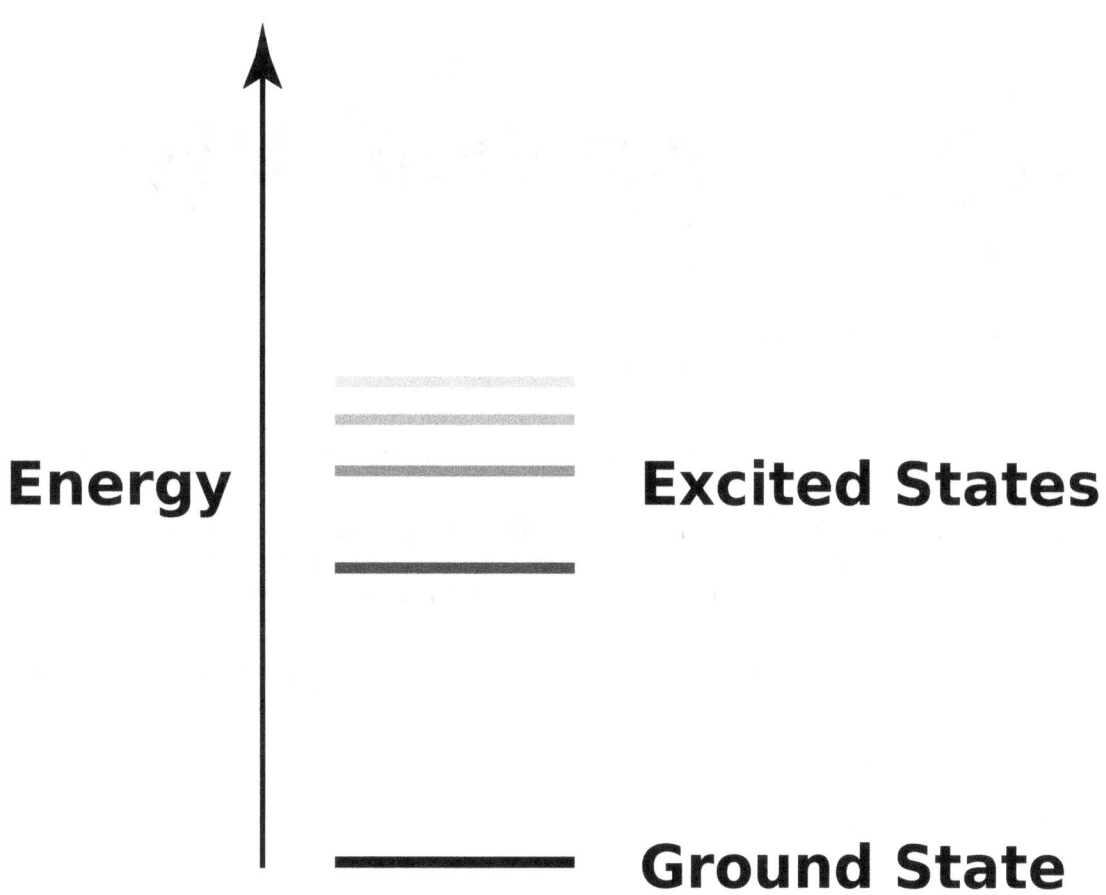

After absorbing energy, an electron may jump from the ground state to a higher energy excited state.

definition for energy level which is often associated with an atom being raised to an excited state.

In quantum mechanics an **excited state** of a system (such as an atom, molecule or nucleus) is any quantum state of the system that has a higher energy than the ground state (that is, more energy than the absolute minimum). The temperature

Excitations of copper 3d orbitals on the CuO2-plane of a high Tc superconductor; The ground state (blue) is x2-y2 orbitals; the excited orbitals are in green; the arrows illustrate inelastic x-ray spectroscopy

of a group of particles is indicative of the level of excitation (with the notable exception of systems that exhibit Negative temperature).

The lifetime of a system in an excited state is usually short: spontaneous or induced emission of a quantum of energy (such as a photon or a phonon) usually occurs shortly after the system is promoted to the excited state, returning the system to a state with lower energy (a less excited state or the ground state). This return to a lower energy level is often loosely described as decay and is the inverse of excitation.

Long-lived excited states are often called metastable. Long-lived nuclear isomers and singlet oxygen are two examples of this.

46.1 Atomic excitation

A simple example of this concept comes by considering the hydrogen atom.

The ground state of the hydrogen atom corresponds to having the atom's single electron in the lowest possible orbit (that is, the spherically symmetric "1s" wavefunction, which has the lowest possible quantum numbers). By giving the atom additional energy (for example, by the absorption of a photon of an appropriate energy), the electron is able to move into an excited state (one with one or more quantum numbers greater than the minimum possible). If the photon has too much energy, the electron will cease to be bound to the atom, and the atom will become ionised.

After excitation the atom may return to the ground state or a lower excited state, by emitting a photon with a characteristic energy. Emission of photons from atoms in various excited states leads to an electromagnetic spectrum showing a series of characteristic emission lines (including, in the case of the hydrogen atom, the Lyman, Balmer, Paschen and Brackett series.)

An atom in a high excited state is termed Rydberg atom. A system of highly excited atoms can form a long-lived condensed excited state e.g. a condensed phase made completely of excited atoms: Rydberg matter. Hydrogen can also be excited by heat or electricity.

46.2   Perturbed gas excitation

A collection of molecules forming a gas can be considered in an excited state if one or more molecules are elevated to kinetic energy levels such that the resulting velocity distribution departs from the equilibrium Boltzmann distribution. This phenomenon has been studied in the case of a two-dimensional gas in some detail, analyzing the time taken to relax to equilibrium.

46.3   Calculation of excited states

Excited states are often calculated using Coupled cluster, Møller–Plesset perturbation theory, Multi-configurational self-consistent field, Configuration interaction,[1] and Time-dependent density functional theory. These calculations are more difficult than non-excited state calculations.[2][3][4][5][6]

46.4   Reaction

A further consequence is reaction of the atom in the excited state, as in photochemistry. Excited states give rise to chemical reaction.

46.5   See also

- Rydberg formula

- Stationary state

- Repulsive state

46.6   References

[1] Hehre, Warren J. (2003). *A Guide to Molecular Mechanics and Quantum Chemical Calculations* (PDF). Irvine, California: Wavefunction, Inc. ISBN 1-890661-06-6.

[2] Glaesemann, Kurt R.; Govind, Niranjan; Krishnamoorthy, Sriram; Kowalski, Karol (2010). "EOMCC, MRPT, and TDDFT Studies of Charge Transfer Processes in Mixed-Valence Compounds: Application to the Spiro Molecule". *The Journal of Physical Chemistry A* **114** (33): 8764–8771. doi:10.1021/jp101761d. PMID 20540550.

[3] Dreuw, Andreas; Head-Gordon, Martin (2005). "Single-Reference ab Initio Methods for the Calculation of Excited States of Large Molecules". *Chemical Reviews* **105** (11): 4009–37. doi:10.1021/cr0505627. PMID 16277369.

[4] Knowles, Peter J.; Werner, Hans-Joachim (1992). "Internally contracted multiconfiguration-reference configuration interaction calculations for excited states". *Theoretica Chimica Acta* **84**: 95. doi:10.1007/BF01117405.

[5] Foresman, James B.; Head-Gordon, Martin; Pople, John A.; Frisch, Michael J. (1992). "Toward a systematic molecular orbital theory for excited states". *The Journal of Physical Chemistry* **96**: 135. doi:10.1021/j100180a030.

[6] Glaesemann, Kurt R.; Gordon, Mark S.; Nakano, Haruyuki (1999). "A study of FeCO+ with correlated wavefunctions". *Physical Chemistry Chemical Physics* **1** (6): 967–975. Bibcode:1999PCCP....1..967G. doi:10.1039/a808518h.

46.7   External links

- NASA background information on ground and excited states

Chapter 47

Resonance (particle physics)

Main article: Relativistic Breit–Wigner distribution

In particle physics, a **resonance** is the peak located around a certain energy found in differential cross sections of scattering experiments. These peaks are associated with subatomic particles (such as nucleons, delta baryons, upsilon mesons) and their excitations. The width of the resonance (Γ) is related to the lifetime (τ) of the particle (or its excited state) by the relation

$$\Gamma = \frac{\hbar}{\tau}$$

where h is the Planck constant.

47.1 See also

- Baryon Resonance Particles

Chapter 48

Particle physics experiments

Particle physics experiments briefly discusses a number of past, present, and proposed experiments with particle accelerators, throughout the world. In addition, some important accelerator interactions are discussed. Also, some notable systems components are discussed, named by project.

48.1 AEGIS (particle physics)

AEGIS is a proposed experiment to be set up at the Antiproton Decelerator at CERN. In addition, *AEGIS* is an acronym for: **A**ntimatter **E**xperiment: **G**ravity, **I**nterferometry, **S**pectroscopy)

The proposed experiment:

It would attempt to determine if gravity affects antimatter in the same way it affects matter by testing its effect on an antihydrogen beam. By sending a stream of antihydrogen through a series of diffraction gratings, the pattern of light and dark patterns would allegedly enable the position of the beam to be pinpointed with up to 1% accuracy.[1]

48.2 Athena

This article is about the CERN project. For the Greek goddess, see Athena. For other uses, see Athena (disambiguation).

ATHENA was an antimatter research project that took place at the AD Ring at CERN. In 2005 ATHENA was disbanded and many of the former members became the ALPHA Collaboration. In August 2002, it was the first experiment to produce 50,000 low-energy antihydrogen atoms, as reported in the journal Nature.[2][3]

For antihydrogen to be created, antiprotons and positrons must first be prepared. Once the antihydrogen is created, a high-resolution detector is needed to confirm that the antihydrogen was created, as well as to look at the spectrum of the antihydrogen in order to compare it to "normal" hydrogen.[4]

The antiprotons are obtained from CERN's Antiproton Decelerator while the positrons are obtained from a positron accumulator. The antiparticles are then led into a recombination trap to create antihydrogen. The trap is surrounded by the ATHENA detector, which detects the annihilation of the antiprotons as well as the positrons.

The ATHENA Collaboration comprised the following institutions[5]

- University of Aarhus, Denmark

- University of Brescia, Italy

- CERN

- University of Genoa, Italy

- University of Pavia, Italy

- RIKEN, Japan

- Federal University of Rio de Janeiro, Brazil

- University of Wales Swansea, United Kingdom

- University of Tokyo, Japan

- University of Zurich, Switzerland

- National Institute for Nuclear Physics, Italy

48.3 ARGUS (experiment)

This article is about the particle physics experiment. For nuclear weapons tests, see Operation Argus.

The **ARGUS** experiment was a particle physics experiment that ran at the electron-positron collider ring *DORIS II* at DESY. It is the first experiment that observed the mixing of the B mesons (in 1987)[6]

The ARGUS detector was a hermetic detector with 90% coverage of the full solid angle. It had drift chambers, a time-of-flight system, an electromagnetic calorimeter and a muon chamber system.[7]

In physics, the **ARGUS distribution**, named after this experiment,[8] is the probability distribution of the reconstructed invariant mass of a decayed particle candidate in continuum background. Its probability density function (not normalized) is:

$$f(x) = x \cdot \sqrt{1 - \left(\frac{x}{c}\right)^2} \exp\left\{-\chi \cdot \left(1 - \left(\frac{x}{c}\right)^2\right)\right\} \text{ for } x > 0.$$

Sometimes a more general form is used to describe a more peaking-like distribution:

$$f(x) = x \cdot \left[1 - \left(\frac{x}{c}\right)^2\right]^p \exp\left\{-\chi \cdot \left(1 - \left(\frac{x}{c}\right)^2\right)\right\}$$

Here parameters c, χ, p represent the cutoff, curvature, and power ($p = 0.5$ gives a regular ARGUS) respectively.

48.4 ATRAP

The **ATRAP** collaboration at CERN developed out of TRAP, a collaboration whose members pioneered cold antiprotons, cold positrons, and first made the ingredients of cold antihydrogen to interact. ATRAP members also pioneered accurate hydrogen spectroscopy and first observed hot antihydrogen atoms. The collaboration includes investigators from Harvard, the University of Bonn, the Max Planck Institute for Quantum Optics, the University of Amsterdam, York University, Seoul National University, NIST, Forschungszentrum Jülich.

48.5   Belle experiment

Main article: Belle experiment

The **Belle experiment** is a particle physics experiment conducted by the Belle Collaboration, an international collaboration of more than 400 physicists and engineers investigating CP-violation effects at the High Energy Accelerator Research Organisation (KEK) in Tsukuba, Ibaraki Prefecture, Japan.

48.6   Systems components

48.6.1   ASTRID particle storage ring

ASTRID is a particle storage ring at the University of Aarhus, Århus, Denmark. It is located in the lower levels of the University of Aarhus Department of Physics and Astronomy.

Its construction was announced on 18 September 1987.[9] By 1998, it had been improved several times, notably increasing its maximum operation time to 30–35 hours.[10] In December 2008, a contract was awarded to design and build ASTRID 2, which will be built adjacent to ASTRID. ASTRID will be used to "top up" the new ring, allowing ASTRID 2 to operate nearly continuously.[11]

ASTRID 2 particle storage ring

ASTRID 2 will be a 46-meter particle storage ring at the University of Aarhus, Århus, Denmark. The contract to build the ring was awarded in December, 2008, and plans are expected to be complete by the end of 2009. It will be built in the lower levels of the University of Aarhus Department of Physics and Astronomy, adjacent to the existing ASTRID particle storage ring. Rather than having an electron beam which decays over time, it will be continually "topped up" by a feed from ASTRID, allowing nearly constant current.[11] It will generate synchrotron radiation to provide a tunable beam of light, expected to be of "remarkable" quality, with wavelengths from the ultraviolet through x-rays.[11]

48.6.2   Anti-proton decelerator

The **antiproton decelerator** (AD) is a storage ring at the CERN laboratory in Geneva. The decelerated antiprotons are ejected to one of the connected experiments.

Current experiments

Former experiments:

48.7   Accelerator interaction overview

48.7.1   Absorber

In high energy physics experiments, an **absorber** is a block of material used to absorb some of the energy of an incident particle. Absorbers can be made of a variety of materials, depending on the purpose; lead and liquid hydrogen are common choices.

Most absorbers are used as part of a detector.

A more recent use for absorbers is for ionization cooling, as in the International Muon Ionization Cooling Experiment.

In solar power, the most important part of the collector takes up the heat of the solar radiation through a medium (water + antifreeze). This is heated and circulates between the collector and the storage tank. A high degree of efficiency is achieved by using black absorbers or, even better, through selective coating.

In sunscreen, ingredients which absorb UVA/UVB rays, such as avobenzone and octyl methoxycinnamate, are known as absorbers. They are contrasted with physical "blockers" of UV radiation such as titanium dioxide and zinc oxide.

48.7.2   Accelerator physics

Main article: Accelerator physics

Accelerator physics is an interdisciplinary topic of applied physics, commonly defined by the intent of designing, building and operating particle accelerators.

The experiments conducted with particle accelerators are not regarded as part of accelerator physics, but belong (according to the objectives of the experiments) to e.g. particle physics, nuclear physics, condensed matter physics or materials physics. The types of experiments done at a particular accelerator facility are determined by characteristics of the generated particle beam such as average energy, particle type, intensity, and dimensions.

48.7.3   Event reconstruction

Main article: Event reconstruction

In a particle detector experiment, event **reconstruction** is the process of interpreting the electronic signals produced by the detector to determine the original particles that passed through, their momenta, directions, and the primary vertex of the event. Thus the initial physical process that occurred at the interaction point of the particle accelerator, whose study is the ultimate goal of the experiment, can be determined. The total event reconstruction is rarely possible (and rarely necessary); usually, only some part of the data described above is obtained and processed.

48.8   See also

- ALPHA Collaboration

- Antimatter

- CERN

- List of synchrotron radiation facilities

- Gravitational interaction of antimatter

- Particle physics

48.9   References

[1] Courtland, R. (12 June 2008). "Would an antimatter apple fall up?". *New Scientist*. Retrieved 2008-10-27.

[2] "Thousands of cold anti-atoms produced at CERN" (Press release). CERN. 19 September 2002. Retrieved 2012-04-10.

[3] Amoretti,M.et al. (2002). "Production and detection of cold antihydrogen atoms".*Nature***419**(6906): 456–9. Bibcode:A. doi:10.1038/nature01096. PMID 12368849.

[4] "How the ATHENA Experiment works". ATHENA. 14 September 2002. Retrieved 2012-04-10.

[5] "The ATHENA Collaboration". ATHENA. 30 January 2006. Retrieved 2012-04-10.

[6] Albrecht, H.; ARGUS Collaboration et al. (1987). "Observation of B^0–B^0 mixing". *Physics Letters B* **192** (1–2): 245. Bibcode:1987PhLB..192..245A. doi:10.1016/0370-2693(87)91177-4.

[7] Albrecht, H.; ARGUS Collaboration et al. (1989). "Argus: A universal detector at DORIS II". *Nuclear Instruments and Methods in Physics Research Section A* **275** (1): 1–48. Bibcode:1989NIMPA.275....1A. doi:10.1016/0168-9002(89)90334-3.

[8] Albretch, H.; ARGUS Collaboration et al. (1990). "Search for hadronic b→u decays". *Physics Letters B* **241** (2): 278. Bibcode:1990PhLB..241..278A. doi:10.1016/0370-2693(90)91293-K. The function has been defined with parameter c representing the beam energy and parameter p set to 0.5. The normalization and the parameter χ have been obtained from data.

[9] Stensgaard, R. (1988). "ASTRID - The Aarhus Storage Ring". *Physica Scripta* **1988** (T22): 315. Bibcode:1988PhST...22..315S. doi:10.1088/0031-8949/1988/T22/051.

[10] Nielsen, J. S.; Møller, S. P. (1998). "New Developments at the ASTRID storage ring" (PDF). *Proceedings from 6th European Particle Accelerator Conference.* Retrieved 2012-04-10.

[11] "ASTRID2 – the ultimate synchrotron radiation source". ASTRID. Retrieved 2012-04-10.

48.10   Further reading

- AEGIS collaboration (8 June 2007). "Proposal for the AEGIS experiment at the CERN Antiproton Decelerator" (PDF). CERN.

- Testera, G. et al. (2008). "Formation of a cold antihydrogen beam in AEGIS for gravity measurements". *AIP Conference Proceedings* **1037**: 5–15. arXiv:0805.4727. Bibcode:2008AIPC.1037....5T. doi:10.1063/1.2977857.

48.11   External links

- ANSTO website

- ANTARES website

- Antiproton Decelerator website

- ARGUS Fest

- ASTRID website

- ASTRID 2 website

- ATHENA website

- CERN's public site

- CERN Antimatter page

- National Medical Cyclotron website

48.12 Text and image sources, contributors, and licenses

48.12.1 Text

- **Hadron** *Source:* https://en.wikipedia.org/wiki/Hadron?oldid=684132151 *Contributors:* Bryan Derksen, Manning Bartlett, Peterlin~enwiki, Edward, Erik Zachte, ESnyder2, Fruge~enwiki, TakuyaMurata, Darkwind, Glenn, Nikai, Ehn, Olya, Phys, Bevo, Topbanana, BenRG, Twang, Donarreiskoffer, Korath, Wjhonson, Merovingian, Ojigiri~enwiki, Sunray, JesseW, Xanzzibar, Giftlite, Xerxes314, Dratman, Physicist, Mikro2nd, LiDaobing, Pthompson, Icairns, Jimaginator, Mike Rosoft, Vsmith, Goochelaar, Sunborn, Livajo, El C, Kwamikagami, Shanes, Fwb22, Jumbuck , Cookiemobsta, Velella, Rebroad, Vuo, Kusma, DV8 2XL, Linas, GrouchyDan, Palica, Marudubshinki, Kbdank71, Mana Excal-ibur, Kinu, Strait, FlaBot, RexNL, Goudzovski, FrankTobia, YurikBot, Radishes, Bambaiah, Hydrargyrum, Salsb, NawlinWiki, Wiki alf, SCZenz, Davemck , Bota47, Scriber~enwiki, Modify, Katieh5584, Eog1916, SmackBot, McGeddon, Gilliam, Benjaminevans82, Dingar, Per-sian Poet Gal, Telempe , DHN-bot~enwiki, Audriusa, Acepectif, Kokot.kokotisko, JorisvS, JarahE, BranStark, SJCrew, Eratticus, Chrumps, Jtuggle, Q43, Epbr123, Wikid77, Headbomb, Escarbot, Deflective, Gcm, NE2, Trapezoidal, Naval Scene, KEKPΩΨ, NeverWorker, Wwmbes, Alexllew, Lvwarren, Jebus0, DariusU, Khalid Mahmood, Adriaan, Rustyfence, Ron2, Leyo, J.delanoy, Maurice Carbonaro, JVersteeg, Rod57, Way2Smart22, Hugh Hudson, Y2H, Ansans, Bobxii, Chris Longley, Useight, Dylan bossart, VolkovBot, TXiKiBoT, Kinkydarkbird, Anony-mous Dissident, Don4of4, Wordsmith, LeaveSleaves, Antixt, Enviroboy, Insanity Incarnate, Nibios, AlleborgoBot, SieBot, Yintan, Lead-SongDog, RadicalOne, Paolo.dL, OKBot, JohnSawyer, Lazarus1907, Pinkadelica, Danthewhale, Martarius, ClueBot, Amaamaddq, Authori-tative Physicist, Wwheaton, Rotational, DragonBot, Sciencedude9998, Tuchomator, El planeto, Kaiba, Thingg, Koshoid, Aitias, Apparition11, Rishi.bedi, TimothyRias, InternetMeme, Jbeans, MystBot, Sgpsaros, Tayste, Addbot, Pkkphysicist, Ehrenkater, Lightbot, Luckas-bot, Yobot, Nallimbot, Dagus2000, Fangfyre, LOLx 9000, Thisaccountwillbebanned, Citation bot, Xqbot, Drilnoth, Br77rino, Wikiedit33, Ajahnjohn, Omnipaedista, RibotBOT, Mashmeister, Tjbright2, My cat's breath smells like catfood, Haeinous, Citation bot 1, Javert, Gil987, I dreamof horses, Jonesey95, Rameshngbot, Thinking of England, Alarichus, SkyMachine, FoxBot, Johnshnappay, Антон Гліністы, Teravolt, Rac-erx11, Naznin farhah, Tommy2010, Rafabaez, Wikipelli, ZéroBot, StringTheory11, Hadron12, Donner60, Bobogoobo, Petrb, ClueBot NG, Gareth Griffith-Jones, Bibcode Bot, BG19bot, Dwightboone, Njavallil, Walterpfeifer, Pfeiferwalter, ChrisGualtieri, Ugog Nizdast, Lithelimbs, RoKo89, Michikohundred, KasparBot, Wwilliam726 and Anonymous: 170

- **Hadronization** *Source:* https://en.wikipedia.org/wiki/Hadronization?oldid=665417782 *Contributors:* Alexwatson, Charles Matthews, Phys, Sam Hocevar, Fwb22, RJFJR, TenOfAllTrades, Isaac Rabinovitch, Ohwilleke, SCZenz, Zwobot, Whobot, SmackBot, Incnis Mrsi, Colonies Chris, Chrumps, A876, Headbomb, Pichote, Sir Link, Moogwrench, Sothisislife101, VVVBot, Addbot, Luckas-bot, Yobot, Citation bot, Double sharp, Naznin farhah, Chaudière, Raktimabir, Jj1236, Bibcode Bot, Prokaryotes and Anonymous: 16

- **Subatomic particle** *Source:* https://en.wikipedia.org/wiki/Subatomic_particle?oldid=684436250 *Contributors:* The Anome, Tarquin, Michael Hardy, FrankH, Ixfd64, CesarB, NuclearWinner, Looxix~enwiki, Ahoerstemeier, LittleDan, Glenn, Kwekubo, Schneelocke, Bevo, Chrisjj, Donarreiskoffer, Baldhur, Romanm, Anthony, Alan Liefting, Giftlite, DocWatson42, Awolf002, Mintleaf~enwiki, Dissident, Xerxes314, Ev-eryking, Bensaccount, Vadmium, Antandrus, Rdsmith4, Kenny TM~~enwiki, Discospinster, ElTyrant, Vsmith, ESkog, RJHall, El C, Bobo192, Shenme, PiccoloNamek, Stephen G. Brown, Alansohn, Ctande, MarkGallagher, Wtshymanski, Egg, DV8 2XL, Ceyockey, Adrian.benko, Oleg Alexandrov, Mindmatrix, JarlaxleArtemis, Duncan.france, Isnow, SeventyThree, Dysepsion, Rjwilmsi, Strait, Klassykittychick, Scor-piuss, Boccobrock, Erkcan, Naraht, DannyWilde, SouthernNights, King of Hearts, Chobot, Wavelength, Bambaiah, Jimp, Phantomsteve, Loom91, Stephenb, PoorLeno, Bachrach44, Spike Wilbury, Syrthiss, Gat0r, Wknight94, Light current, KGasso, GraemeL, Rlove, Sitenl, As-terion, SmackBot, Incnis Mrsi, JohnRussell, Darkgod, Jordan.ambra, Chris the speller, Bluebot, Persian Poet Gal, DHN-bot~enwiki, Sbharris, Can't sleep, clown will eat me, Drkirkby, V1adis1av, Voyajer, Pax85, Radagast83, Edwtie, Drphilharmonic, Thinkingman, Lambiam, Doug Bell, Rigadoun, Ortho, 041744, Ckatz, RandomCritic, Ginkgo100, Esurnir, Tawkerbot2, JForget, Megaboz, Johnlogic, Myasuda, Christian75, Thijs!bot, Epbr123, Guyla, Mbell, Nonagonal Spider, Headbomb, Mjollnir783, Weasel5i2, Escarbot, Ssr, Mentifisto, AntiVandalBot, Luna Santin, Refried, JAnDbot, Instinct, Acroterion, Bongwarrior, VoABot II, Ling.Nut, Glen, Geboy, Mike6271, Davburns, J.delanoy, Yonide-bot, Acalamari, Jakey665, McSly, TomasBat, Benito001, Juliancolton, Logic20, Idioma-bot, ArchetypeRyan, VolkovBot, Philip Trueman, Amother, TXiKiBoT, Geht, Jhannah, GcSwRhIc, Seraphim, Martin451, Optigan13, Krazywrath, Lamro, Insanity Incarnate, Monty845, Lk-leinow, AlleborgoBot, Freependulum, Borne nocker, Hazel77, S8333631, Doclecticwiki, SieBot, Gerakibot, 4RM0~enwiki, Keilana, Flyer22, Tiptoety, The Evil Spartan, Sohelpme, AlexWaelde, Wombatcat, Lisatwo, Antman123, Nimbusania, DarkCatalyst, ClueBot, The Thing That Should Not Be, Drugieuk, Arakunem, Drmies, DragonBot, Djr32, Excirial, Monobi, WikiZorro, SpikeToronto, Peter.C, Darren23, Little Mountain 5, Mifter, Garycompugeek, PicoGils, Addbot, Willking1979, Ronhjones, Knowledgesupreme, Download, Chamal N, Bassbonerocks, CosmiCarl, Barak Sh, IOLJeff, Ehrenkater, VASANTH S.N., Tide rolls, Lightbot, Megaman en m, Luckas-bot, Yobot, Amirobot, Eric-Wester, Kulmalukko, AnomieBOT, Ciphers, ^musaz, Jim1138, Icalanise, Ulric1313, Materialscientist, Citation bot, Maxis ftw, ArthurBot, Xqbot, Phazvmk, Gopal81, Addihockey10, Turk oğlan, Dubravko49, GrouchoBot, АлександрВв, TR4YH4N, Mitraunodo, Sushiflinger, Dave3457, Mark Renier, Doremo, Steve Quinn, Gigigogo, Citation bot 1, Pinethicket, Tom.Reding, Maude Frickert, Teamspoad, TobeBot, Mptb3, Koyae, MartinHiggs, Diannaa, Harrasser, Tbhotch, RjwilmsiBot, Narayanan20092009, EmausBot, Ajraddatz, Heyimawesome, Wikipelli, Hhhippo, JSquish, Harddk, Shuipzv3, Permenent, Anir1uph, Access Denied, Maschen, Epicstonemason, VictorianMutant, CharlieEchoTango, ClueBot NG, Preon, IfYouDoIfYouDon't, Widr, MerlIwBot, Helpful Pixie Bot, Novusuna, IrishStephen, BG19bot, Northamerica1000, Neutral cur-rent, Kord Kakurios, Zombiecat181, Jethro B, Ducknish, MadGuy7023, Jmeg82, Webclient101, Mogism, Reatlas, Faizan, Bigdaddysound, Frankalbertson, The Herald, Ginsuloft, AddWittyNameHere, Colecharb, 123gogeta, TheRapeTrainBackFromTheDead, Again, CraigyDavi, Robdistasio, Darkenergydesigner, Tetra quark, Jesus.Like.For.Real, Jennifer1122, Harshita1999, ProprioMe OW and Anonymous: 368

- **Standard Model** *Source:* https://en.wikipedia.org/wiki/Standard_Model?oldid=686594371 *Contributors:* AxelBoldt, Derek Ross, CYD, Bryan Derksen, The Anome, Ed Poor, Andre Engels, Roadrunner, David spector, Isis~enwiki, Youandme, Ram-Man, Stevertigo, Edward, Patrick, Boud, Michael Hardy, SebastianHelm, Looxix~enwiki, Julesd, Glenn, AugPi, Mxn, Raven in Orbit, Reddi, Phr, Tpbradbury, Populus, Hao-herb428, Phys, Floydian, Bevo, Pierre Boreal, AnonMoos, BenRG, Jeffq, Dmytro, Drxenocide, Robbot, Nurg, Securiger, Texture, Roscoe x, Fuelbottle, Mattflaschen, Tobias Bergemann, Alan Liefting, Ancheta Wis, Giftlite, Dbenbenn, Harp, Herbee, Monedula, LeYaYa, Xerxes314, Dratman, Alison, JeffBobFrank, Dmmaus, Pharotic, Brockert, Bodhitha, Andycjp, Sonjaaa, HorsePunchKid, APH, Icairns, AmarChandra, Gscshoyru, Kate, Arivero, FT2, Rama, Vsmith, David Schaich, Xezbeth, D-Notice, Dfan, Bender235, Pt, El C, Laurascudder, Shanes, Drhex, Fogger~enwiki, Brim, Rbj, Jeodesic, Jumbuck, Alansohn, Gary, ChristopherWillis, Guy Harris, Axl, Sligocki, Kocio, Stillnotelf, Alinor,

Wtmitchell, Egg, TenOfAllTrades, H2g2bob, Killing Vector, Linas, Mindmatrix, Benbest, Dodiad, Mpatel, Faethon, TPickup, Faethon34, Palica, Dysepsion, Faethon36, Qwertyca, Drbogdan, Rjwilmsi, Zbxgscqf, Macumba, Strangethingintheland, Dstudent, R.e.b., Bubba73, Drrngrvy, Agasicles, FlaBot, Naraht, Agasides, DannyWilde, Dave1g, Itinerant1, Gparker, Jrtayloriv, Goudzovski, Chobot, Bgwhite, FrankTobia, YurikBot, Bambaiah, Ohwilleke, VoxMoose, Bhny, JabberWok, Bovineone, Krbabu, SCZenz, JulesH, Davemck, Lomn, E2mb0t~enwiki, Dna-webmaster, Jrf, Dv82matt, Tetracube, Hirak 99, Arthur Rubin, Netrapt, JLaTondre, Caco de vidro, RG2, GrinBot~enwiki, That Guy, From That Show!, Hal peridol, SmackBot, YellowMonkey, Tom Lougheed, Melchoir, Bazza 7, KocjoBot~enwiki, Jagged 85, Thunderboltz, Setanta747 (locked), Skizzik, Dauto, Chris the speller, Bluebot, TimBentley, Sirex98, Silly rabbit, Complexica, Metacomet, DHN-bot~enwiki, MovGP0, QFT, Kittybrewster, Addshore, Jmnbatista, Cybercobra, Jgwacker, BullRangifer, Soarhead77, Daniel.Cardenas, Yevgeny Kats, Byelf2007, TriTertButoxy, Craig Bolon, Ajnosek, Ekjon Lok, Bjankuloski06, Tarcieri, Waggers, JarahE, Michaelbusch, Lottamiata, Newone, Twas Now, IanOfNorwich, Srain, Patrickwooldridge, J Milburn, Mosaffa, Gatortpk, Vessels42, Geremia, Van helsing, Harrigan, Phatom87, Cydebot, David edwards, Verdy p, Michael C Price, Xantharius, Crum375, JamesAM, Thijs!bot, Epbr123, Headbomb, Phy1729, Stannered, Tariqhada, Seaphoto, Orionus, Voyaging, Gnixon, Jbaranao, Jrw@pobox.com, Len Raymond, Narssarssuaq, Bakken, CattleGirl, Davidoaf, Vanished user ty12k189jq10, Lvwarren, Taborgate, Leyo, HEL, J.delanoy, Hans Dunkelberg, Stephanwehner, Wbellido, Aoosten, Jacksonwalters, The Transliterator, DadaNeem, Student7, Joshmt, WJBscribe, Jozwolf, Hexane2000, BernardZ, Awren, Sheliak, Physicist brazuca, Schucker, Goop Goop, Fences and windows, Dextrose, Mcewan, Swamy g, TXiKiBoT, Sharikkamur, Thrawn562, Voorlandt, Escalona, Setreset, PDFbot, Pleroma, UnitedStatesian, Piyush Sriva, Kacser, Billinghurst, Francis Flinch, Moose-32, Ptrslv72, David Barnard, SieBot, ShiftFn, Robdunst, Jim E. Black, SheepNotGoats, Gerakibot, Nozzer42, Mr swordfish, Wing gundam, Bamkin, Likebox, Arthur Smart, HungarianBarbarian, Commutator, KathrynLybarger, Iomesus, C0nanPayne, Crazz bug 5, ClueBot, Superwj5, Wwheaton, Garyzx, SuperHamster, Elsweyn, Maldmac, DragonBot, Djr32, Diagramma Della Verita, Nymf, Eeekster, Brews ohare, NuclearWarfare, PhySusie, Ordovico, Mastertek, DumZiBoT, BodhisattvaBot, Guarracino, Mitch Ames, Truthnlove, Stephen Poppitt, Tayste, Addbot, Deepmath, Eric Drexler, DWHalliday, Mjamja, Leszek Jańczuk, NjardarBot, Mwoldin, Bassbonerocks, Barak Sh, AgadaUrbanit, Lightbot, Smeagol 17, Abjiklam, Ve744, Luckas-bot, Yobot, Orion11M87, AnomieBOT, JackieBot, Icalanise, Citation bot, ArthurBot, Northryde, LilHelpa, Xqbot, Sionus, Professor J Lawrence, Tomwsulcer, Edsegal, GrouchoBot, Trongphu, QMarion II, Ernsts, A. di M., Bytbox, FrescoBot, Paine Ellsworth, Aliotra, Steve Quinn, Citation bot 1, Rameshngbot, MJ94, RedBot, MastiBot, Aknochel, Sijothankam, Puzl bustr, Beta Orionis, Physics therapist, Bj norge, Innotata, Jesse V., RjwilmsiBot, Mathewsyriac, Afteread, EmausBot, Bookalign, WikitanvirBot, Wilhelm-physiker, Bdijkstra, DerNeedle, Kenmint, Dbraize, Tanner Swett, HeptishHotik, مهریار شیخشمن, Suslindisambiguator, Quondum, Webbeh, UniversumExNihilo, Vanished user fijw983kjaslkekfhj45, Maschen, RockMagnetist, Stormymountain, Ζeta ζ, Whoop whoop pull up, Isocliff, ClueBot NG, Smtchahal, Snotbot, Tonypak, O.Koslowski, CharleyQuinton, Dsperlich, Theopolisme, ZakMarksbury, Helpful Pixie Bot, Bibcode Bot, BG19bot, Tirebiter78, AvocatoBot, Lukys~enwiki, Stapletongrey, Ownedroad9, Chip123456, ChrisGualtieri, Khazar2, Billyfesh399, Rhlozier, JYBot, Dexbot, Doom636, Rongended, Cerabot~enwiki, CuriousMind01, Cjean42, Jayanta mallick, Joeinwiki, Kowtje, JPaestpreornJeolhlna, Eyesnore, Euan Richard, Nigstomper, Particle physicist, Prokaryotes, Jernahthern, Ginsuloft, Dimension10, JNrgbKLM, Krabaey, 1codesterS, FelixRosch, Monkbot, Delbert7, BradNorton1979, Lathamboyle, Tetra quark, KasparBot, Buckbill10, Huritisho, S3rr8s and Anonymous: 358

- **List of particles** *Source:* https://en.wikipedia.org/wiki/List_of_particles?oldid=682746251 *Contributors:* AxelBoldt, Danny, Rmhermen, Stevertigo, Bdesham, Ahoerstemeier, Stan Shebs, Docu, Salsa Shark, Nikai, Evercat, Schneelocke, Charles Matthews, Jitse Niesen, CBDunkerson, Bevo, Raul654, Donarreiskoffer, Robbot, Sanders muc, Merovingian, Pengo, Giftlite, Herbee, Xerxes314, Dratman, Jeremy Henty, Alensha, Bodhitha, Physicist, Hayne, Quadell, RetiredUser2, Mysidia, Icairns, Asbestos, D6, Urvabara, Discospinster, Rich Farmbrough, FT2, Qutezuce, ArnoldReinhold, Neko-chan, El C, Laurascudder, Susvolans, EmilJ, Physicistjedi, Minghong, Gbrandt, Eddideigel, Axl, Mac Davis, David Ko, Radical Mallard, RJFJR, Count Iblis, Dirac1933, TenOfAllTrades, LFaraone, Oleg Alexandrov, Linas, JarlaxleArtemis, Duncan.france, GregorB, Cedrus-Libani, Karam.Anthony.K, Palica, Rjwilmsi, Zbxgscqf, JLM~enwiki, Strait, Ems57fcva, Krash, Dan Guan, DannyWilde, Lmatt, Goudzovski, Chobot, YurikBot, Bambaiah, Vuvar1, Madkayaker, Hydrargyrum, Presscorr, Chaos, Salsb, Tavilis, SCZenz, Lexicon, TUSHANT JHA, Dna-webmaster, Tomvds, Poulpy, Cstmoore, TLSuda, NeilN, MacsBug, Tom Lougheed, McGeddon, Bazza 7, WookieInHeat, Derdeib, Yamaguchi先生, Betacommand, Bluebot, Master of Puppets, DHN-bot~enwiki, Raistuumum, Juancnuno, Kittybrewster, Acepectif, Ligulembot, TriTertButoxy, ArglebargleIV, Khazar, John, FrozenMan, JorisvS, 041744, Dr Greg, Slakr, Mets501, Scorpion0422, Cbuckley, Iridescent, TwistOfCain, Happy-melon, JRSpriggs, Flickboy, Van helsing, Lithium6, Neelix, Rotiro, Cydebot, Quibik, Christian75, Omicronpersei8, Thijs!bot, Qwyrxian, TauLibrus, Headbomb, Inner Earth, 49, Guptasuneet, Scottmsg, WinBot, Elmoosecapitan, Tyco.skinner, AubreyEllenShomo, Arch dude, Johnman239, Mwarren us, TheEditrix2, CalamusFortis, MartinBot, Sadisticsuburbanite, Bissinger, Anaxial, CommonsDelinker, Maurice Carbonaro, Zojj, OliverHarris, Joshmt, Adanadhel, Lseixas, Graphite Elbow, VolkovBot, Jmrowland, Quilbert, Anonymous Dissident, Dstary, Escalona, JPMasseo, Figureskatingfan, Inx272, Meters, Antixt, Hamish a e fowler, GoddersUK, Bluetryst, SieBot, Ishvara7, WereSpielChequers, Audrius u, VovanA, Paolo.dL, RSStockdale, Anchor Link Bot, StewartMH, Explicit, ClueBot, Unbuttered Parsnip, Nolimitownass, DragonBot, Atomic7732, TimothyRias, SkyLined, Addbot, DOI bot, Jojhutton, Favonian, LinkFA-Bot, OlEnglish, Teles, Legobot, Luckas-bot, Yobot, Dov Henis, Azcolvin429, AnomieBOT, Götz, Icalanise, Flewis, Materialscientist, OllieFury, Vuerqex, ArthurBot, Vulcan Hephaestus, Blennow, Reality006, Coretheapple, Jcimorra, RibotBOT, Ernsts, A. di M., Axelfoley12, Zosterops, FrescoBot, Paine Ellsworth, Citation bot 1, JIK1975, Tom.Reding, Diffequa, WikitanvirBot, Racerx11, 112358sam, Aegnor.erar, Hops Splurt, HESUPERMAN, Hhhippo, AvicBot, JSquish, StringTheory11, Waperkins, Bamyers99, Suslindisambiguator, L Kensington, DennisIsMe, RockMagnetist, ClueBot NG, Snotbot, Primergrey, Vio45lin, Widr, MsFionnuala, Oklahoma3477, Bibcode Bot, CityOfSilver, Cap'n G, BML0309, Dan653, Twocount, Penguinstorm300, Dexbot, LightandDark2000, Ohiggy, TwoTwoHello, Andyhowlett, Printersmoke, Orion 2013, ARUNEEK, Seino van Breugel, AspaasBekkelund, TheMagikCow, Vyom27, ParkersComments, Selva Ganapathy and Anonymous: 290

- **Fundamental interaction** *Source:* https://en.wikipedia.org/wiki/Fundamental_interaction?oldid=686669427 *Contributors:* AxelBoldt, Zundark, The Anome, Tarquin, AstroNomer~enwiki, William Avery, Roadrunner, Ellmist, Robert Foley, Heron, Isis~enwiki, Stevertigo, Patrick, Michael Hardy, Gdarin, CesarB, Looxix~enwiki, Cyp, William M. Connolley, Theresa knott, Mxn, Bemoeial, Reddi, Zoicon5, Finlay McWalter, Robbot, Lowellian, Brjaga, Roscoe x, Seth Ilys, Ancheta Wis, Giftlite, Christopher Parham, Herbee, Monedula, Xerxes314, Alison, Pcarbonn, Beland, Melikamp, Karol Langner, AmarChandra, Mike Rosoft, Jørgen Friis Bak, JimJast, Discospinster, Guanabot, FT2, Harriv, Quietly, GoldenRing, Clement Cherlin, El C, Lycurgus, Joanjoc~enwiki, Alereon, Euyyn, Kanzure, Army1987, Rbj, Haham hanuka, Nsaa, Jumbuck, Foant, Dachannien, Kdau, ReyBrujo, Reaverdrop, BDD, Someoneinmyheadbutit'snotme, DV8 2XL, Kazvorpal, Woohookitty, Linas, Mindmatrix, Sabejias, StradivariusTV, Mpatel, Miss Madeline, Isnow, Elvey, Chun-hian, Koavf, Strait, Jmcc150, RE, Gadha, FlaBot, DClement, ZoneSeek, Alfred Centauri, Lmatt, Rell Canis, Mstroeck, Chobot, Subtractive, Visor, GangofOne, Mysekurity, YurikBot, Ashleyisachild, Bambaiah, Lucinos~enwiki, Wavesmikey, Chaos, FFLaguna, Dbfirs, Trigger hippie77, Enormousdude, Shimei, RG2, Bweenie,

Phr en, GrinBot~enwiki, SmackBot, Unyoyega, Andy M. Wang, Vvarkey, Jjalexand, Mithaca, Acipsen, DHN-bot~enwiki, Colonies Chris, Andy120290, Addshore, SundarBot, Jgwacker, LeoNomis, Sadi Carnot, TTE, SashatoBot, FrozenMan, Philosophus, A. Parrot, Fangfufu, GDallimore, Avanishsharma, CRGreathouse, Green caterpillar, McVities, MaxEnt, A. Exeunt, Scott.medling, LouisBB, Thijs!bot, Martin Hogbin, Mojo Hand, Headbomb, Dfrg.msc, Dodecahedron~enwiki, JAnDbot, The penfool, Fordskydog, MER-C, TheEditrix2, Fabrictramp, Leyo, Trusilver, Joshuaali, Idioma-bot, VolkovBot, TXiKiBoT, Anonymous Dissident, MackSalmon, Praveen pillay, BotKung, Gnomon13, Lamro, RMW42, EmxBot, Neparis, SieBot, WereSpielChequers, ToePeu.bot, Avargasm, RadicalOne, Dhatfield, SuperSpy00bob, Sbowers3, Beast of traal, Lightmouse, Nskillen, Sunrise, OKBot, Bpeps, C0nanPayne, StewartMH, Sfan00 IMG, ClueBot, MichaelVernonDavis, Super-Hamster, Djr32, Sadiqsaleem09, PixelBot, Eeekster, Zamis45, Yonskii, 1ForTheMoney, Noctibus, Truthnlove, Addbot, Mabdul, LinkFA-Bot, F Notebook, Lightbot, Legobot, Clay Juicer, Luckas-bot, Yobot, II MusLiM HyBRiD II, Rifter0x0000, AnomieBOT, Glen Dillon, Girl Scout cookie, Cleroth, JackieBot, Piano non troppo, Flewis, AthenaO, Xqbot, Omnipaedista, RibotBOT, A. di M., Ironboy11, Goodbye Galaxy, Jmbenham, Unkownkid2400, Rameshngbot, Jschnur, RedBot, Σ, Frankjohnson123, IVAN3MAN, Right-wing genius, Lokentaren, Setsuna29, EngineerFromVega, RjwilmsiBot, Anuandraj, Beyond My Ken, Deadlyops, Carbo1200, Kbasford, ClueBot NG, Greedohun, MelbourneStar, Grannis3, Kasirbot, Kaos Magician, Einsteiner900, Widr, Shelbylv, CasualVisitor, Hz.tiang, JimmyMachineGunHand, Cengime, Batty-Bot, Prof. Squirrel, Dexbot, Mogism, Makecat-bot, Ryan.laff, Jamesx12345, Ttitts, CsDix, Kenanwang, GregRos, Prokaryotes, Robertpb97, Occurring, Basedrawnz, Learnerktm, Julietvbarbara, Barbarousbunch815, Brendapallister, Nicholaspurcellstudio, Pickleslover, Tetra quark, Claudio.nahmad.arcaraz, Trumpet21, KasparBot, The oracle 2015, Mitzionne, Huritisho and Anonymous: 255

- **Strong interaction** *Source:* https://en.wikipedia.org/wiki/Strong_interaction?oldid=681830498 *Contributors:* AxelBoldt, Sodium, Bryan Derksen, RK, Andre Engels, Danny, Peterlin~enwiki, Heron, Xavic69, Tim Starling, EddEdmondson, Gdarin, Ixfd64, Wintran, Salsa Shark, AugPi, Timwi, Bemoeial, Wikiborg, Fuzheado, ElusiveByte, Populus, Phys, Omegatron, Bevo, PuzzletChung, Robbot, Mayooranathan, Henrygb, Giftlite, DocWatson42, Sj, Harp, Monedula, Xerxes314, Remy B, Jason Quinn, Utcursch, Zfr, AmarChandra, Sam Hocevar, Nicobn~enwiki, Jørgen Friis Bak, Discospinster, FT2, Vsmith, ArnoldReinhold, Trekie8472, Roybb95~enwiki, David Schaich, Gianluigi, Marknewlyn~enwiki, Drhex, CDN99, Army1987, Matt McIrvin, Danski14, VivaEmilyDavies, TenOfAllTrades, Vuo, DV8 2XL, Kazvorpal, Oleg Alexandrov, Su-perstring, Spettro9, BillC, Eras-mus, Tevatron~enwiki, Mendaliv, Zbxgscqf, Strait, Loudenvier, Donotresus, Yamamoto Ichiro, Siv0r, Lmatt, Srleffler, Chobot, DVdm, Wavelength, Borgx, Bambaiah, Phmer, Jimp, Supasheep, Limulus, Salsb, RazorICE, Expensivehat, Dhollm, Lob-wedge, Superiority, Tetracube, Willtron, GrinBot~enwiki, Asterion, Finell, AndrewWTaylor, SmackBot, RDBury, Tom Lougheed, Trojo~enwiki, Jrockley, Dauto, Chris the speller, Jjalexand, Complexica, Sbharris, Colonies Chris, Richard001, TTE, Spiritia, Titus III, FrozenMan, Newone, Happy-melon, Conrad.Irwin, Rowellcf, Chrisahn, Cydebot, Danny Bierek, Mtpaley, Wannabe Runny, Irigi, Headbomb, Niduzzi, KP Botany, Tlabshier, Hanzoro 5, Dougher, Steelpillow, JAnDbot, Supertheman, Roleplayer, Magioladitis, Kopovoi, Vssun, Hoverfish, Khalid Mahmood, Pan Dan, Robin S, Vortimer, Sujaybhu, Natsirtguy, Peter Chastain, Cpiral, Tygrrr, Treisijs, Alpvax, VolkovBot, JohnBlackburne, TXiKiBoT, Marskuzz, Muro de Aguas, Qxz, Sintaku, SieBot, Gerakibot, Escape Artist Swyer, Proton666, ObfuscatePenguin, ClueBot, GorillaWarfare, Jackey0105, DragonBot, Jefflayman, Nownownow, Cenarium, Razorflame, Zahnrad, Silvercromagnon, InternetMeme, Rreagan007, WikHead, Drogs630, Addbot, Guoguo12, Omega Squad, ThisIsMyWikipediaName, Seratna, CarsracBot, Purple Emu, CosmiCarl, AgadaUrbanit, Tiderolls, Lightbot, Luckas-bot, Timeroot, Donthedev, Rifter0x0000, Umnum, AnomieBOT, VanishedUser sdu9aya9fasdsopa, Orange Knightof Passion, Piano non troppo, Citation bot, Obersachsebot, Xqbot, DSisyphBot, Barelistido, Almabot, GrouchoBot, RibotBOT, SassoBot, Mnnngb, CES1596, Gummer85, Citation bot1, Boulaur, RedBot, Jauhienij, Surf5270, ElPeste, Slon02, EmausBot, John of Reading, Mnky-man, JSquish, Cogiati, Bamyers99, Rexprimoris, Donner60, ChuispastonBot, ClueBot NG, Jj1236, Helpful Pixie Bot, Bolatbek, ElphiBot, J.wong.wiki, Glevum, Zedtwitz, Zedshort, Nishantkumar19, Kisokj, YFdyh-bot, Andyhowlett, Reatlas, CsDix, EvergreenFir, Aurelianjh, Jwratner1, Diggerh, Kshitizarora2993, Tetra quark, KasparBot and Anonymous: 171

- **Free particle** *Source:* https://en.wikipedia.org/wiki/Free_particle?oldid=686137688 *Contributors:* Sverdrup, Lupin, MathKnight, Fastfission, Karol Langner, PAR, Linas, B-Con, SmackBot, Chris the speller, Kcordina, John, ShelfSkewed, Bmk, Cydebot, Thijs!bot, Lapsiman, JAnDbot, Lseixas, PixelBot, Lisalima, Addbot, Amirobot, Yngvadottir, Erik9bot, Rausch, Grondilu, ZéroBot, Timetraveler3.14, ReecyBoy42, Helpful Pixie Bot, F=q(E+v^B), X1320x, Len loker and Anonymous: 16

- **Quark** *Source:* https://en.wikipedia.org/wiki/Quark?oldid=684869775 *Contributors:* AxelBoldt, Derek Ross, Vicki Rosenzweig, Mav, Bryan Derksen, The Anome, Gareth Owen, Andre Engels, PierreAbbat, Peterlin~enwiki, Ben-Zin~enwiki, Zoe, Heron, Montrealais, Hfastedge, Edward, Dante Alighieri, Ixfd64, CesarB, Card~enwiki, NuclearWinner, Looxix~enwiki, Ahoerstemeier, Elliot100, Docu, J-Wiki, Nanobug, Aarchiba, Julesd, Glenn, Schneelocke, Jengod, A5, Timwi, Dysprosia, DJ Clayworth, Phys, Ed g2s, Bevo, Olathe, MD87, Jni, Phil Boswell, Sjorford, Donarreiskoffer, Robbot, Sanders muc, Moncrief, Merovingian, PxT, Texture, Bkell, UtherSRG, Widsith, Ancheta Wis, Giftlite, ShaunMacPherson, Harp, Nunh-huh, Lupin, Herbee, Leflyman, Monedula, 0x6D667061, Xerxes314, Anville, Hoho~enwiki, Alison, Beardo, Moogle10000, Wronkiew, Jackol, Bobblewik, Bodhitha, Piotrus, Kaldari, Elroch, Icairns, Zfr, TonyW, Ukexpat, BrianWilloughby, Grunt, O'Dea, Jiy, Discospinster, Rich Farmbrough, Guanabot, T Long, Vsmith, Saintswithin, SocratesJedi, Mani1, Bender235, Lancer, RJHall, Mr. Billion, El C, Kwamikagami, Laurascudder, Susvolans, Triona, Axezz, Bobo192, Army1987, C S, Ziggurat, Rangelov, Matt McIrvin, Jojit fb, Nk, Pentalis, Obradovic Goran, Fwb22, Lysdexia, Benjonson, Alansohn, Gary, Gintautasm, Guy Harris, Keenan Pepper, MonkeyFoo, Lectonar, Mac Davis, Wdfarmer, Snowolf, Schapel, Knowledge Seeker, Evil Monkey, VivaEmilyDavies, CloudNine, Kusma, Kazvorpal, Kay Dekker, Crosbiesmith, Mogigoma, Linas, Mindmatrix, JarlaxleArtemis, ScottDavis, LOL, Wdyoung, Before My Ken, Tylerni7, Jwanders, Dataphiliac, AndriyK, Noetica, Wayward, Wisq, Palica, Marudubshinki, Calréfa Wéná, GSlicer, Graham87, Deltabeignet, Kbdank71, Yurik, Crzrussian, Rjwilmsi, Bremen, Marasama, SpNeo, Mike Peel, Bubba73, DoubleBlue, Matt Deres, Yamamoto Ichiro, Algebra, Dsnow75, RobertG, Nihiltres, Jeff02, RexNL, TeaDrinker, Chobot, DVdm, Jpacold, Gwernol, Elfguy, Roboto de Ajvol, YurikBot, Wavelength, Bambaiah, Sceptre, Hairy Dude, Jimp, Phantomsteve, TheDoober, Dobromila, JabberWok, CambridgeBayWeather, Chaos, Salsb, Wimt, Ugur Basak, NawlinWiki, Spike Wilbury, Bossrat, SCZenz, Randolf Richardson, Danlaycock, Tony1, DRosenbach, Robertbyrne, Dna-webmaster, WAS 4.250, Closedmouth, Pietdesomere, Heathhunnicutt, Kevin, Banus, RG2, Kamickalo, That Guy, From That Show!, Veinor, MacsBug, Smack-Bot, Aigarius, BBandHB, Incnis Mrsi, InverseHypercube, C.Fred, Bazza 7, Ikip, Anastrophe, Jrockley, Eskimbot, AnOddName, Jonathan Karlsson, Edgar181, Gilliam, Dauto, NickGarvey, Vvarkey, Bluebot, KaragouniS, Keegan, Dahn, Bigfun, Miquonranger03, OrangeDog, Silly rabbit, Metacomet, Tripledot, Nbarth, DHN-bot~enwiki, Sbharris, Colonies Chris, Hallenrm, Scwlong, Gsp8181, Can't sleep, clown will eat me, Mallorn, Jeff DLB, TKD, Addshore, Mqjjb30e, Cybercobra, Khukri, B jonas, Jdlambert, Lpgeffen, Nrcprm2026, Akriasas, Zadignose, Jóna Þórunn, Bdushaw, Beyazid, TriTertButoxy, SashatoBot, SciBrad, Doug Bell, Soap, Richard L. Peterson, John, Mgiganteus1, SpyMagician, Edconrad, Loadmaster, 2T, Waggers, SandyGeorgia, Ravi12346, Dbzfrk15146, Peyre, Newone, GDallimore, Happy-melon, Majora4, Chovain, Tawkerbot2, Cryptic C62, JForget, Vaughan Pratt, Hello789, ZICO, SUPRATIM DEY, Ruslik0, CuriousEric, Paulfriedman7, Logical2u, Myasuda, RoddyYoung, Typewritten, Cydebot, Abeg92, Mike Christie, Grahamec, Gogo Dodo, Jayen466, 879(CoDe), Michael C Price, Tawkerbot4, Ameliorate!, Akcarver, Gimmetrow, SallyScot, Casliber, Thijs!bot, Epbr123, NeoPhyteRep, LeBofSportif, Markus Pössel, Anupam, Sopranosmob781, Headbomb, Marek69, John254, KJBurns, MichaelMaggs, Escarbot, Eleuther, Ice Ardor, Aadal, AntiVandalBot, SmokeyTheCat, Tyco.skinner, Exteray, RobJ1981, Rsocol, Ke garne, Deflective, Husond, MER-C, CosineKitty, Andonic, East718, Pkoppenb, DanPMK, Magioladitis, WolfmanSF, Thasaidon, Bongwarrior, VoABot II, باسم, Inertiatic076, Kevinmon, Christoph Scholz~enwiki, Aka042, Giggy, Tanvirzaman, Johnbibby, Cyktsui, ArchStanton69, Ace42, Allstarecho, Shijualex, DerHexer, Elandra, Denis tarasov, MartinBot, Poeloq, Dorvaq, CommonsDelinker, HEL, J.delanoy, Nev1, Ops101ex, DrKay, Hgpot, Ferdyshenko, Jigesh, DJ1AM, Tarotcards, Coppertwig, TomasBat, Nikbuz, SJP, FJPB, Vainamainien, Tiggydong, Robprain, Sheliak, Cuzkatzimhut, Lights, X!, VolkovBot, Off-shell, CWii, ABF, John Darrow, Holme053, Nousernamesleft, Ryan032, GimmeBot, Davehi1, A4bot, Captain Courageous, Guillaume2303, Anonymous Dissident, Drestros power, Qxz, Anna Lincoln, Eldaran~enwiki, Leafyplant, Don4of4, PaulTanenbaum, Abdullais4u, Jbryancoop, Mbalelo, Gilisa, Eubulides, Chronitis, Seresin, Dustybunny, Insanity Incarnate, Upquark, Edge1212, Ollieho, AOEU Warrior, SieBot, Graham Beards, WereSpielChequers, Csmart287, Guguma5, Winchelsea, Jbmurray, Caltas, Vanished User 8a9b4725f8376, Keilana, Bentogoa, Aillema, RadicalOne, Arbor to SJ, Elcobbola, Physics one, Dhatfield, RSStockdale, Son of the right hand, Ngexpert5, Ngexpert6, Ngexpert7, Psycherevolt, Sean.hoyland, Mygerardromance, Dabomb87, Nergaal, Muhends, Romit3, SallyForth123, Atif.t2, ClueBot, The Thing That Should Not Be, Wwheaton, Xeno malleus, Harland1, Piledhigheranddeeper, Maxtitan, DragonBot, Glopso, Choonkiat.lee, Himynameisdumb, Worth my salt, Arthur Quark, Estirabot, Brews ohare, Jotterbot, PhySusie, Brianboulton, Dekisugi, ANOMALY-117, Sallicio, Yomangan, Jtle515, Katanada, DumZiBoT, TimothyRias, XLinkBot, Vayalir, Oldnoah, Saintlucifer2008, Nathanwesley3, Dragonfiremage, Devilist666, Mancune2001, Jbeans, Wiki-Dao, SkyLined, Truthnlove, Airplaneman, Eklipse, Addbot, Eric Drexler, AVand, Some jerk on the Internet, Captain-tucker, Giants2008, Iceblock, Ronhjones, Quarksci, Mseanbrown, Looie496, LaaknorBot, Peti610botH, AgadaUrbanit, Tide rolls, Vicki breazeale, Gail, Extruder~enwiki, Abduallah mohammed, Dealer77, Luckas-bot, Yobot, Fraggle81, Cflm001, Legobot II, Amble, Mmxx, Superpenguin1984, Worm That Turned, The Vector Kid, Planlips, Fangfyre, TestEditBot, Azcolvin429, Vroo, Synchronism, Bility, Orion11M87, AnomieBOT, Xi rho, Rubinbot, Jim1138, Bookaneer, Yotcmdr, Crystal whacker, Sonic h, Materialscientist, Citation bot, Pitke, Vuerqex, Bci2, ArthurBot, LilHelpa, Xqbot, Jeffrey Mall, AbigailAbernathy, Srich32977, Alex2510, Almabot, Uscbino, Pmlineditor, RibotBOT, Shmomuffin, Gunjan verma81, Chotarocket, Ernsts, Renverse, A. di M., Weekendpartier, FrescoBot, Paine Ellsworth, DelphinidaeZeta, Steve Quinn, Citation bot 1, AstaBOTh15, Pinethicket, Jonesey95, Calmer Waters, Skyerise, Pmokeefe, Jschnur, Searsshoesales, Jrobbinz123, Lissajous, Turian, Lando Calrissian, Wotnow, Ansumang, Reaper Eternal, 564dude, Jackvancs, Bobotast, MINTOPOINT, TjBot, DexDor, Антон Гліністы, Daggersteel10, Chiechiecheist, EmausBot, John of Reading, WikitanvirBot, Duskbrood, FergalG, Slightsmile, Barak90, Wikipelli, TheLemon1234, Manofgrass, Brazmyth, H3llBot, Stoneymufc29, GeorgeBarnick, Brandmeister, Ego White Tray, RockMagnetist, TYelliot, ClueBot NG, Gilderien, A520, Cheeseequalsyum, Timothy jordan, 123Hedgehog456, Maplelanefarm, 336, Helpful Pixie Bot, Jeffreyts11, 123456789malm, Bibcode Bot, BG19bot, Hurricanefan25, MusikAnimal, Davidiad, MosquitoBird11, Mydogpwnsall, MrBill3, Njavallil, Glacialfox, Walterpfeifer, Thebannana, CE9958, Marioedesouza, Mediran, Dexbot, Rishab021, TwoTwoHello, Cjean42, Sriharsh1234, Sam boron100, Wankybanky, Wikitroll12345, RojoEsLardo, Jwratner1, NottNott, Saebre, JNrgbKLM, KheltonHeadley, AspaasBekkelund, HectorCabreraJr, Hazinho93, Quadrupedi, QuantumMatt101, Philipphilip0001, Monkbot, RiderDB, Egfraley, Tetra quark, Weed305, KasparBot and Anonymous: 705

- **Quark model** *Source:* https://en.wikipedia.org/wiki/Quark_model?oldid=685637327 *Contributors:* Rmhermen, Stevertigo, AugPi, Chrisjj, Donarreiskoffer, Rorro, Xerxes314, Doshell, Rich Farmbrough, Masudr, David Schaich, Laurascudder, Andrew Gray, Feezo, Linas, Wdyoung, Tabletop, Kbdank71, Marasama, Chekaz, Nigosh, Srleffler, Commander Nemet, Elfguy, Bambaiah, Hairy Dude, Jmauro2000, Salsb, Zwobot, Ilmari Karonen, Finell, SmackBot, C.Fred, Kevin Ryde, RProgrammer, Voyajer, VMS Mosaic, Radagast83, Eassin, Drinibot, ShelfSkewed, Michael C Price, Headbomb, OrenBochman, AntiVandalBot, Magioladitis, Ludvikus, Tarotcards, Coppertwig, Cuzkatzimhut, LokiClock, The Stickler, Muhends, ClueBot, Rotational, Glopso, PixelBot, SchreiberBike, Addbot, Luckas-bot, Yobot, Citation bot, ArthurBot, Zhividya, TechBot, Omnipaedista, Cekli829, Kwiki, Citation bot 1, DEm, WikitanvirBot, Hesana, Jjspinorfield1, Suslindisambiguator, AManWithNo-Plan, Brandmeister, ClueBot NG, Bibcode Bot, WikiDenvah, Dexbot, Tony Mach, Arlene47, Truocled, Alakzi and Anonymous: 35

- **Electric charge** *Source:* https://en.wikipedia.org/wiki/Electric_charge?oldid=685481752 *Contributors:* AxelBoldt, Mav, Andre Engels, Roadrunner, Peterlin~enwiki, Heron, JohnOwens, Michael Hardy, Ixfd64, Delirium, Looxix~enwiki, Ellywa, Mdebets, Glenn, Rossami, Nikai, Andres, Raven in Orbit, Reddi, Omegatron, Gakrivas, Lumos3, Rogper~enwiki, Gentgeen, Robbot, Fredrik, Dukeofomnium, Wikibot, Fuelbottle, Wjbeaty, Giftlite, DavidCary, Herbee, Snowdog, Dratman, Valen~enwiki, RScheiber, Jason Quinn, Brockert, OldakQuill, Manuel Anastácio, LiDaobing, Karol Langner, Icairns, Iantresman, GNU, Vincom2, Discospinster, Guanabot, Jpk, Dbachmann, ZeroOne, Laurascudder, Bobo192, Rbj, Giraffedata, Kjkolb, Scentoni, Mdd, Alansohn, Atlant, ABCD, Velella, Wtshymanski, HenkvD, Mikeo, DV8 2XL, Gene Nygaard, HenryLi, Oleg Alexandrov, Nuno Tavares, Cimex, Rocastelo, StradivariusTV, Oliphaunt, BillC, Eleassar777, Cyberman, Palica, BD2412, Demonuk, Edison, SMC, Krash, Dougluce, FlaBot, Psyphen, Nivix, Alfred Centauri, Gurch, Kri, Gdrbot, Manscher, YurikBot, Bambaiah, Lucinos~enwiki, Stephenb, Manop, Pseudomonas, JDoorjam, TDogg310, Chichui, Kkmurray, Wknight94, Light current, Enormousdude, Johndburger, Tcsetattr, Pinikas, Reyk, Canley, Geoffrey.landis, JDspeeder1, GrinBot~enwiki, Mejor Los Indios, Sbyrnes321, Marquez~enwiki, Moeron, Vald, Thunderboltz, Dmitry sychov, HalfShadow, Gilliam, Oscarthecat, Andy M. Wang, Chris the speller, Lenko, DHN-bot~enwiki, Dual Freq, Hallenrm, Rrburke, The tooth, MichaelBillington, Hgilbert, Drphilharmonic, Daniel.Cardenas, Springnuts, Yevgeny Kats, Andrei Stroe, DJIndica, Naui~enwiki, Nmnogueira, SashatoBot, Richard L. Peterson, Slowmover, Cronholm144, Mgiganteus1, Bjankuloski06en~enwiki, Nonsuch, Ben Moore, RandomCritic, MarkSutton, Stikonas, Dicklyon, Levineps, Igoldste, Tawkerbot2, Chetvorno, JForget, CmdrObot, Kehrli, Jsd, Myasuda, Cydebot, Fl, Bvcrist, Meno25, Gogo Dodo, WISo, Christian75, Ssilvers, Thijs!bot, Epbr123, Barticus88, N5iln, Mojo Hand, Headbomb, Gerry Ashton, Escarbot, Aadal, AntiVandalBot, Seaphoto, Prolog, DarkAudit, Lyricmac, Tim Shuba, WikifingHelper, Asgrrr, JAnDbot, Acroterion, Bongwarrior, VoABot II, J2thawiki, Sstolper, Jjurik, Bubba hotep, User A1, DerHexer, InvertRect, Robin S, MartinBot, M. Bilal Shafiq, LedgendGamer, Pharaoh of the Wizards, Numbo3, Hans Dunkelberg, NightFalcon90909, Uncle Dick, Ginsengbomb, Katalaveno, DarkFalls, NewEnglandYankee, QuickClown, Juliancolton, ACBest, Treisijs, Lseixas, Jefferson Anderson, Sheliak, Philip Trueman, TXiKiBoT, The Original Wildbear, Ayan2289, Nickipedia 008, LuizBalloti, Monty845, Jpalpant, Biscuittin, Demmy100, SieBot, Gerakibot, Caltas, Gastin, Wing gundam, Msadaghd, JerrySteal, Jojalozzo, Oxymoron83, Faradayplank, Avnjay, Anchor Link Bot, Neo., Loren.wilton, ClueBot, The Thing That Should Not Be, Arakunem, Termine, Mild Bill Hiccup, Stephaninator, LeoFrank, Excirial, Kocher2006, Jusdafax, Brews ohare, Cenarium, Jotterbot, PhySusie, SchreiberBike, Wuzur, JDPhD, Versus22, Thinking Stone, Rror, Cernms, Truthnlove, Addbot, Some jerk on the Internet, CanadianLinuxUser, NjardarBot, LaaknorBot, Scottyferguson, LinkFA-Bot, Naidevinci, Ocwaldron, Tide rolls, Lightbot, JDSperling, Legobot, Luckas-bot, Yobot, CinchBug, Duping Man, AnomieBOT, DemocraticLuntz, Sertion, Jim1138, IRP, Pyrrhus16, Kingpin13, Bluerasberry, Materialscientist, Geek1337~enwiki, ImperatorExercitus,

Xqbot, TheAMmollusc, Phazvmk, Addihockey10, Capricorn42, Nnivi, ProtectionTaggingBot, RibotBOT, Srr712, A. di M., Constructive editor, Frozenevolution, Ryryrules100, Jc3s5h, Drunauthorized, Mithrandir, Steve Quinn, Davidteng, Fast kartwheels, BenzolBot, DivineAlpha, AstaBOTh15, Pinethicket, I dream of horses, Jivee Blau, Calmer Waters, Tinton5, MastiBot, Serols, Meaghan, Lalrang2007, Logical Gentleman, FoxBot, TobeBot, SchreyP, Jonkerz, Ndkartik, Vrenator, Taytaylisious09, Ammodramus, Jamietw, DARTH SIDIOUS 2, Eshmate, Irfanyousufdar, EmausBot, John of Reading, GoingBatty, K6ka, Darkfight, Hhhippo, JSquish, Harddk, Stephen C Wells, Liam McM, Sonygal, L Kensington, Donner60, Peter Karlsen, Sven Manguard, Planetscared, ClueBot NG, Jack Greenmaven, Cking1414, Ihwood, Ulflund, CocuBot, MelbourneStar, O.Koslowski, Brickmack, AvocatoBot, Ushakaron, Rm1271, Altaïr, F=q(E+v^B), Snow Blizzard, Brad7777, Bhaskarandpm, Eduardofeld, GoShow, Dexbot, JoshyyP, Brandonsmacgregor, Reeceyboii, Frosty, Reatlas, I am One of Many, Eyesnore, Tentinator, Germeten, Nablacdy, Spyglasses, Freddyboi69, 20M030810, SpecialPiggy, Marizperoj, Peterfreed, Rigid hexagon, Jiteshkumar727464, Dyeith, Podayeruma, Oleaster, Layfi, BlueDecker, GeneralizationsAreBad, Pritam kumar Barik, KasparBot, Ramprakashsfc and Anonymous: 428

- **Mass** *Source:* https://en.wikipedia.org/wiki/Mass?oldid=686749080 *Contributors:* CYD, Mav, Zundark, The Anome, Tarquin, Drj, Claudine-Chionh, Andre Engels, Ben-Zin~enwiki, Heron, Isis~enwiki, Stevertigo, Ubiquity, Patrick, Michael Hardy, Tim Starling, Erik Zachte, Ixfd64, TakuyaMurata, Delirium, Looxix~enwiki, Ihcoyc, Ahoerstemeier, Cyp, Stevenj, Snoyes, JWSchmidt, Julesd, Glenn, RadRafe, Mxn, Denny, RodC, Charles Matthews, RickK, Reddi, Dysprosia, Kbk, DJ Clayworth, SEWilco, Fibonacci, Omegatron, Head, Raul654, Pakaran, Donarreiskoffer, Gentgeen, Robbot, Jredmond, Mayooranathan, Merovingian, Lsy098~enwiki, Texture, Sunray, Bkell, Papadopc, Aomarks, Giftlite, DocWatson42, Harp, Lethe, Herbee, Fropuff, Wwoods, Wouterhagens, Bensaccount, Pharotic, Jrdioko, Pavelfilo, Gadfium, Pcarbonn, Antandrus, Beland, Mako098765, Jossi, Rdsmith4, Maximaximax, Icairns, Zfr, Sam Hocevar, Asbestos, Lindberg G Williams Jr, Joyous!, Klemen Kocjancic, Trevor MacInnis, Mike Rosoft, Brianjd, D6, JimJast, Discospinster, Rich Farmbrough, FT2, Hidaspal, Pjacobi, Vsmith, Arthur Holland, Dbachmann, Aardark, Bender235, ESkog, Neko-chan, RJHall, Lycurgus, Shanes, RoyBoy, Spoon!, Dannown, Adambro, Bobo192, Nigelj, Longhair, Redlentil, I9Q79oL78KiL0QTFHgyc, Juzeris, Nicop (Usurp), Bert Hickman, Nk, Deryck Chan, Timsheridan, MPerel, Haham hanuka, Nsaa, Alansohn, Gary, Arthena, Keenan Pepper, Calton, Axl, Lee S. Svoboda, Snowolf, Wtmitchell, Max Naylor, Count Iblis, RainbowOfLight, Rhialto, Bsadowski1, Gene Nygaard, MIT Trekkie, Iustinus, Woohookitty, GrouchyDan, Camw, Ukulele~enwiki, StradivariusTV, Benhocking, Kurzon, Sengkang, Maidden, Macaddct1984, Shanedidona, Wgmleslie, Allen3, Dysepsion, Mandarax, Ashmoo, Rpeblack, DePiep, Josh Parris, Sjö, Sjakkalle, Rjwilmsi, Саша Стефановиђ, Lordkinbote, SMC, SeanMack, FlaBot, Anskas, Mathbot, Nihiltres, RexNL, Fresheneesz, Carrionluggage, Srleffler, Physchim62, King of Hearts, Chobot, Antilived, DVdm, Athello9, Sanpaz, Bgwhite, PointedEars, Roboto de Ajvol, The Rambling Man, Zaidpjd~enwiki, RobotE, Sceptre, Phmer, Bhny, Chaser, NawlinWiki, Wiktionary4Prez!, Borbrav, Długosz, Irishguy, Dhollm, Moe Epsilon, Tony1, Syrthiss, Dbfirs, DeadEyeArrow, Derek.cashman, Wknight94, WAS 4.250, FF2010, Fibula, Light current, Gossja, Enormousdude, Lt-wiki-bot, E Wing, KGasso, Tetigit, Cassini83, Kevin, Willtron, Staxringold, MagneticFlux, Cmglee, DVD R W, Finell, CIreland, Tom Morris, ChemGardener, Itub, SmackBot, MegamanXplosion, David Kernow, Slashme, InverseHypercube, KnowledgeOfSelf, Hydrogen Iodide, Kenny56, Piroteknix, Bomac, Perico~enwiki, ZerodEgo, Srnec, Xaosflux, Yamaguchi⬚⬚, Gilliam, Ohnoitsjamie, Hmains, Rajeevmass~enwiki, MK8, Miquonranger03, MalafayaBot, George Rodney Maruri Game, Silly rabbit, Complexica, Nervestaple, Jfsamper, DHN-bot~enwiki, Sbharris, Darth Panda, Conorchurch, Shalom Yechiel, Grover cleveland, Cybercobra, Nakon, IrisKawling, LoveMonkey, Jbergquist, DMacks, Henning Makholm, Bejnar, Yevgeny Kats, Ace ETP, Wvbailey, Xaminmo, Vanished user 9i39j3, Soap, Kuru, Scientizzle, Gobonobo, Shadowlynk, Accurizer, Mgiganteus1, Scetoaus, Joshua Scott, PseudoSudo, Ckatz, A. Parrot, Slakr, Dicklyon, Ryulong, KJS77, Emx~enwiki, Iridescent, Spinnick597, Lottamiata, Igoldste, Blehfu, Civil Engineer III, Courcelles, Tawkerbot2, JForget, Hilmarz, CRGreathouse, Ale jrb, The Font, Runningonbrains, Dgw, MarsRover, Logical2u, Colostomyexplosion, Cydebot, Danrok, Bvcrist, Vanished user vjhsduheuiui4t5hjri, Michaelas10, Gogo Dodo, JFreeman, Anonymous 198736, Chasingsol, Edgerck, Shirulashem, Christian75, Satori Son, Thijs!bot, Epbr123, GigaAndy, Thedarxide, O, Sagaciousuk, N5iln, Andyjsmith, Bethan 182, Mojo Hand, Headbomb, Marek69, NorwegianBlue, I do not exist, X201, Guptasuneet, Ryanl91, Thljcl, Greg L, MichaelMaggs, Futurebird, Escarbot, Dzubint, Mentifisto, AntiVandalBot, Luna Santin, Opelio, EarthPerson, Lyricmac, Jj137, Yanickborg, Naveen Sankar, IrishFlukey, MikeLynch, Husond, MER-C, PhilKnight, Kirrages, Rothorpe, Jameskeates, Kborland, Pervect, Bongwarrior, VoABot II, Meredyth, Mgmirkin, Fusionmix, Wikidudeman, Hasek is the best, Brusegadi, Animum, Allstarecho, Vssun, Esanchez7587, Hbent, Hdt83, MartinBot, Intesvensk, Mårten Berglund, Armanalp, Autocratique, CommonsDelinker, AlexiusHoratius, Um3k, Tgeairn, JCarlos, J.delanoy, Trusilver, Maurice Carbonaro, Tdadamemd, Acalamari, Katalaveno, Enuja, Ncmvocalist, McSly, Brickc1, Snowfisher, Bailo26, Tarotcards, WaiteDavid137, WHeimbigner, TomasBat, Fylwind, Joshua Issac, Juliancolton, Azrak, Bonadea, JavierMC, Useight, Lseixas, Idioma-bot, Meiskam, VolkovBot, Zeoblast, Enderminh, Lear's Fool, Fences and windows, Al.locke, Semidimes, Philip Trueman, TXiKiBoT, Oshwah, GroveGuy, Red Act, HarryAlffa, Ask123, XRaine, Clarince63, Corvus cornix, Martin451, JakeTM, LeaveSleaves, Natg 19, Mr.Kennedy1, Meters, Synthebot, Falcon8765, Burntsauce, Nwh5305, Rdengler, Brianga, HeirloomGardener, AlleborgoBot, Symane, The Mad Genius, NHRHS2010, EmxBot, SieBot, StAnselm, Timb66, Ttony21, Nubiatech, Tresiden, YonaBot, Tiddly Tom, Moonriddengirl, Scarian, Winchelsea, Rockstone35, Dawn Bard, Lexiphyxia, Yintan, LeadSongDog, Xxdark vampxx, Keilana, Quest for Truth, Flyer22, Alexfusco5, The Evil Spartan, SweetCarmen, JSpung, Oxymoron83, Antonio Lopez, Faradayplank, Hello71, Lightmouse, BenoniBot~enwiki, OKBot, Stfg, Escape Orbit, Kanonkas, Quinling, Krishnashiva, Atif.t2, Deavenger, Martarius, De728631, ClueBot, Andrew Nutter, Avenged Eightfold, Snigbrook, The Thing That Should Not Be, Ariadacapo, IceUnshattered, RYNORT, Mild Bill Hiccup, Ianchristie, Unitfreak, S234432, Komet lover 39, EggMan247, Neverquick, Puchiko, Gakusha, DragonBot, Naerii, Alexbot, Amanman78, Smallie11, Themaab, Lartoven, Humanino, Brews ohare, Mechidiot, Tyler, Pot, Cenarium, Jotterbot, PhySusie, Vboo-belarus, Hans Adler, Razorflame, Noosentaal, Dekisugi, La Pianista, Thingg, Aitias, Daxxisback, Seviolia, Versus22, T71024, Johnuniq, SoxBot III, Homocion, Forbes72, Fastily, Dthomsen8, Nepenthes, Facts707, SilvonenBot, ZooFari, Truthnlove, Addbot, Some jerk on the Internet, Jojhutton, Tcncv, Fyrael, Bagzawi, Njaelkies Lea, Fieldday-sunday, CanadianLinuxUser, Leszek Jańczuk, Fluffernutter, Cst17, Download, SoSaysChappy, Ld100, SpBot, LinkFA-Bot, Numbo3-bot, Robert The Rebuilder, Tide rolls, Lightbot, Ivancurtisivancurtis, Superboy112233, Quantumobserver, Ben Ben, Legobot, Luckas-bot, Yobot, Tohd8BohaithuGh1, Fraggle81, II MusLiM HyBRiD II, Aldebaran66, Gobbleswoggler, QueenCake, Duping Man, SwisterTwister, AnakngAraw, AnomieBOT, DemocraticLuntz, Kristen Eriksen, Piano non troppo, NickK, Materialscientist, Cgrrp, ImperatorExercitus, Citation bot, OllieFury, StephenPCook, Bob Burkhardt, GB fan, Neurolysis, Ariessantua, Rightly, Badboy321, Xqbot, Wavgfkl, Cureden, JimVC3, Capricorn42, TechBot, Nasnema, Mononomic, The Evil IP address, Aa77zz, NOrbeck, Programming gecko, Omnipaedista, RibotBOT, Doulos Christos, Twin Hills student, Paul.rogers1, Bubbles1996, Shadowjams, Darkest tree, PM800, A. di M., Dougofborg, Nagualdesign, FrescoBot, Cyrus3195, Rufusleo, Wikipe-tan, Defribrillation Nation, Sky Attacker, PhysicsExplorer, Galorr, Dger, Jmrkool, Steve Quinn, PeterEastern, DivineAlpha, HamburgerRadio, Citation bot 1, Amplitude101, Busukxuan, Pinethicket, I dream of horses, Elockid, Rameshngbot, Tom.Reding, Tinton5, Share and Enjoy, RedBot, Chezz444, Île flottante, Σ, Rotalumis, Andieda, ActivExpression, IVAN3MAN, Lemmiwinks2, Tim1357, FoxBot, TobeBot, Trappist the monk, Knox59, Gallusgallus, Isaac909, 777player, Euandrew, JV Smithy, Stroppolo, Minimac, Stringence, Mean as custard,

Luckas-bot, Yobot, Citation bot, ArthurBot, Xqbot, DSisyphBot, Paine Ellsworth, Citation bot 1, Tim1357, Trappist the monk, EmausBot, ZéroBot, Quondum, Rezabot, Helpful Pixie Bot, Bibcode Bot, TheMan4000, 786b6364, Monkbot and Anonymous: 22

- **Strange quark***Source:* https://en.wikipedia.org/wiki/Strange_quark?oldid=663285537*Contributors:* Bryan Derksen, Alfio, Jni, Owain, Xerxes 314,Soman, Kjoonlee, Kwamikagami, Rsholmes, Esb82, Neonumbers, Rjwilmsi, Mike Peel, Gurch, Erik4, Chobot, YurikBot, Jimp, Salsb, SCZenz,Poulpy, SmackBot, Bluebot, NCurse, Vina-iwbot~enwiki, Yevgeny Kats, Zzzzzzzzzzz, Laplace's Demon, MightyWarrior, Myasuda, Thijs!bot,Headbomb, Chillysnow, JAnDbot, Abyssoft, Bongwarrior, Albmont, McSly, I310342~enwiki, Pdcook, Sheliak, VolkovBot, SieBot, Muhends,Auntof6, Iohannes Animosus, TimothyRias, IngerAlHaosului, Addbot, ProbablyAmbiguous, Luckas-bot, Yobot, AnomieBOT, Citation bot,Sarah12sarah, Erik9bot, Thehelpfulbot, Paine Ellsworth, Rkr1991, Citation bot 1, Skyerise, Johann137, Trappist the monk, Puzl bustr, Agrasa,Wikiborg4711, EmausBot, Hhhippo, ZéroBot, Quondum, CocuBot, Helpful Pixie Bot, Bibcode Bot, Vkpd11, P76837, Matthew gib, Glaisher,RhinoMind and Anonymous: 38

- **Charm quark** *Source:* https://en.wikipedia.org/wiki/Charm_quark?oldid=663286006 *Contributors:* Bryan Derksen, Alfio, Bogdangiusca, Xerxes314, Bodhitha, Perey, Kjoonlee, Rjwilmsi, Mike Peel, Chobot, YurikBot, Bambaiah, Conscious, Salsb, SCZenz, Scottfisher, Poulpy, SmackBot, Delldot, Warhol13, Rezecib, Vina-iwbot~enwiki, Happy-melon, Laplace's Demon, CRGreathouse, Michael C Price, Thijs!bot, Headbomb, Nisselua, JAnDbot, Abyssoft, Uncle.wink, Bryanhiggs, HEL, I310342~enwiki, Qoou.Anonimu, Idioma-bot, Sheliak, Anonymous Dissident, Kumorifox, BeIsKr, AlleborgoBot, SieBot, Muhends, TimothyRias, Addbot, Mjamja, Lightbot, Luckas-bot, Yobot, Nallimbot, Citation bot, ArthurBot, Quebec99, Xqbot, DSisyphBot, GrouchoBot, RibotBOT, SassoBot, A. di M., Paine Ellsworth, Dogposter, D'ohBot, Citation bot 1, Citation bot 4, RedBot, MastiBot, Trappist the monk, EarthCom1000, Alph Bot, EmausBot, ZéroBot, Quondum, Anita5192, CocuBot, Rezabot, Helpful Pixie Bot, Bibcode Bot, Penguinstorm300, Hoppeduppeanut, Leowestland and Anonymous: 35

- **Top quark** *Source:* https://en.wikipedia.org/wiki/Top_quark?oldid=679102575 *Contributors:* Damian Yerrick, Bryan Derksen, HPA, Haryo, Bkell, Giftlite, Xerxes314, Edcolins, Bodhitha, David Schaich, Kjoonlee, Axl, Woohookitty, Rjwilmsi, Strait, Mike Peel, Vegaswikian, Wikiliki, Goudzovski, Chobot, YurikBot, Bambaiah, JabberWok, Gaius Cornelius, Salsb, Howcheng, SCZenz, Emijrp, Physicsdavid, SmackBot, Incnis Mrsi, ZerodEgo, Mr.Z-man, Jgwacker, Pulu, Stikonas, Mets501, Peyre, RekishiEJ, Banedon, הסרפד, Headbomb, Davidhorman, Oreo Priest, AntiVandalBot, JAnDbot, Abyssoft, Maliz, HEL, Fatka, I310342~enwiki, Idioma-bot, Sheliak, Biggus Dictus, TXiKiBoT, Reibot, Kachuak, Ptrslv72, SieBot, Hatster301, Muhends, ClueBot, Niceguyedc, Noca2plus, Choonkiat.lee, Brews ohare, Kakofonous, Jtle515, TimothyRias, Prostarplayer321, SkyLined, Cockatoot, Addbot, Mr0t1633, Mjamja, ChenzwBot, Ginosbot, Zorrobot, Luckas-bot, Naudefjbot~enwiki, Dreamer08, AnomieBOT, Icalanise, Citation bot, ArthurBot, LilHelpa, DSisyphBot, Unready, GrouchoBot, RibotBOT, Soandos, Paine Ellsworth, Citation bot 1, Jonesey95, Thinking of England, Nomis2k, Higgshunter, RjwilmsiBot, Mophoplz, EmausBot, John of Reading, WikitanvirBot, Barak90, StringTheory11, Peter.poier, Quondum, Samlever, Whoop whoop pull up, Reify-tech, Helpful Pixie Bot, Bibcode Bot, Glevum, Kephir, Mmitchell10, Quadrupedi, Monkbot, BrunoUbaldo and Anonymous: 64

- **Bottom quark** *Source:* https://en.wikipedia.org/wiki/Bottom_quark?oldid=676070719 *Contributors:* Bryan Derksen, Xerxes314, Bodhitha, Icairns, Kjoonlee, Bobo192, Pinar, WadeSimMiser, Rjwilmsi, Mike Peel, Erkcan, FlaBot, Itinerant1, Chobot, YurikBot, Bambaiah, Jimp, Conscious, Ozabluda, SpuriousQ, Salsb, SCZenz, Lexicon, Poulpy, Physicsdavid, SmackBot, Hmains, Luís Felipe Braga, Laplace's Demon, CmdrObot, Outriggr (2006-2009), Niubrad, הסרפד, Thijs!bot, Headbomb, JAnDbot, Abyssoft, Pkoppenb, Dr. Morbius, I310342~enwiki, Joshmt, Idioma-bot, Sheliak, VolkovBot, Antixt, AlleborgoBot, BartekChom, Muhends, Auntof6, TimothyRias, Lockalbot, Addbot, Mr Sme, Luckas-bot, THEN WHO WAS PHONE?, Citation bot, ArthurBot, Xqbot, GrouchoBot, StevenVerstoep, Thehelpfulbot, Paine Ellsworth, Citation bot 1, Jonesey95, Double sharp, TjBot, EmausBot, Barak90, TuHan-Bot, ZéroBot, StringTheory11, Quondum, Chris857, ChuispastonBot, Widr, Helpful Pixie Bot, Bibcode Bot, P76837, ChrisGualtieri, Ajd268, Mfb, Monkbot, Axel Azzopardi, Kenijr and Anonymous: 33

- **Baryon** *Source:* https://en.wikipedia.org/wiki/Baryon?oldid=681412960 *Contributors:* AxelBoldt, Tobias Hoevekamp, Bryan Derksen, Ben-Zin~enwiki, Heron, Tim Starling, Alan Peakall, Paul A, Salsa Shark, Glenn, Mxn, Charles Matthews, The Anomebot, ElusiveByte, Phys, Bevo, Traroth, Donarreiskoffer, Robbot, Korath, Kristof vt, Merovingian, Ojigiri~enwiki, Sunray, Wikibot, Giftlite, DocWatson42, Shaun-MacPherson, Herbee, Xerxes314, Dratman, DÅ,ugosz, Kaldari, OwenBlacker, Icairns, JohnArmagh, Rich Farmbrough, Guanabot, Mani1, E2m, Tompw, El C, Bobo192, I9Q79oL78KiL0QTFHgyc, Giraffedata, Physicistjedi, Jumbuck, Gary, ABCD, Oleg Alexandrov, Woohookitty, Tevatron~enwiki, BD2412, Kbdank71, Nightscream, Ae77, MZMcBride, Chekaz, R.e.b., Erkcan, Maxim Razin, Oo64eva, Chobot, Roboto de Ajvol, YurikBot, Bambaiah, Jimp, Salsb, Ergzay, DragonHawk, SCZenz, E2mb0t~enwiki, Bota47, Simen, Sbyrnes321, Lainagier, Timotheus Canens, Bluebot, Colonies Chris, Kingdon, Shadow1, Bigmantonyd, Drphilharmonic, Kseferovic, Wierdw123, Physicsdog, Torrazzo, Verdy p, Michael C Price, Thijs!bot, Headbomb, Hcobb, Orionus, QuiteUnusual, Spartaz, Plantsurfer, Amateria1121, Diamond2, Swpb, BatteryIncluded, Hveziris, Saxophlute, Gwern, Ben MacDui, R'n'B, Ash, Tgeairn, Maurice Carbonaro, STBotD, VolkovBot, GimmeBot, NoiseEHC, Tearmeapart, BotKung, BrianADesmond, Antixt, AlleborgoBot, Lou427, SieBot, VVVBot, Gerakibot, LeadSongDog, Keilana, Paolo.dL, Doctorfluffy, TrufflesTheLamb, OKBot, Hamiltondaniel, TubularWorld, ClueBot, Artichoker, ChandlerMapBot, CalumH93, Addbot, LaaknorBot, CarsracBot, Jonhstone12, Legobot, Luckas-bot, Bugbrain_04, AnomieBOT, JackieBot, Materialscientist, Citation bot, ArthurBot, Xqbot, Omnipaedista, SassoBot, Spellage, WaysToEscape, FrescoBot, Citation bot 1, FoxBot, Noommos, EmausBot, John of Reading, JSquish, ZéroBot, StringTheory11, Stibu, Ethaniel, Markinvancouver, ClueBot NG, Koornti, Kasirbot, Rezabot, Bibcode Bot, Atomician, Zedshort, Marioedesouza, ChrisGualtieri, WorldWideJuan, CoolHandLouis, Monkbot, KasparBot and Anonymous: 106

- **Meson** *Source:* https://en.wikipedia.org/wiki/Meson?oldid=672777660 *Contributors:* AxelBoldt, Bryan Derksen, Josh Grosse, PierreAbbat, Ben-Zin~enwiki, Xavic69, TakuyaMurata, Fwappler, Ahoerstemeier, Ping, Phys, Bcorr, Jeffq, Donarreiskoffer, Robbot, Fredrik, Sanders muc, Merovingian, Rursus, Ojigiri~enwiki, David19999, DocWatson42, Harp, Marcika, Xerxes314, Niteowlneils, Eequor, Physicist, Eroica, Icairns, Sam Hocevar, Lehi, Rich Farmbrough, Pjacobi, Tjic, Robotje, Nicke Lilltroll~enwiki, Pearle, Jumbuck, Jérôme, Bucephalus, Falcorian, Palica, Tevatron~enwiki, Mandarax, Kbdank71, Strait, Titoxd, FlaBot, Jeremygbyrne, Chobot, YurikBot, Wavelength, Bambaiah, Phmer, Jimp, Ozabluda, JabberWok, Salsb, Leutha, Długosz, SCZenz, Ravedave, Gadget850, Antiduh, Tetracube, SmackBot, Melchoir, Eskimbot, Chris the speller, DHN-bot~enwiki, Sbharris, Kevinpurcell, Mesons, DMacks, Jashank, JorisvS, Mgiganteus1, Geologyguy, Ryulong, JarahE, Myasuda, ChrisKennedy, Michael C Price, Thijs!bot, Headbomb, Escarbot, Orionus, Spartaz, Gökhan, Deflective, Magioladitis, Swpb, Khalid Mahmood, Tercer, Kostisl, Hans Dunkelberg, Tarotcards, Xiahou, JeffreyRMiles, VolkovBot, Prizrak, TXiKiBoT, Muro de Aguas, Martin451, LeaveSleaves, Antixt, SieBot, Majeston, Gerakibot, Graf Von Crayola, Humanityisthedisease, Mimihitam, Fratrep, OKBot, ClueBot, Terrorist96, Diagramma Della Verita, Brews ohare, Neville35, RMFan1, WikHead, Stephen Poppitt, Addbot, Gtakanis, Chzz, Debresser, CosmiCarl, AgadaUrbanit, Dickdock, Magog the Ogre, AnomieBOT, StratoWiki, Altruism2010, Citation bot, ArthurBot, Xqbot, Omnipaedista,

WaysToEscape, FrescoBot, Paine Ellsworth, Ironboy11, Steve Quinn, 000ojjo000, Yehoshua2, Citation bot 1, Wdcf, Thinking of England, Puzl bustr, Ale And Quail, Discovery4, Mean as custard, Dkzico007, John of Reading, WikitanvirBot, GoingBatty, Hanretty, ZéroBot, StringTheory11, Markinvancouver, ClueBot NG, Christian.kolen, Wallace Kneeland, Helpful Pixie Bot, Bibcode Bot, Glevum, DerekWinters, Mark viking, Justin567Hicks, Prokaryotes, SJ Defender, Monkbot, KasparBot and Anonymous: 87

- **Exotic meson** *Source:* https://en.wikipedia.org/wiki/Exotic_meson?oldid=671307183 *Contributors:* Xavic69, Michael Hardy, Glenn, Phys, Xerxes314, Gzornenplatz, Physicist, Icairns, D6, Bender235, Gauge, Apyule, Pearle, Fwb22, Kbdank71, Strait, Erkcan, YurikBot, Bambaiah, Salsb, Garion96, Banus, SmackBot, Bluebot, V1adis1av, Ziusudra, ChrisCork, Physic sox, Difluoroethene, Thijs!bot, Headbomb, Fallschirmjäger, VolkovBot, RedAndr, Antixt, Addbot, TaBOT-zerem, FrescoBot, Jennyxie, ClueBot NG, Parcly Taxel, Rezabot, Bibcode Bot and Anonymous: 14

- **Proton** *Source:* https://en.wikipedia.org/wiki/Proton?oldid=686592084 *Contributors:* Mav, Bryan Derksen, Zundark, Manning Bartlett, Ap, Andre Engels, Josh Grosse, Danny, XJaM, Ellmist, Heron, Jaknouse, Stevertigo, Dwmyers, Bdesham, Patrick, Kku, Stewacide, TakuyaMurata, Egil, NuclearWinner, Looxix~enwiki, Mkweise, Ahoerstemeier, Sobekhotep, Salsa Shark, Glenn, Kaihsu, Jordi Burguet Castell, Mxn, Denny, Timwi, Paul-L~enwiki, Omegatron, Geraki, David.Monniaux, Donarreiskoffer, Robbot, Josh Cherry, Jakohn, RedWolf, Altenmann, Yelyos, Merovingian, Flauto Dolce, Hadal, Giftlite, JamesMLane, Mikez, Tom harrison, Lupin, Herbee, Xerxes314, Fleminra, Bensaccount, Foobar, PlatinumX, Utcursch, Knutux, Antandrus, Beland, Eroica, Melikamp, Rdsmith4, DragonflySixtyseven, Icairns, Cglassey, Deglr6328, Grunt, Thorwald, Jenlight, Mike Rosoft, Diagonalfish, Discospinster, Cacycle, Vsmith, Jpk, Wikiacc, Mani1, Bender235, Kjoonlee, RJHall, El C, Shrike, Femto, Bobo192, O18, Army1987, Smalljim, GTubio, Vortexrealm, Elipongo, Foobaz, Kjkolb, Obradovic Goran, Nsaa, Eddideigel, Anthony Appleyard, Mattpickman, Apoc2400, Carmelbuck, Spangineer, Wtmitchell, Saga City, Uucp, Crobzub, Vcelloho, RainbowOfLight, TenOfAllTrades, Computerjoe, Kusma, Itsmine, Falcorian, Richard Arthur Norton (1958-), Firien, GregorB, Macaddct1984, Mayz, Karam.Anthony.K, Marudubshinki, Bebenko, Rtcpenguin, Graham87, Kbdank71, Ketiltrout, Drbogdan, Rjwilmsi, Strait, NeonMerlin, RadicalJester, Bubba73, Yamamoto Ichiro, Rangek, FlaBot, Nivix, RexNL, Fresheneesz, Srleffler, Imnotminkus, King of Hearts, CiaPan, Chobot, Deyyaz, Roboto de Ajvol, The Rambling Man, YurikBot, Bambaiah, JWB, Anuran, Pip2andahalf, RussBot, Wigie, Jumbo Snails, Raquel Baranow, Hellbus, Salsb, Oni Lukos, Anomalocaris, NawlinWiki, Injinera, Welsh, Długosz, Martin Ulfvik, Moe Epsilon, BOT-Superzerocool, DeadEyeArrow, Bota47, D-Day, Mtu, Pooryorick~enwiki, J S Ayer, Theodolite, Bayerischermann, Closedmouth, Reyk, Petri Krohn, DGaw, Paul D. Anderson, Katieh5584, RG2, SDS, GrinBot~enwiki, Nekura, Orii, Luk, Sycthos, Itub, Attilios, SmackBot, Moeron, Incnis Mrsi, Melchoir, CyclePat, Edgar181, HalfShadow, Dhochron, Munky2, Gilliam, Chris the speller, Rajeevmass~enwiki, Rkitko, AndrewBuck, Bethling, SchfiftyThree, Complexica, Sbharris, Rogermw, Can't sleep, clown will eat me, Shalom Yechiel, PeteShanosky, Writtenright, Homestarmy, Wikiwikiwiki3~enwiki, SundarBot, COMPFUNK2, Dreadstar, Orczar, Drphilharmonic, Dvorak729, DMacks, Mion, Vinaiwbot~enwiki, Bdushaw, SashatoBot, ArglebargleIV, Khazar, Cholerashot, Rijkbenik, Spacecadethailey, Herr apa, Deathcakes, Noah Salzman, Aeluwas, Waggers, Mozzura, Mattabat, Elb2000, Newone, MOBle, Igoldste, Rhetth, Frank Lofaro Jr., Tawkerbot2, CmdrObot, Ale jrb, Sir Vicious, KyraVixen, Ruslik0, McVities, TheTito, Cydebot, Nick Y., Gogo Dodo, Red Director, Umdunno, Difluoroethene, Odie5533, Q43, Tawkerbot4, Dwool99f, Narayanese, Rasheedy, Zalgo, Lo2u, Jenswort, Thijs!bot, Epbr123, Tsogo3, Headbomb, Marek69, Electron9, Mnemeson, Dfrg.msc, Philippe, Aadal, AntiVandalBot, Seaphoto, HairyDan, Shirt58, EarthPerson, Gregnx, Jj137, Naturalnumber, Myanw, Ellissound, Leuko, MER-C, CosineKitty, Fetchcomms, Andonic, Kerotan, .anacondabot, Acroterion, Plynn9, Casmith 789, Bongwarrior, VoABot II, Astrangequark, Swpb, WODUP, Recurring dreams, Avicennasis, Catgut, Dirac66, 28421u2232nfenfcenc, Hveziris, Fang 23, The Real Marauder, Oddworth, JaGa, MartinBot, Rettetast, Pbroks13, Artaxiad, J.delanoy, WeglarczykJ, Silverxxx, C.A.T.S. CEO, Maurice Carbonaro, 12dstring, WarthogDemon, Acalamari, Exdejesus, TomasBat, Antony-22, Potatoswatter, KylieTastic, Vanished user 39948282, Treisijs, S, SoCalSuperEagle, Xiahou, Specter01010, Idioma-bot, Ciju, 28bytes, VolkovBot, Doc7777777777, Jeff G., Soliloquial, Tuffcarrot, Philip Trueman, TXiKiBoT, Oshwah, The Original Wildbear, Vipinhari, Bjman, Bigyaks, Alexalexalex123~enwiki, Meters, Antixt, Spinningspark, Insanity Incarnate, Upquark, AlleborgoBot, Vitalikk, B41988, Petergans, Demmy100, SieBot, Accounting4Taste, Jauerback, Studnic12, Xe1881, Yintan, GlassCobra, Keilana, RadicalOne, Flyer22, Sbowers3, Prestonmag, Oxymoron83, BenoniBot~enwiki, Jacob.jose, Mygerardromance, Rajbboy69, ClueBot, The Thing That Should Not Be, RODERICKMOLASAR, Tigerboy1966, Regibox, ChandlerMapBot, Mr blabla, Excirial, Alexbot, Robbie098, Poopmister91191, Ploft, NuclearWarfare, Lunchscale, PhySusie, SoxBot, El pobre Pedro, Thehelpfulone, La Pianista, Thingg, Kanxkawii, Aitias, Subash.chandran007, Johnuniq, XLinkBot, Avoided, WikHead, SilvonenBot, SkyLined, Mls1492, Weletahoozyzog, Addbot, Zrules, Arcturus87, Ronhjones, Cst17, Glane23, AndersBot, Wandering Traveler, Omnipedian, LinkFA-Bot, Numbo3-bot, Tide rolls, Thermalimage, Luckas-bot, Yobot, Велетень, Wickedwizardofoz, Newportm, Kilom691, Heart of a Lion, Eric-Wester, Jay0205, AnomieBOT, LeftyAce, Götz, Jim1138, Sp eloc, Judoc, Materialscientist, The High Fin Sperm Whale, Citation bot, OllieFury, Apollo, Xqbot, Phazvmk, Blennow, Cureden, Wyklety, Aa77zz, Squishywushy123, Srich32977, Rueyfgugdtj, RibotBOT, PM800, A. di M., ꧁꧂, Rain bowell, CES1596, FrescoBot, Paine Ellsworth, Tobby72, Citation bot 1, Pinethicket, HRoestBot, Calmer Waters, Bejinhan, Impala2009, Nicklcms, Blckmgc, SkyMachine, Gryllida, Double sharp, TobeBot, Ilovefatchicks, Keegscee, DARTH SIDIOUS 2, TjBot, StudentDoc73, Nachos0123, EmausBot, WikitanvirBot, ANDREVV, Bencbartlett, Zues zeus kratos, Pcorty, Sterrettc, K6ka, JSquish, Stuffness12, John Cline, Harddk, Fæ, StringTheory11, Brazmyth, Suslindisambiguator, Wayne Slam, Zach444, Rcsprinter123, Wiggles007, Brandmeister, Donner60, Sarthak 94, RockMagnetist, Xonqnopp, ClueBot NG, Timelord360, This lousy T-shirt, IHopeThisNameIsntTaken, Corusant, Cntras, Dictabeard, Rezabot, Helpful Pixie Bot, Wbm1058, Bibcode Bot, BG19bot, ArthropodOfDoom, RadioActiveKitKat, AvocatoBot, Metricopolus, JacobTrue, Mlkamitso, Toccata quarta, Blaspie55, Mhutchison43, 220 of Borg, Anbu121, Hitheresir, RudolfRed, Knodir, BattyBot, ChrisGualtieri, Hower64, Ducknish, Stephen Glass, BrightStarSky, Dexbot, Mogism, Sheehan Cein14, Frosty, Pidotclan, Marcoapc.84, Delnium strex, Faizan, Huddydakota, Prof.Professer, Dustin V. S., Borreswafflertron, Ugog Nizdast, Jwratner1, Javierha, JaconaFrere, AspaasBekkelund, Bballbro62, Melcous, Monkbot, Jayakumar RG, Scorpion1045, Krebs49, Yollowswagger19, Maddie005, Junchuann, Chrisbrownthathoe, Orgasam069, Tktobykerby, Interpuncts, Tetra quark, Cjohnson2020, KasparBot, Muzammil Alam Baig, Rambunctious Racoon, Teo boruch and Anonymous: 642

- **Neutron** *Source:* https://en.wikipedia.org/wiki/Neutron?oldid=682674206 *Contributors:* AxelBoldt, Tobias Hoevekamp, Chenyu, Trelvis, Calypso, Mav, Bryan Derksen, The Anome, AstroNomer~enwiki, Malcolm Farmer, Andre Engels, Xaonon, Danny, XJaM, Roadrunner, Jaknouse, Olivier, Patrick, Michael Hardy, Valery Beaud, Ixfd64, TakuyaMurata, NuclearWinner, Looxix~enwiki, ArnoLagrange, Mkweise, Ellywa, Ahoerstemeier, Cyp, Andrewa, Aarchiba, Julesd, Glenn, Nikai, Andres, Stone, Denni, Kbk, Tarosan~enwiki, Maximus Rex, Donarreiskoffer, Gentgeen, Robbot, Fredrik, Romanm, Merovingian, Rursus, Wikibot, Alan Liefting, Dave6, Giftlite, Mikez, Art Carlson, Herbee, Xerxes314, Everyking, Dratman, NeoJustin, Bensaccount, Poupoune5, Jorge Stolfi, Christofurio, Knutux, Karol Langner, Aecarol, Icairns, Zfr, Cglassey, Peter bertok, Frau Holle, M1ss1ontomars2k4, Sparky2002b, Mike Rosoft, Guanabot, Vsmith, Dbachmann, Bender235, Kjoonlee, AlDragon, Geoking66, Neko-chan, RJHall, CanisRufus, El C, Susvolans, Femto, CDN99, Bobo192, O18, Smalljim, SpeedyGonsales, Kjkolb, Obradovic Goran, Sam Korn, Nsaa, Jakew, Eddideigel, Jumbuck, Patsw, Alansohn, Interiot, Riana, Wtmitchell, BRW, NickMartin, Vuo, DV8 2XL, HenryLi, Tchaika, Forteblast, Falcorian, Richard Arthur Norton (1958-), JarlaxleArtemis, WadeSimMiser, Sega381, SDC, Jon Harald Søby, Prashanthns, Abd, LexCorp, Graham87, Magister Mathematicae, Doughboy, Ketiltrout, Rjwilmsi, Nightscream, Zbxgscqf, Strait, AySz88, Oo64eva, Rangek, FlaBot, Nihiltres, Goudzovski, Srleffler, Ronebofh, King of Hearts, Chobot, DVdm, YurikBot, RobotE, Bambaiah, JWB, TSO1D, Jimp, Phantomsteve, KyleDantarin, Stephenb, Gaius Cornelius, Yyy, Salsb, NawlinWiki, Tupungato, Wiki alf, Complainer, Grafen, Długosz, Voidxor, Scottfisher, Kkmurray, Spute, Dna-webmaster, Wknight94, Stefan Udrea, Mike Serfas, Closedmouth, Reyk, Modify, Alchie1, CWenger, RG2, Paul Erik, Triple333, Attilios, SmackBot, Caiyern, Melchoir, Wiki Tiki God, Unyoyega, Jrockley, Dr.Science, Edgar181, Yamaguchi⬛⬛, Kdliss, Wigren, Chris the speller, Rajeevmass~enwiki, Persian Poet Gal, SchfiftyThree, Complex-ica, DHN-bot~enwiki, Sbharris, Colonies Chris, Brainblaster52, Can't sleep, clown will eat me, DéRahier, Juancnuno, SundarBot, DFriend, Aldaron, KunalKathuria, Nakon, Mwtoews, DMacks, Soarhead77, Bdushaw, Pilotguy, Renafaye77, SashatoBot, Demicx, Tim bates, Mgi-ganteus1, Slakr, Citicat, Asyndeton, BranStark, Shoeofdeath, Newone, Tawkerbot2, Atomobot, Mosaffa, CmdrObot, Wafulz, Dycedarg, Rwflammang, Joelholdsworth, Lokal Profil, Karenjc, Myasuda, Safalra, Icek~enwiki, Badseed, Nick Y., Gogo Dodo, Chasingsol, Phydend, Gimmetrow, Thijs!bot, Epbr123, Montazmeahii, Goods21, Tsogo3, N5iln, Oerjan, Headbomb, Marek69, SouthernMan, RoboServien, Escar-bot, Aadal, WikiSlasher, AntiVandalBot, Seaphoto, Naturalnumber, Spencer, Astavats, Husond, CosineKitty, Medconn, TheEditrix2, Bong-warrior, VoABot II, Kuyabribri, JamesBWatson, WODUP, Mother.earth, Animum, BatteryIncluded, Dirac66, LorenzoB, DerHexer, JaGa, Hans Moravec, Hyray, Patstuart, MartinBot, Church of emacs, Gnuarm, Mennoblaauw, Andre.holzner, Rettetast, J.delanoy, Dbiel, Extransit, Acalamari, Ncmvocalist, TomasBat, MetsFan76, Joshmt, Heavens is the world, Scott Illini, TraceyR, Idioma-bot, Mviduka4197, VolkovBot, Tourbillon, Thedjatclubrock, Jeff G., Mocirne, Seattle Skier, TXiKiBoT, DoctorPiouk, Dev 176, Martin451, ABigGreenHippo, Abdullais4u, FreeFull, Wikiisawesome, Scarymaryfwfc, RadiantRay, Roomyt, W1k13rh3nry, Antixt, Deanslinclair, Enviroboy, Burntsauce, Brianga, Alle-borgoBot, EmxBot, Neparis, D. Recorder, Ponyo, YohanN7, SieBot, Cwkmail, Yintan, Agesworth, JerrySteal, Keilana, RadicalOne, Toddst1, Tiptoety, JetLover, Arjen Dijksman, Sbowers3, Aruton, Oxymoron83, AnonGuy, Beej175560, Techman224, Anyeverybody, Nergaal, Denis-arona, Lord Shivan, Naturespace, ClueBot, RudolfSchmidt, PipepBot, Fasettle, Fyyer, The Thing That Should Not Be, Starkiller88, Indus-trieman, Mild Bill Hiccup, Polyamorph, Shjacks45, ChandlerMapBot, DragonBot, Gnome de plume, Jusdafax, Ju7kik8ol568r, Cenarium, Jotterbot, Vboo-belarus, Subash.chandran007, Plasmic Physics, Versus22, XLinkBot, Dark Mage, PL290, SkyLined, Addbot, Taschna, DOI bot, Ronhjones, Mr. Wheely Guy, LaaknorBot, CarsracBot, JBukon, Favonian, LinkFA-Bot, 5 albert square, AgadaUrbanit, Morgrimm, Numbo3-bot, Ehrenkater, LarryFrank, Tide rolls, Lightbot, Teles, Legobot, Luckas-bot, Yobot, Велетень, 2D, Tohd8BohaithuGh1, Cabb99, AnakngAraw, AnomieBOT, Bsimmons666, Jim1138, AdjustShift, Bluerasberry, Materialscientist, Hdehuer, The High Fin Sperm Whale, Citation bot, Satan's Kitchen, Maxis ftw, Raven1977, ArthurBot, Marshallsumter, Xqbot, Gopal81, Capricorn42, Drilnoth, DSisyphBot, Gilo1969, Paula Pilcher, Faatoafe90, Goostyyy, WaveEtherSniffer, GrouchoBot, Abce2, Amaury, Doulos Christos, Gordonrox24, Shadow-jams, A. di M., Samwb123, R8R Gtrs, FrescoBot, LucienBOT, Paine Ellsworth, Cannolis, Citation bot 1, Ecko15, Biker Biker, Pinethicket, HRoestBot, Jonesey95, Nicklcms, Seattle Jörg, Abhinav paulite, Double sharp, Darrell cosare, كاشف عقیلى, Mr.98, Diannaa, Ironnickel, Andrea105, Onel5969, TjBot, MagnInd, Jackehammond, Jimmy be, Robert Johnson 10, EmausBot, Green Day143, WikitanvirBot, Unken-ruf, GoingBatty, Illdz, Psturm~enwiki, Pcorty, Wikipelli, Hhhippo, ZéroBot, John Cline, Brazmyth, Quondum, GianniG46, Copper.nanotube, Brandmeister, L Kensington, Epicstonemason, Sjkimminau, Chris857, VictorianMutant, DASHBotAV, Whoop whoop pull up, ClueBot NG, Nebulosus, CocuBot, Satellizer, Letoya123, TruPepitoM, OverQuantum, Heyheyheyhohoho, Rezabot, Android1188, Widr, Diyar se, Ieditpa-gesincorrectly, Bibcode Bot, Neutronscattering, Wiki13, Metricopolus, Contact '97, Universuminkeisari, Nathanrohler, Zedshort, Hamish59, Nitrobutane, Hobos-r-us12, Oznitecki, BattyBot, MeowMeowArf, Dansalmo, ChrisGualtieri, GeorgEhlers, Ducknish, Gladiator222, Dexbot, Mogism, 331dot, TwoTwoHello, Lugia2453, Graphium, FaerieChilde, Fossilsnout, Morg00, Cldorian, Xuanmingzi, DihllonJessie, Jesse.johns, The Herald, Zenibus, Darkch2, Jwratner1, Javierha, My name is not dave, Cytokinetics, EtymAesthete, DudeWithAFeud, Abitslow, Aspaas-Bekkelund, Bballbro62, Mahusha, Light on the wall, Monkbot, Profesionalpretzels, Jayakumar RG, Haftswinch532, Selmatoed50, Istillcant, HMSLavender, Petahr, Orduin, Kethrus, Pulkit 4325, DiscantX, TSchonfeldt, Matan Kovac, KasparBot, Kafishabbir, Lord Wingus The Third and Anonymous: 554

- **Antiparticle** *Source:* https://en.wikipedia.org/wiki/Antiparticle?oldid=686699606 *Contributors:* AxelBoldt, CYD, Mav, Bryan Derksen, An-dre Engels, Josh Grosse, Stevertigo, Mrwojo, Patrick, RTC, Paddu, CesarB, Nikai, Nikola Smolenski, Charles Matthews, The Anomebot, Wik, Omegatron, Bevo, Altenmann, Merovingian, Intangir, Wikibot, Martinwguy, Giftlite, Bogdanb, Harp, BenFrantzDale, Herbee, Spencer195, Fleminra, Jason Quinn, Zeimusu, Mako098765, Karol Langner, Mike Rosoft, Helohe, Rich Farmbrough, Guanabot, Pjacobi, Guanabot2, Mr. Billion, Joanjoc~enwiki, Kghose, Cmdrjameson, Giraffedata, Matt McIrvin, HasharBot~enwiki, Pediddle, Deror avi, Woohookitty, Mindma-trix, Wdyoung, GregorB, SeventyThree, Justin Ormont, Palica, Marudubshinki, Tevatron~enwiki, Rjwilmsi, Ae77, MZMcBride, KaiMartin, FlaBot, Krackpipe, Commander Nemet, Roboto de Ajvol, YurikBot, Borgx, Bambaiah, Zhaladshar, Spike Wilbury, Bota47, Terbospeed, Mkossick, Tim314, ⬛⬛⬛⬛ robot, SmackBot, FocalPoint, Alsandro, Srnec, Dauto, Octahedron80, Drphilharmonic, Marcus Brute, Vina-iwbot~enwiki, Jake-helliwell, Grumpyyoungman01, Newone, Mellery, Van helsing, Tim1988, Myasuda, Gogo Dodo, Goldencako, Thijs!bot, Headbomb, Tyco.skinner, JAnDbot, Steveprutz, Ferritecore, Jpod2, Singularity, Dbiel, TomasBat, Eternalmatt, Joshmt, DorganBot, Cuck-ooman4, VolkovBot, TXiKiBoT, Red Act, Anonymous Dissident, AlleborgoBot, SieBot, Likebox, RadicalOne, Flyer22, KoenDelaere, Thomega, RW Marloe, BrightRoundCircle, Davidmosen, Jacob.jose, Anyeverybody, ClueBot, Diagramma Della Verita, Alexbot, Eeekster, Rishi.bedi, SilvonenBot, NellieBly, Lilaspastia, SkyLined, AkhtaBot, CarsracBot, Lightbot, Legobot, Luckas-bot, Yobot, Planlips, Csmallw, AnomieBOT, Citation bot, Vuerqex, ArthurBot, Xqbot, Omnipaedista, RibotBOT, Muhwang, EmausBot, John of Reading, L Kensington, Benazhack, Clue-Bot NG, Geekingreen, Mesoderm, Bibcode Bot, B wik, Mark Arsten, Rm1271, Penguinstorm300, Robotsheepboy, YFdyh-bot, 77Mike77, संजीव कुमार, Dert567, Monkbot, Dhm4444, Nazo!nin, KasparBot and Anonymous: 100

- **Exotic hadron** *Source:* https://en.wikipedia.org/wiki/Exotic_hadron?oldid=675789447 *Contributors:* AxelBoldt, Glenn, Xerxes314, Rich Farmbrough, YUL89YYZ, Keenan Pepper, Count Iblis, April Arcus, Linas, Bambaiah, Wogsland, Vladislav, Acjohnson55, Happy-melon, JRSpriggs, Postmodern Beatnik, ZICO, Thijs!bot, Whatever1111, Headbomb, Stannered, Leyo, VolkovBot, Antixt, Muhends, Curtis95112, MystBot, SkyLined, Addbot, Tide rolls, Luckas-bot, Yobot, Carlog3, Tarsilia, Acather96, Carbosi, ZéroBot, StringTheory11, Ethaniel, Chuis-pastonBot, Comicboy1996, Trompedo and Anonymous: 16

- **Tetraquark** *Source:* https://en.wikipedia.org/wiki/Tetraquark?oldid=676856701 *Contributors:* Bryan Derksen, Phys, Phil Boswell, Merovin-gian, Herbee, Varlaam, Physicist, Setokaiba, Icairns, SeaDour, Drbogdan, Rjwilmsi, Bubba73, Nhussein, Bambaiah, Todd Vierling, Antiduh, 2over0, Teply, SmackBot, Vladislav, Wiki me, Yevgeny Kats, Newone, Headbomb, Sobreira, Hcobb, Ron2, VolkovBot, Antixt, Muhends,

Voidxor, Deane@gooroos.com, FyzixFighter, SmackBot, Incnis Mrsi, Chris the speller, DHN-bot~enwiki, Croquant, Sbharris, Voyajer, Jgwacker, Bdushaw, Safalra, Kanags, Cricketgirl, Headbomb, Bobblehead, Mhaitham.shammaa, Magioladitis, Bongwarrior, Cgingold, JJ Harrison, Robin S, J.delanoy, VolkovBot, JohnBlackburne, Clarince63, Tamorlan, Biasoli, Hoopssheaffer, Abortz, SieBot, Sanya3, Proton666, Dolphin51, Bschaeffer~enwiki, ClueBot, Binksternet, Chris Illert, Manishearth, Djr32, RexxS, Ladsgroup, Karthik bala2009, SilvonenBot, Addbot, Tcncv, Mjamja, WFPM, CarsracBot, Tide rolls, Luckas-bot, Yobot, VanishedUser sdu9aya9fasdsopa, Orange Knight of Passion, Kingpin13, Bci2, Δζ, Pinethicket, RedBot, Ripchip Bot, Mema mema12, Gfoley4, XinaNicole, Wikipelli, ZéroBot, Quondum, Brandmeister, ClueBot NG, Helpful Pixie Bot, Bibcode Bot, Krenair, Vkpd11, Mdahir, Kylegodbey, Ginsuloft, Marj2117, JellyPatotie, Tetra quark and Anonymous: 71

- **Glueball** *Source:* https://en.wikipedia.org/wiki/Glueball?oldid=685875288 *Contributors:* Paul A, Loren Rosen, Phys, Sanders muc, Xerxes314, Mennonot, MuDavid, Jeodesic, Bambaiah, Hairy Dude, Ohwilleke, Xaxafrad, Smurrayinchester, Triple333, Saravask, Kmarinas86, V1adis1av, Sasata, Zaphody3k, Thijs!bot, Headbomb, Magioladitis, Nyq, Idioma-bot, Anonymous Dissident, Antixt, YonaBot, Avidallred, Boemmels, Alexbot, SchreiberBike, Addbot, Mpfiz, Lightbot, Luckas-bot, Dreamer08, AnomieBOT, Archon 2488, Pra1998, Tom.Reding, Loqueelvientoajuarez, RjwilmsiBot, Carbosi, Drummermean, JSquish, ZéroBot, Suslindisambiguator, Whoop whoop pull up, Bibcode Bot, Vkpd11, Retnuh66, ChrisGualtieri, Richardbernstein and Anonymous: 10

- **Quantum number** *Source:* https://en.wikipedia.org/wiki/Quantum_number?oldid=686328109 *Contributors:* The Anome, Stevertigo, Xavic69, Tim Starling, EddEdmondson, Ellywa, Mxn, Aliekens, Donarreiskoffer, Sverdrup, Hadal, Syntax~enwiki, Decumanus, Giftlite, Suspekt~enwiki, Dmmaus, Christopherlin, H Padleckas, RetiredUser2, Tsemii, Edsanville, Kareeser, Waza, Trevor MacInnis, Hidaspal, Vsmith, Spoon!, SpeedyGonsales, Nsaa, Arthena, BryanD, Pol098, Mpatel, MaximH, SeventyThree, Brownsteve, DaveTheRed, BD2412, DePiep, Happy-Camper, Fred Bradstadt, Azure8472, FlaBot, Margosbot~enwiki, Fresheneesz, Albrozdude, Bubbachuck, YurikBot, Wavelength, Mushin, Bambaiah, JWB, Hairy Dude, Huw Powell, JabberWok, Phantombantam, Jengelh, Chaos, ManoaChild, Bota47, Wknight94, Arthur Rubin, Modify, Teply, That Guy, From That Show!, Itub, SmackBot, Eskimbot, Complexica, Colonies Chris, Voyajer, Cubbi, Wybot, Yevgeny Kats, Yoshigev, Maatghandi, DJIndica, John, Sadeq, Davemcarlson, Mets501, Dan Gluck, Charles Baynham, Kushal one, Im.a.lumberjack, Imnotoneofyou, Myasuda, Cydebot, Kanags, Matrix61312, Waxigloo, SpK, Barticus88, JAnDbot, Trapezoidal, .anacondabot, Magioladitis, J.delanoy, Choihei, Acalamari, Somdebg, NewEnglandYankee, Fylwind, TraceyR, Larryisgood, AlnoktaBOT, Katoa, Go2slash, Venny85, Cobaro, EmxBot, SieBot, Gerakibot, Yintan, Allmightyduck, Agur bar Jacé, ClueBot, Keraunoscopia, Niceguyedc, Jotterbot, Versus22, Wake chaser, Joyonicity, SkyLined, Ivy martin08, Addbot, Zahd, WikiUserPedia, Jasper Deng, Drova, PV=nRT, WikiDreamer Bot, Luckas-bot, Yobot, AnomieBOT, Qmonkey, Law, Materialscientist, Citation bot, DirlBot, Melmann, Br77rino, Sahehco, A.amitkumar, FrescoBot, ProgramadorCCCP, Craig Pemberton, Redrose64, Pinethicket, Adlerbot, RobinK, Musicality213, Jordgette, Reach Out to the Truth, DARTH SIDIOUS 2, EarthCom1000, EmausBot, Gfoley4, Ravilovefriends, Demeza13, Tommy2010, Akshanshshrivastava, JSquish, ZéroBot, Mrfair, Sealbock, AManWithNoPlan, L0ngpar1sh, Just granpa, Sp4cetiger, ClueBot NG, Accelerometer, Frietjes, El.vegaro, DerekRobinson, Theopolisme, Helpful Pixie Bot, Shivsagardharam, Titodutta, Petermahlzahn, Mark Arsten, F=q(E+v^B), MrBill3, Fylbecatulous, Troller Hi, Dexbot, Webclient101, Fox2k11, Mpov, Vanamonde93, Avdhesh avistein, JaconaFrere, Septate, Tomasz59 and Anonymous: 225

- **Poincaré group** *Source:* https://en.wikipedia.org/wiki/Poincar%C3%A9_group?oldid=685453530 *Contributors:* AxelBoldt, Zundark, The Anome, XJaM, Mbecker, Stevertigo, Patrick, Michael Hardy, Marco Krohn, AugPi, Stupidmoron, Charles Matthews, Phys, Anupamsr, Giftlite, Lethe, Fropuff, DefLog~enwiki, Jossi, Rich Farmbrough, Bender235, Rgdboer, Aronbeekman, Danski14, Keenan Pepper, Gene Nygaard, Oleg Alexandrov, JFG, Mpatel, Allen3, Rjwilmsi, DVdm, YurikBot, That Guy, From That Show!, SmackBot, Incnis Mrsi, Nbarth, Tsca.bot, Kcordina, Cybercobra, JRSpriggs, Cydebot, Headbomb, Nearyan, JAnDbot, Fetchcomms, Yill577, SHCarter, Sullivan.t.j, Cuzkatzimhut, XCelam, Drschawrz, YohanN7, VVVBot, Phe-bot, Addbot, Luckas-bot, Ptbotgourou, AnomieBOT, Omnipaedista, Thinking of England, Meaghan, Jowa fan, Skater00, ZéroBot, Quondum, Git2010, Maschen, JFB80, Dexbot, CsDix, Prokaryotes, Kfitzell29, Ryanexler and Anonymous: 32

- **Parity(physics)** *Source:* https://en.wikipedia.org/wiki/Parity_(physics)?oldid=685934580 *Contributors:* Patrick, TakuyaMurata, Charles Matthews, Phys, SoLando, Tobias Bergemann, Giftlite, Xerxes314, Beland, Karol Langner, Lumidek, CALR, Pak21, Nvj, Cmdrjameson, Eruantalon, Sergio Macías, Wtmitchell, Knowledge Seeker, Count Iblis, Oleg Alexandrov, Joriki, Marudubshinki, Ae77, Nihiltres, Thecurran, Wave-length, Bambaiah, Archelon, Pseudomonas, Kabirramola, E2mb0t~enwiki, Elkman, GrinBot~enwiki, SmackBot, Incnis Mrsi, Tom Lougheed, Lei fisme, QFT, Wiki me, Akriasas, WhiteHatLurker, Erwin, JarahE, JRSpriggs, Raghunathan, Usgnus, Cydebot, Michael C Price, Thijs!bot, Barticus 88, Mbell, Headbomb, Pjvpjv, Dougher, Magioladitis, Thasaidon, Dirac66, HEL, Tarotcards, Idioma-bot, Gerrit C. Groenenboom, Cuzkatzimhut, Red Act, Pamputt, Antixt, SieBot, BotMultichill, Paolo.dL, Anchor Link Bot, ClueBot, Sun Creator, DumZiBoT, Lazyrussian, Rror, TravisAF, Addbot, Luckas-bot, Yobot, Tonyrex, PianoDan, Citation bot, ArthurBot, Omnipaedista, Theaucitron, Sławomir Biały, CraigPemberton, Merongb10, RedBot, TobeBot, Heurisko, Linguisticgeek, Queller69, RjwilmsiBot, EmausBot, Albear-And, ZéroBot, Quondum, Kmva, ClueBot NG, Greedohun, Tamila Shalumova, Helpful Pixie Bot, Bibcode Bot, Vkpd11, Slumdog2011, Goodbear3, MuonRay, Abitslow, JellyPatotie, Are you freaking kidding me and Anonymous: 64

- **C parity** *Source:* https://en.wikipedia.org/wiki/C_parity?oldid=667472921 *Contributors:* Xavic69, Michael Hardy, RJFJR, JYOuyang, SmackBot, Prateek.agrawal, Alan.ca, JoeBot, CmdrObot, Michael C Price, Headbomb, WinBot, Alphachimpbot, Magioladitis, Maliz, Bryan A, Venny85, Yobot, Omnipaedista, BenzolBot, Hhhippo, Ego White Tray, Bibcode Bot, ChrisGualtieri, Gabobaby, Monkbot, Choi koun, Kasuga and Anonymous: 7

- **Flavour (particle physics)** *Source:* https://en.wikipedia.org/wiki/Flavour_(particle_physics)?oldid=681888935 *Contributors:* Schewek, MichaelHardy, Nurg, Xerxes314, Varlaam, Andycjp, R. fiend, DragonflySixtyseven, CALR, STGM, Andrew Gray, Knowledge Seeker, Egg, Alai, Sylvain Mielot, Linas, Mindmatrix, SpNeo, Drrngrvy, YurikBot, Bambaiah, Hairy Dude, NTBot~enwiki, Bhny, Cossy, Długosz, SCZenz, Nick, Karl Andrews, SmackBot, Incnis Mrsi, Dauto, Doug Bell, Zero sharp, Ompty, BFD1, Ruslik0, Cydebot, Hydraton31, Xxanthippe, Michael C Price, Thijs!bot, Headbomb, FelixP~enwiki, Rompe, Hayesgm, Knotwork, CosineKitty, Robin S, Askielboe, Yonidebot, Choihei, I310342~enwiki, Thecinimod, VolkovBot, A4bot, Kresadlo, Maxim, Odellus, Ptrslv72, SieBot, VVVBot, The Stickler, Muhends, PixelBot, Jtle515, Count Truthstein, DumZiBoT, MystBot, SkyLined, Addbot, ZeroOmega, SpBot, Ehrenkater, HerculeBot, Luckas-bot, Ptbotgourou, Magog the Ogre, Icalanise, Omnipaedista, Citation bot 1, Xtermin8R645, B2NVB2, Jrobbinz123, 777sms, Bizzurp, EmausBot, VinculumMan, AvocatoBot, Drift chambers, Skynden, Isambard Kingdom and Anonymous: 42

- **Isospin** *Source:* https://en.wikipedia.org/wiki/Isospin?oldid=682122579 *Contributors:* Stone, Giftlite, Xerxes314, Michael Devore, RScheiber, Jason Quinn, AmarChandra, Lumidek, Perey, Rich Farmbrough, Hidaspal, V79, Cmdrjameson, RJFJR, Linas, Robert K S, Jwanders, TPickup,

Ddn2, FreplySpang, Rjwilmsi, Strait, Mike Peel, Margosbot~enwiki, Goudzovski, M7bot, Bambaiah, Bhny, Archelon, Welsh, Thiseye, Smack-Bot, Incnis Mrsi, Sue Anne, Colonies Chris, Sawran~enwiki, KI, Iridescent, Cydebot, Michael C Price, My Flatley, Zalgo, Thijs!bot, Headbomb, Knotwork, JAnDbot, Madmarigold, Avicennasis, Lilac Soul, KIAaze, Tarotcards, Fylwind, VolkovBot, Quilbert, Anonymous Dissident, Antixt, OlekG, PaddyLeahy, SieBot, Likebox, OsamaBinLogin, Uzdzislaw, Albambot, Addbot, Luckas-bot, Citation bot, ArthurBot, Bozzochet, Obersachsebot, Glenmark, Br77rino, J04n, Ernsts, RedAcer, Citation bot 1, Minivip, FoxBot, WikitanvirBot, Helpful Pixie Bot, Bibcode Bot, BG19bot, Jamisonsloan, Monkbot, Kfitzell29, GioComitini and Anonymous: 45

- **G-parity***Source:* https://en.wikipedia.org/wiki/G-parity?oldid=541031378*Contributors:* Xavic69, Michael Hardy, Charles Matthews, Xerxes 314,Hidaspal, Strait, Bambaiah, SmackBot, Dauto, Bluebot, Nberger, BWDuncan, CapitalR, Brichcja, MystBot, Addbot, Luckas-bot, Erik9bot,Carlog3, EleferenBot, QuantumSquirrel, Bibcode Bot and Anonymous: 5

- **Strangeness** *Source:* https://en.wikipedia.org/wiki/Strangeness?oldid=674819737 *Contributors:* Xavic69, Ahoerstemeier, Timwi, Herbee, Xerxes314, JeffBobFrank, RScheiber, Icairns, Xeroc, Mike Rosoft, Jkl, Jag123, LostLeviathan, Fred Condo, Mel Etitis, Tevatron~enwiki, Eyu100, Donotresus, FlaBot, Who, Fresheneesz, Srleffler, Roboto de Ajvol, Bambaiah, Hairy Dude, Conscious, Shawn81, Kyorosuke, SCZenz, 99 Willys on Wheels on the wall, 99 Willys on Wheels..., SmackBot, Stepa, JSpudeman, Complexica, Richard L. Peterson, ZICO, ShelfSkewed, Cydebot, Dchristle, Mbell, Headbomb, AntiVandalBot, NE2, The sage, I310342~enwiki, Pernogr~enwiki, Anonymous Dissident, Pamputt, Riwnodennyk, Callie.hoon, SilvonenBot, Addbot, Mr0t1633, Zorrobot, Citation bot, Wnme, Ernsts, A. di M., Qwarx, Yutsi, Johann137, Turian, Alarichus, Dinamik-bot, EmausBot, Vacation9, 図図図, Furkhaocean, JamesMoose, Ibnbaja and Anonymous: 33

- **Baryon number** *Source:* https://en.wikipedia.org/wiki/Baryon_number?oldid=675888853 *Contributors:* Andre Engels, Stevertigo, Delirium, Phys, Sanders muc, Securiger, Herbee, Xerxes314, Dratman, RScheiber, Jason Quinn, Discospinster, Pjacobi, Brim, Guy Harris, H2g2bob, Linas, Ted BJ, Isnow, Ddn2, BD2412, Raymond Hill, Bubba73, Margosbot~enwiki, Fresheneesz, Cannywizard, PointedEars, Roboto de Ajvol, YurikBot, Bambaiah, Tom Lougheed, V1adis1av, QFT, Doug Bell, Dan Gluck, Cydebot, Thijs!bot, Headbomb, Barakitty, Richard n, JAnDbot, CosineKitty, Bbi5291, Siryendor, FaTTshady74, STBotD, TXiKiBoT, A4bot, Venny85, SieBot, Muhends, Erodium, Addbot, WikiDreamer Bot, Legobot, Luckas-bot, Amirobot, ArthurBot, XZeroBot, Ernsts, MastiBot, Ttsush, Sahimrobot, Ernest3.141 and Anonymous: 17

- **Ground state** *Source:* https://en.wikipedia.org/wiki/Ground_state?oldid=666336191 *Contributors:* Marj Tiefert, Samw, Charles Matthews, Grendelkhan, Bevo, Topbanana, Robbot, Rfc1394, Wile E. Heresiarch, Giftlite, Mintleaf~enwiki, Fropuff, COMPATT, Karol Langner, Rdsmith4, H Padleckas, Canterbury Tail, Mwanner, La goutte de pluie, Kjkolb, SeventyThree, FlaBot, Fresheneesz, Chobot, Bambaiah, CWenger, Chris the speller, Pen of bushido, Robofish, JoseREMY, JorisvS, Mets501, BrainMagMo, MaxEnt, Hazmat2, Headbomb, Second Quantization, CosineKitty, R'n'B, VolkovBot, Thurth, SieBot, OlliffeObscurity, Greenpickle, Auntof6, Djr32, Addbot, PV=nRT, CiasoMs, Omino di carta, AnomieBOT, Eumolpo, GrouchoBot, Mnmngb, Jc3s5h, Udifuchs, Ysyoon, MastiBot, Double sharp, AvicBot, Tin2019, AManWithNoPlan, ClueBot NG, Art and Muscle, Curb Chain, Dr. Whooves, PeterBmftjg, Draughor and Anonymous: 29

- **Excited state** *Source:* https://en.wikipedia.org/wiki/Excited_state?oldid=675596412 *Contributors:* XJaM, Andres, Smack, Giftlite, Bensaccount, Vogon, RetiredUser2, Icairns, Vsmith, Jag123, Passw0rd, Alansohn, Andrewwall, Andrew Gray, Rjwilmsi, Phantom784, Fresheneesz, Mushin, Conscious, Grafen, Andreaskem, Wknight94, Pb30, The Photon, Xaosflux, Armeria, Pieter Kuiper, Bduke, OrangeDog, Behaafarid, Voyajer, DMacks, Spiritia, Anlace, John, Buchanan-Hermit, Atomobot, Rozzychan~enwiki, ShelfSkewed, Thijs!bot, Hazmat2, Headbomb, RogueNinja, Alphachimpbot, JAnDbot, Loonymonkey, Dell9300, Uvainio, Phasechange, Warut, Bieusinh92, ARTE, Jorfer, Equazcion, Bovineboy2008, Nomaan8, Bartosik, TXiKiBoT, A4bot, Ronningt, Spinningspark, AlleborgoBot, YonaBot, Kropotkine 113, Behtis, Cm176, Djr32, Pifreak94, Boleyn, SkyLined, Prowikipedians, Addbot, Chzz, Frozenguild, Numbo3-bot, Zorrobot, Luckas-bot, Yobot, Jnivekk, Materialscientist, Citation bot, JohnnyB256, Acebulf, Djalover, Mnmngb, Rushbugled13, Armando-Martin, Enchirion, Hezimmerman, Tommy2010, Chemprofguy, ZéroBot, AManWithNoPlan, SBaker43, Wwwmrzkwww, ClueBot NG, Bibcode Bot, Snow Blizzard, Johnhgagon, Thepoopinator, Vinoodle, Mpj7 and Anonymous: 66

- **Resonance (particle physics)** *Source:* https://en.wikipedia.org/wiki/Resonance_(particle_physics)?oldid=676034557 *Contributors:* Dratman, Rich Farmbrough, Jerome Charles Potts, Rogermw, Alavirad~enwiki, Headbomb, Davidhorman, Magioladitis, Cuzkatzimhut, VolkovBot, BartekChom, Addbot, Luckas-bot, Yobot, Obersachsebot, FrescoBot, B2NVB2, Enedrox, Coronium, TobeBot, Guerillero, Hhhippo, Loew Galitz, DaltonCastle and Anonymous: 8

- **Particle physics experiments***Source:* https://en.wikipedia.org/wiki/Particle_physics_experiments?oldid=679007924*Contributors:* The Anome,Michael Hardy, Robbot, Hidaspal, Oleg Alexandrov, Rjwilmsi, Erkcan, Kolbasz, Ilmari Karonen, Chris the speller, Headbomb, CosineKitty,The Anomebot2, Sheppa28, Vanished User 1004, XLinkBot, Addbot, Yobot, Citation bot, Kikuyu3, Steve Quinn, Trappist the monk, MiraclePen, H3llBot, BR84, BG19bot, Monkbot and Anonymous: 2

48.12.2 Images

- **File:3gluon.png** *Source:* https://upload.wikimedia.org/wikipedia/commons/a/a8/3gluon.png *License:* Public domain *Contributors:* Transferred from en.wikipedia by SreeBot *Original artist:* Bambaiah at en.wikipedia

- **File:8foldway.png** *Source:* https://upload.wikimedia.org/wikipedia/commons/9/9f/8foldway.png *License:* CC-BY-SA-3.0 *Contributors:* Transferred from en.wikipedia to Commons. *Original artist:* Bambaiah

- **File:Acap.svg** *Source:* https://upload.wikimedia.org/wikipedia/commons/5/52/Acap.svg *License:* Public domain *Contributors:* Own work *Original artist:* F l a n k e r

- **File:Ambox_important.svg** *Source:* https://upload.wikimedia.org/wikipedia/commons/b/b4/Ambox_important.svg *License:* Public domain *Contributors:* Own work, based off of Image:Ambox scales.svg *Original artist:* Dsmurat (talk · contribs)

- **File:Asymmetricwave2.png** *Source:* https://upload.wikimedia.org/wikipedia/commons/0/0d/Asymmetricwave2.png *License:* CC BY 3.0 *Contributors:* Own work *Original artist:* TimothyRias

- **File:Baryon-decuplet-small.svg** *Source:* https://upload.wikimedia.org/wikipedia/commons/7/78/Baryon-decuplet-small.svg *License:* Public domain *Contributors:* Own work *Original artist:* Trassiorf

48.12.3 Content license